OCEANOGRAPHY

OCEANOGRAPHY

Editor

Alfred Eckstein

scitus
academics

Oceanography
Edited by **Alfred Eckstein**

ISBN: 978-1-68117-671-0
Library of Congress Control Number: 2016946537

SCITUS Academics LLC,
616, Corporate Way, Suite 2, 4766,
Valley Cottage, NY 10989

www.scitusacademics.com

Notice

Printed in United States of America on Acid Free Paper

PREFACE

Humans have been studying the ocean since pre-historic times when people would venture out from their homes along the coast in rafts. Around 2000-3000 years ago, naturalists and philosophers, including Aristotle, began to try to understand the vast bodies of water. While over 70% of the globe is covered by the ocean, much of it remains a mystery to us. Its immense size, depth, and power entice scientists to uncover the mysteries the ocean holds. Oceanography is the study of the world ocean, including aspects of biology, chemistry, physics, geology, meteorology, and astronomy, among many others.

Oceanography provides a solid integration of physical, chemical, geological, and biological topics. Organized to allow instructors to advance methodically, this book presents latest coverage of the marine world and its organization, giving a complete outline of how and why the fluid earth system acts the way it does. The developments in the field of ocean acoustics over recent years make this book an important tool for specialists in acoustics, oceanography, marine biology, and related fields.

Oceanography covers a wide range of topics, from marine life and ecosystems, to currents and waves, to the movement of sediments, to seafloor geology. As humans have come to populate most corners of the globe, our impact on the oceans is stressing their ability to continue operating normally. Healthy oceans are critically significant to maintaining a healthy planet. Perhaps the most important observation is that oceanography gives us a world view, an understanding, of the global system that is our environment, which can inspire our work, wherever it leads.

TABLE OF CONTENTS

1

COMPARATIVE ANALYSIS OF SEA SURFACE TEMPERATURE PATTERN IN THE EASTERN AND WESTERN GULFS OF ARABIAN SEA AND THE RED SEA IN RECENT PAST USING SATELLITE DATA

Neha Nandkeolyar,[1] Mini Raman,[2] G. Sandhya Kiran,[1] and Ajai[2]

[1]Department of Botany, Faculty of Science, M.S. University, Baroda, Vadodara 390002, India

[2]Marine Optics Division, Marine and Planetary Sciences Group, Space Application Centre, Ahmedabad 380015, India

ABSTRACT

With unprecedented rate of development in the countries surrounding the gulfs of the Arabian Sea, there has been a rapid warming of these gulfs. In this regard, using Advanced Very High Resolution Radiometer (AVHRR) data from 1985 to 2009, a climatological study of Sea Surface Temperature (SST) and its inter annual variability in the Persian Gulf (PG), Gulf of Oman (GO), Gulf of Aden (GA), Gulf of Kutch (KTCH), Gulf of Khambhat (KMBT), and Red Sea (RS) was carried out using the normalized SST anomaly index. KTCH, KMBT, and GA pursued the typical Arabian Sea basin bimodal SST pattern, whereas PG, GO, and RS followed unimodal SST curve. In the western gulfs and RS, from 1985 to 1991-1992, cooling was observed followed by rapid warming phase from 1993 onwards, whereas in the eastern gulfs, the phase of sharp rise of SST was observed from 1995 onwards. Strong influence of the El Niño and La Niña and the Indian Ocean Dipole on interannual variability of SST of gulfs was observed. Annual and seasonal increase of SST was lower in the eastern gulfs than the western gulfs. RS showed the highest annual increase of normalized SST anomaly (+0.64/decade) followed by PG (+0.4/decade).

1. INTRODUCTION

In today's era, one of the greatest challenges faced by humankind is "global warming." There has been a 0.6°C increase in global temperature in the last century, and it is projected to further increase by 1.8°C to 4°C in the 21st century [1], posing a serious threat to the socioeconomic sector worldwide. With oceans covering approximately 72% of the earth's surface, any discussion on climate change and global warming would remain incomplete without including the role of the oceans. Studies have revealed that because of the increase in greenhouse gases, the oceans are warming significantly, which is affecting the marine ecosystem [2–5], resulting in shifting of habitats and dwindling of marine biota. Most of the studies with respect to climate change have focussed on global scale changes [5–15]. However, an exhaustive analysis at the regional and local levels is needed for framing adaptive and mitigative strategies to alleviate the effect of global warming.

In the Indian context, the recent finding, that amongst the oceans, the warming of the Indian Ocean is second highest [15], is a cause of worry and calls for immediate attention. Even though climatological studies have been done in the past in the Indian Ocean encompassing the Arabian Sea and the Bay of Bengal [16–20], a comparative analysis of the changing Sea Surface Temperature (SST) pattern the gulfs of Arabian Sea (Persian Gulf, Gulf of Oman, Gulf of Aden, Gulf of Kutch, and Gulf of Khambhat) and the Red sea has not been done. The gulfs of the Arabian Sea are not just strategically important with rich sources of oil and natural gas but are also the hot spots of the marine biodiversity [21–28]. However, in the recent years, there has been an expeditious change in the marine ecosystem of the eastern and western gulfs of the Arabian Sea and the Red Sea, owing to anthropogenic interference [29–32]. Since 1990s, 40% of the coasts of the Persian Gulf have been modified [33]. The Persian Gulf and the Red Sea areas have been reported to be warming rapidly owing to the developmental projects undertaken in the surrounding coastal countries [30–32]. The Gulf of Kutch on the Indian coast is being aggressively developed as oil importing bases because of its proximity to the Middle East countries

[29]. In general, the issues of common concern in the eastern and western gulfs of the Arabian Sea and the Red Sea include various anthropogenic activities like industrialization, coastal infrastructure development projects, setting up of new ports and oil terminals, oil pollution from shipping industry, overfishing, dredging, and increase in tourism and recreational activities. [29–32, 34], resulting in habitat destruction, changes in temperature and salinity profile, and causing a significant loss of biodiversity [35].

In this regard, the present work was taken up to study the monthly, seasonal, and annual pattern of Sea Surface Temperature (SST) and to analyse its changing pattern with emphasis on interannual variability in the eastern (Gulf of Kutch and Gulf of Khambhat) and the western gulfs (Persian Gulf, Gulf of Oman, and Gulf of Aden) of the Arabian Sea and the Red sea.

2. MATERIALS AND METHODS

2.1. Area of Study

The gulfs of the Arabian Sea basin (Figure 1(a)) can be grouped into the eastern and the western branches. The eastern branch includes the Gulf of Kutch and the Gulf of Khambhat on the Indian coast, whereas the western branch includes the Gulf of Oman and the Persian Gulf in the northwestern part of Arabian Sea, and the Gulf of Aden in the southwestern part of Arabian Sea, connecting it with the Red Sea.

2.1.1. Persian Gulf

The Persian Gulf (24–3°N; 48–56.5°E), (Figure 1(b)) located in an arid subtropical zone, is a semienclosed shallow, marginal sea with an average water depth of 36 m. It is spread over an area of 239,000 km^2 and is 990 km in length. It is connected to the Gulf of Oman through the 56 km wide Strait of Hormuz [36].

In the Persian Gulf, the seasonal differences of insolation, along with cold winds from the nearby highlands, result in extreme temperature (ranging from 16°C to 35°C) [36] and salinity (ranging from 36 to 43 parts per thousand) conditions [37]. However, despite the extreme conditions, the offshore waters of the Persian Gulf are rich in nutrients and support a variety of marine photoautotroph like the macroalgae, phytoplankton, mangroves, sea grass beds, and intertidal vegetation [25].

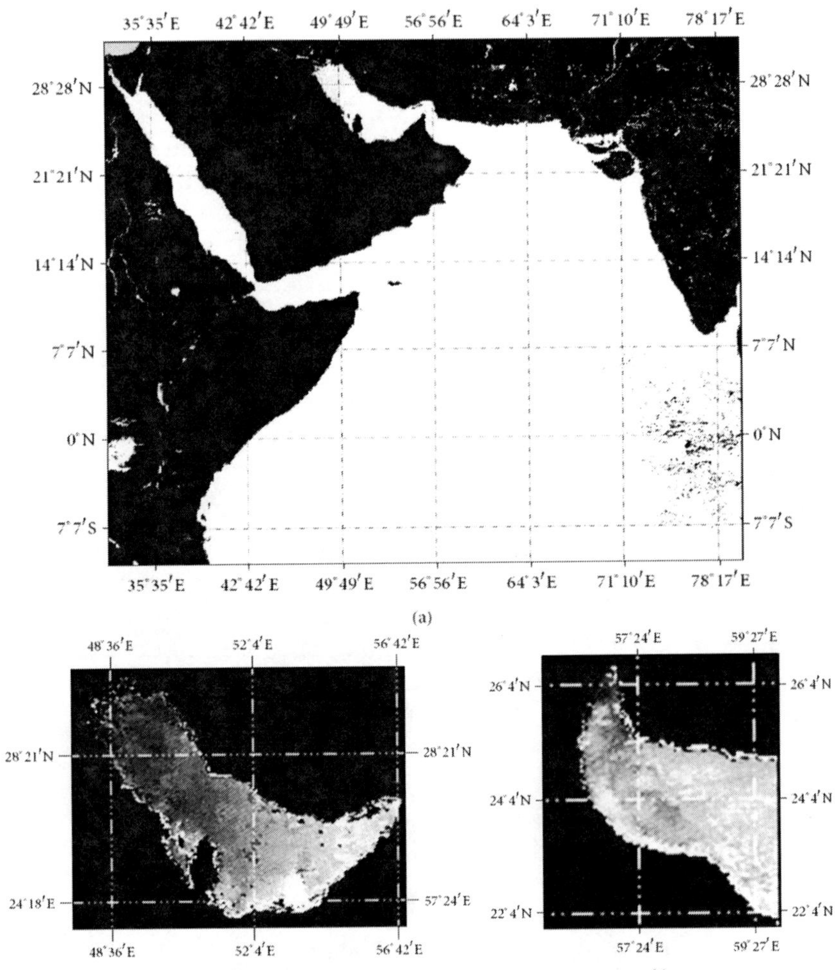

Figure 1: (a) Area of Study showing Arabian Sea basin. (b) Persian Gulf. (c) Gulf of Oman. (d) Gulf of Aden. (e) Red Sea. (f) Gulf of Kutch and Gulf of Khambhat.

2.1.2. Gulf of Oman

The Gulf of Oman (22°3–26°5′N; 56.5–61°43′E) is a strait connecting the Arabian Sea with the Persian Gulf (Figure 1(c)). The Gulf of Oman is 320 km wide and 560 km long.

Some of the most powerful eddy currents of the Arabian Sea are located close to the Omani coast. The coastal areas of the Gulf of Oman support varied habitats including mangrove swamps, lagoons, and mudflats.

2.1.3. Gulf of Aden

The Gulf of Aden (GA) (10°–15°N; 43°–52°E), connecting the Red Sea with the Arabian Sea, is part of the Suez Canal shipping route between the Mediterranean Sea and the Indian Ocean (Figure 1(d)). It is about 900 km long, with average depth of 1800 m, and covers an area of about 220×10^3 km^2. Its importance lies in the fact that it provides an outlet to the saline water masses of the Red Sea into the Arabian Sea [38].

2.1.4. Red Sea

The Red Sea (RS) (12°29′N–27°57′N; 34°36′E–43°30′E) is approximately 2,100 km long and 280 km wide (Figure 1(e)). The average depth is about 500 m, but at places it is more than 2,000 m deep. The Red Sea is linked to the Mediterranean Sea by the Suez Canal in the north, whereas in the south, it is connected to the Arabian Sea through the Gulf of Aden.

The waters of the Red Sea are warm and saline. The temperature ranges from 21°C–28°C in the north and 26°C to 32°C in the south. Salinities in the Red Sea range from 37% in the south to 42% in the north. A consequence of these extreme conditions is that some species within the Red Sea (like mangroves, shallow sea grasses, etc.) probably exist at the limits of their physiological tolerance [30].

2.1.5. Gulf of Kutch

The Gulf of Kutch (22°15′–23°4′N; 68°20′–70°40′E) encloses an area of 7350 km^2 (Figure 1(f)). Towards the western end, the Gulf is about 75 km wide and 60 m deep, while in the eastern end it is 18 km wide and less than 20 m deep. It is under the influence of strong tidal currents. High rate of evaporation, along with the release of salty water from the adjoining saltpans of Rann of Kutch, makes the eastern part of the Gulf of Kutch more saline (40%) than the western part (35%). The temperature of the Gulf waters varies between 24°C to 30°C.

The Gulf of Kutch is one of India's only coastal areas endowed with coral reefs. It provides a platform for different habitats like coral reefs, mangroves, creeks, mud flats, islands, rocky shore, sandy shore, and so forth and hence is enriched in biodiversity [29].

2.1.6. Gulf of Khambhat

Gulf of Khambhat (between 72°2′E to 72°6′E and 21° to 22°2′N) is one of the major fishing areas along western coast of India (Figure 1(f)). It is about 80 km wide at mouth and tapers to 25 km along the coast. It is about 140 km in length and is characterised by several inlets and creeks formed by the confluence of rivers. The Gulf of Khambhat is shallow with about 30 m average depth and abounds in shoals and sandbanks. It is known for its extreme tides. The tidal range at Gulf of Khambhat is known to be one of the largest along the Indian coastline [39].

2.2. Data Used

The regional effect of the global warming in the oceanic water masses needs to be assessed and quantified. However, the lack of uniform data and methodology hampers such regional analysis [1]. With the availability of National Oceanic and Atmospheric Administration's (NOAA) Advanced Very High Resolution Radiometer (AVHRR) satellite data set, for 25 years of span (1985–2009), to some extent the problem of uniformity of data has been solved. Besides, the high resolution satellite derived SST data gives better results than in situ data in terms of its continuity and spatial coverage. Even though Reynolds SST and National Center for Environmental Prediction (NCEP) reanalyzed data sets [40] are available for longer periods, but their low spatial resolution (1° and 2.5°, resp.) and interpolation and reanalysis methods do not accurately capture the coastal areas and the upwelling regions [41].

In this study the monthly NOAA AVHRR Pathfinder (version 5.0) SST data at 4 km resolution was obtained from NASA'S Jet Propulsion Laboratory's Physical Oceanographic Centre (http://podaac.jpl.nasa.gov/), for the period 1985–2009, for climatological study of the eastern and western gulfs of the Arabian Sea and the Red Sea. SST fields were derived through the Multichannel Sea Surface Temperature (MCSST) algorithm, as given by McClain [42].

2.3. Methodology

For each of the defined areas of study, the monthly images were masked to avoid the influence of the land and clouds using the image processing software ENVI 4.1 and ERDAS 9.0. For analysis all the pixels of the study area were included. Care was taken to exclude the pixels with zero values while averaging. Following Joint Global Ocean Flux Study (JGOFS) [43], the four climatological seasons described here are(a)northeast monsoon (December–March) (NEM),(b)spring intermonsoon (April-May) (SIM),(c)southwest monsoon (June–September) (SWM), and(d)fall intermonsoon (October-November) (FIM).The climatological mean (CM_{25}) of 25 years (1985–2009) for each month was calculated by averaging the monthly mean (Myi). The interannual variability was analyzed using the monthly normalized anomalies, computed by subtracting the monthly climatological mean (CM_{25}) from the monthly mean (Myi) of each year, and normalized to the standard deviation for that month (SD_{25}) and given asThe seasonal and annual normalized anomalies were computed by averaging the monthly normalized anomalies over appropriate seasons and years. Coefficient of variation (CV) was used to express the magnitude of interannual variability in the annual and the seasonal SST for the eastern and the western gulfs of the Arabian Sea and the Red sea [44] and is calculated asTo study the effect of El Niño and La Niña, comparison with ENSO (El Niño-Southern Oscillation) was carried out using the multivariate ENSO index, that is, MEI index, provided by the climate diagnostic centre (http://www. cdc.noaa.gov/).

3. RESULTS

3.1. Comparison of the Monthly Climatological Mean SST (1985–2009)

Figure 2 shows the comparative analysis of the monthly climatological mean SST (1985–2009) of the eastern and western gulfs of the Arabian Sea and the Red Sea. The SST in the Arabian Sea open ocean varies in a typical bimodal pattern with warming during spring inter monsoon (SIM) (April-May) and fall intermonsoon (FIM) (October-November) and cooling during the southwest monsoon (SWM) (June–September) and northeast monsoon (NEM) (December–March) seasons [45–47].

The Gulf of Aden and the eastern gulfs of the Arabian Sea, namely, Gulf of Kutch and Gulf of Khambhat followed the typical bimodal SST pattern as in the Arabian Sea. However, unlike the Arabian Sea open ocean where the minimum SST is observed during the months of July-August (SWM season), the minimum SST in these gulfs was found in the month of January (NEM season). In the bimodal pattern, the observed second peak of SST in the Gulf of Kutch and Gulf of Khambhat was in the month of October, similar to the Arabian Sea open ocean pattern, but in the Gulf of Aden, the second peak of SST was observed in the month of September. Climatologically, from November to April, Gulf of Aden was cooler than the Arabian Sea open ocean (by an average 1°C), while from May to October, it was warmer than the Arabian Sea open ocean (by an average 2.02°C). Climatologically, both the eastern gulfs were cooler than the Arabian Sea open ocean (by an average difference of 1.32°C and 0.38°C, resp.), for all the months, except for September and October. Besides, in the months of July and August Gulf of Kutch was found to be cooler than Arabian Sea open ocean months by more than 2°C, whereas in Gulf of Khambhat no such significant difference was found.

The Persian Gulf, Gulf of Oman, and the Red Sea exhibited a distinct unimodal SST curve instead of the typical bimodal pattern as in the Arabian Sea open ocean, with minimum temperature during the NEM season and maximum temperature during the SWM season. Whereas the minimum temperature for these three water bodies was in February, the maximum temperature in the Persian Gulf and the Red sea was in August, whereas in the Gulf of Oman, the maximum temperature was in the month of June. Climatologically from November to May, the Persian Gulf, Gulf of Oman, and the Red sea were found to be cooler than the Arabian Sea open ocean (by an average 4.9°C, 2.2°C, and 1.9°C, resp.), while from June to October they were warmer than the

Arabian Sea open ocean (by an average 3.7°C, 2.7°C and 2.08°C, resp.). Amongst the eastern and western gulfs of the Arabian Sea and the Red Sea, the Persian Gulf exhibited the highest range of SST (11.97°C).

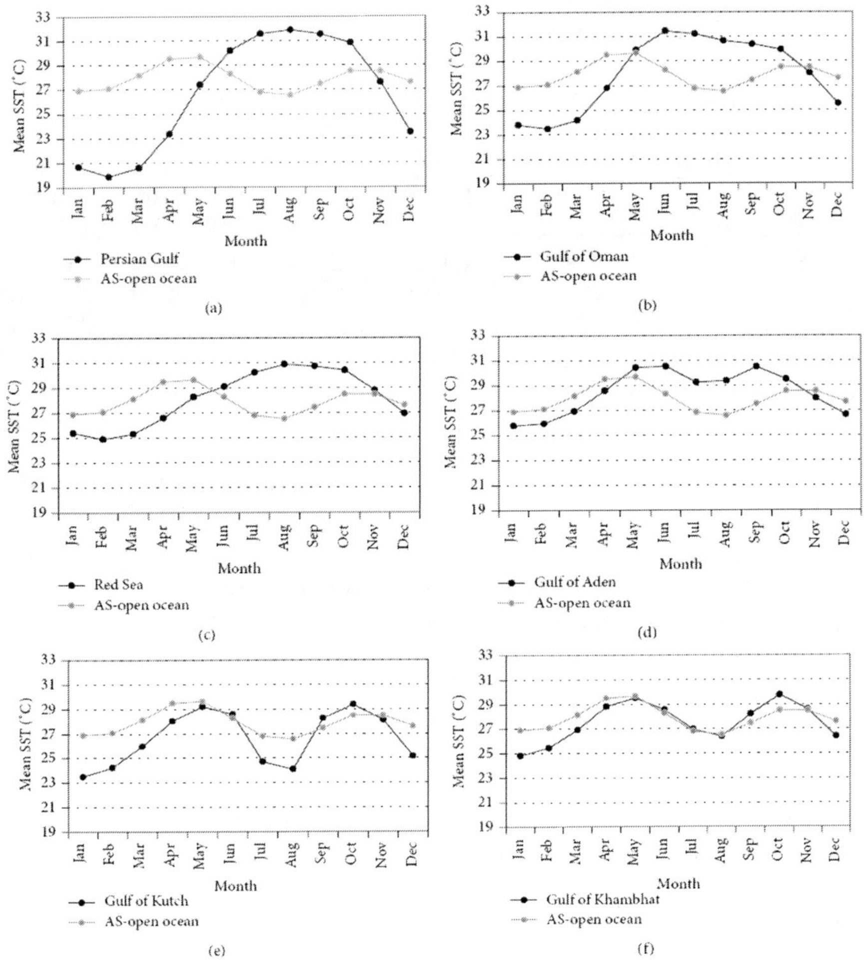

Figure 2: Comparison of the monthly climatological mean SST (1985–2009) of the Persian Gulf, Gulf of Oman, Red Sea, Gulf of Aden, Gulf of Kutch, and Gulf of Khambhat with Arabian Sea open ocean.

3.2. Interannual Variability of SST in the Eastern and Western Gulfs of Arabian Sea and Red Sea from 1985 to 2009

3.2.1. Comparison of the Annual Normalized SST Anomaly from 1985 to 2009

A time series study of the interannual variability of SST can be seen from Figure 3. A warming trend was observed in all the gulfs of the Arabian Sea in terms of the increasing annual normalized SST anomalies from 1985 to 2009. Whereas in the eastern gulfs of the Arabian Sea, the annual increase of normalised SST anomaly from 1985 to 2009 was insignificant with the Gulf of Kutch and Gulf of Khambhat showing an increase of 0.19/decade and 0.1/decade, respectively, in the western gulfs of the Arabian Sea and the Red sea, a sharp increase in SST was observed with the maximum rate of increase of normalized SST anomaly in the Red Sea (0.64/decade) followed by the Persian Gulf (0.4/decade).

A remarkable time lag in the rise of annual SST was found between the eastern and the western gulfs of the Arabian Sea and the Red Sea. For the western gulfs of the Arabian Sea and the Red Sea, a sharp increase in the annual SST was observed from 1992-1993 to 2009. However, for the eastern gulfs, the warming was delayed till 1995, and hence instead of 1992-1993, the sharp rise of annual SST was found from 1995 onwards.

For the Persian Gulf, Gulf of Oman, and Gulf of Aden, a decrease in annual normalized anomaly (cooling) was observed from 1985 to 1991, and thereafter from 1992 to 2009, a sharp increase (warming) was found. In the Persian Gulf, the normalized anomaly decreased (by −1.61), from −0.25 in the 1985 to −1.36 in 1991 and thereafter increased (by +1.48) to +0.12 in 2009. In the Gulf of Oman normalized anomaly decreased (by −1.0) from −0.44 (1985) to −1.44 (1991) and later increased (by +1.0) to +0.45 (2009). However, in the Red Sea, the normalized anomaly decreased from 1985 to 1992, (instead of 1991), from −0.76 (1985) to −1.53 (1992), and thereafter increased to +0.64 (2009).

In the Gulf of Kutch and Gulf of Khambhat (eastern gulfs of the Arabian Sea), a significant decrease in normalized anomaly (cooling) was observed from 1985 to 1994, and thereafter from 1995 to 2009 a sharp increase in normalized anomaly (warming) was found. In the Gulf of Kutch and Gulf of Khambhat, the normalized anomaly decreased from

−0.10 in 1985 to −0.92 in 1994 and subsequently increased to +0.75 in 2009. Similarly, in the Gulf of Khambhat the normalized anomaly decreased from −0.22 in 1985 to −0.75 in 1994 and later increased to +0.58 in 2009.

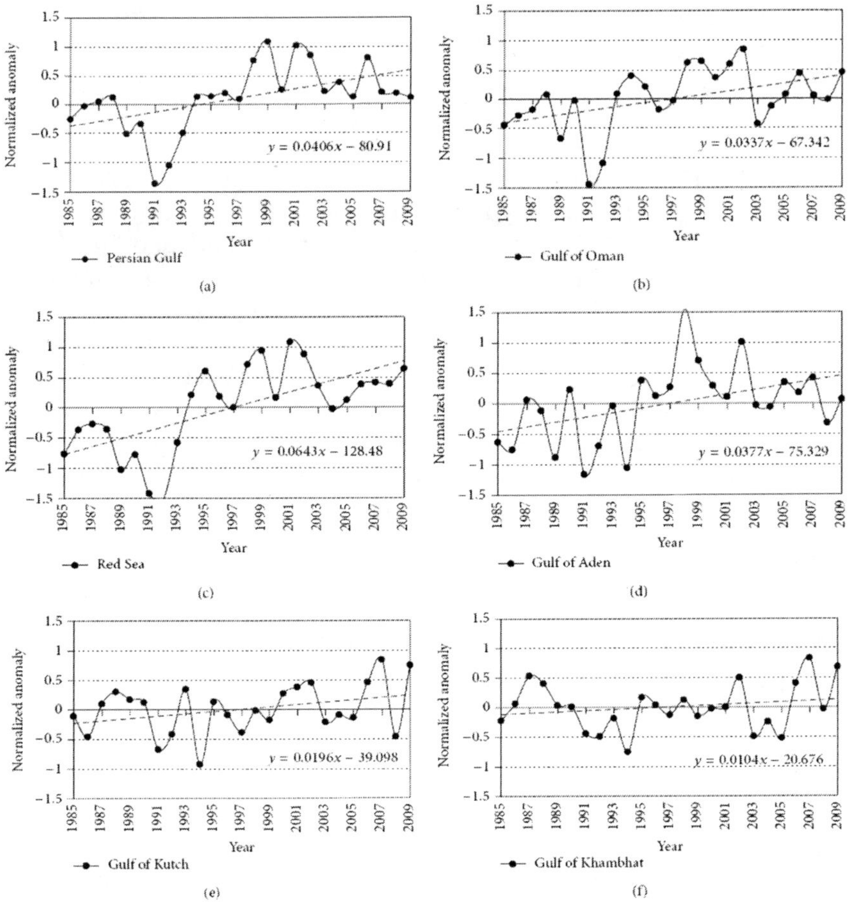

Figure 3: Comparison of the annual normalised SST anomaly from 1985 to 2009 of the Persian Gulf, Gulf of Oman, Red Sea, Gulf of Aden, Gulf of Kutch, and Gulf of Khambhat.

In the three western gulfs of the Arabian Sea, namely, Persian Gulf, Gulf of Oman, and Gulf of Aden, the maximum negative normalized SST anomaly was observed in 1991 (−1.36, −1.44, and −1.16, resp.). In the Red Sea the maximum negative normalized SST anomaly was found in 1992 (−1.5). On the other hand in both the eastern gulfs of the Arabian Sea, maximum negative normalized SST anomaly was observed in 1994 (−0.93 in Gulf of Kutch and −0.75 in Gulf of Khambhat). The maximum positive normalized SST anomaly was found to be in 1998 in Gulf of Aden (+1.54), in 1999 in Persian Gulf (+1.1), in 2001 in Red Sea (+1.1), and in 2002 in Gulf of Gulf of Oman, whereas in both the eastern gulfs of the Arabian Sea, maximum positive normalized anomaly was observed in 2007.

In the western gulfs of Arabian Sea and the Red Sea, the influence of the El Niño and the La Niña events can be seen from the crest and the troughs of the graph depicting the annual SST anomaly. For instance the La Niña years like 1985, 1986, 1989, 1996, 1999, 2000, and 2008 were found to have a high negative anomalies (cooling), and the El Niño years like 1986, 1988, 1991–1995, 1997-1998, 2002, 2006, and 2009 were observed to have a high positive normalized anomalies (warming). However, the influence of the El Niño and La Niña varied in terms of the difference in magnitude of the positive and negative anomalies. Besides, in the eastern gulfs of the Arabian Sea, only few of the El Niño (namely, 1987, 1988, 2002, and 2006) or La Niña (namely, 1989 and 2008) events influenced the warming or cooling of the gulfs.

3.2.2. Comparison of Seasonal Normalized SST Anomaly from 1985–2009

The interannual variation of SST during the four seasons, namely, northeast monsoon (NEM) (December–March), spring intermonsoon (SIM) (April-May), southwest Monsoon (SWM) (June–September), and fall inter Monsoon (FIM) (October-November), affecting the gulfs of Arabian Sea and the Red sea, was analyzed using the normalized SST anomaly indices as in Figure 4.

As seen from Figures 4(a), 4(b), 4(c), and 4(d), a similar interannual variation of SST pattern for the western gulfs of the Arabian Sea (namely, Persian Gulf, Gulf of Oman, and Gulf of Aden) and the Red Sea was observed with an increasing trend of SST in all the four seasons. However, in the eastern gulfs (Gulf of Kutch and Gulf of Khambhat), an increasing pattern of SST was found during the NEM, SIM, and FIM, but during the SWM season, cooling was observed.

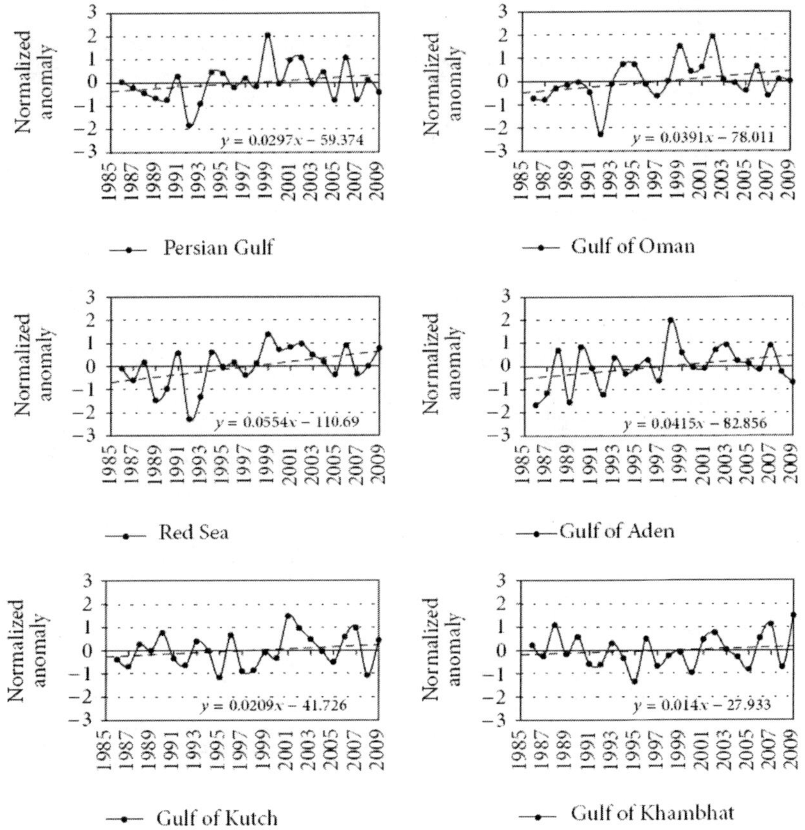
Normalized anomaly
y = 0.0297x − 59.374
Persian Gulf
y = 0.0391x − 78.011
Gulf of Oman
y = 0.0554x − 110.69
Red Sea
y = 0.0415x − 82.856
Gulf of Aden
y = 0.0209x − 41.726
Gulf of Kutch
y = 0.014x − 27.933
Gulf of Khambhat

(a) NEM

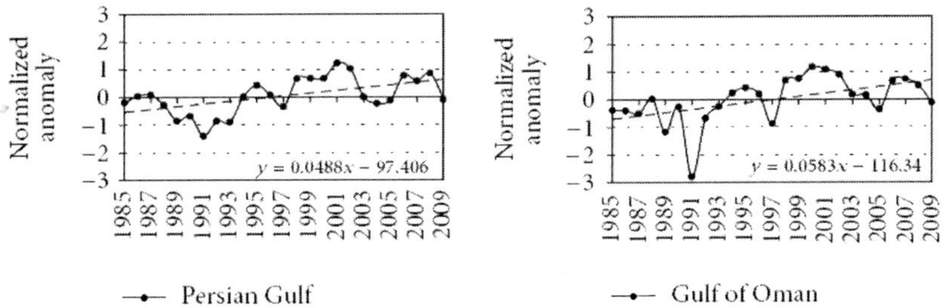
Normalized anomaly
y = 0.0488x − 97.406
Persian Gulf
y = 0.0583x − 116.34
Gulf of Oman

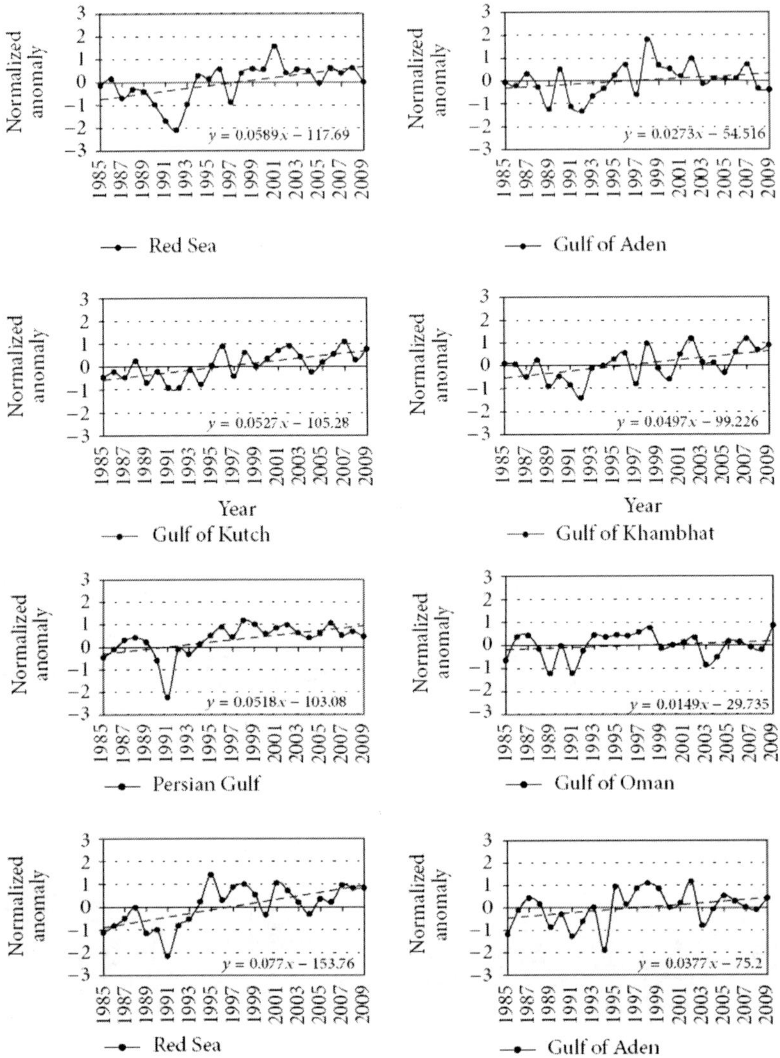
Normalized anomaly
y = 0.0589x − 117.69
Red Sea
y = 0.0273x − 54.516
Gulf of Aden
y = 0.0527x − 105.28
Year
Gulf of Kutch
y = 0.0497x − 99.226
Year
Gulf of Khambhat
y = 0.0518x − 103.08
Persian Gulf
y = 0.0149x − 29.735
Gulf of Oman
y = 0.077x − 153.76
Red Sea
y = 0.0377x − 75.2
Gulf of Aden

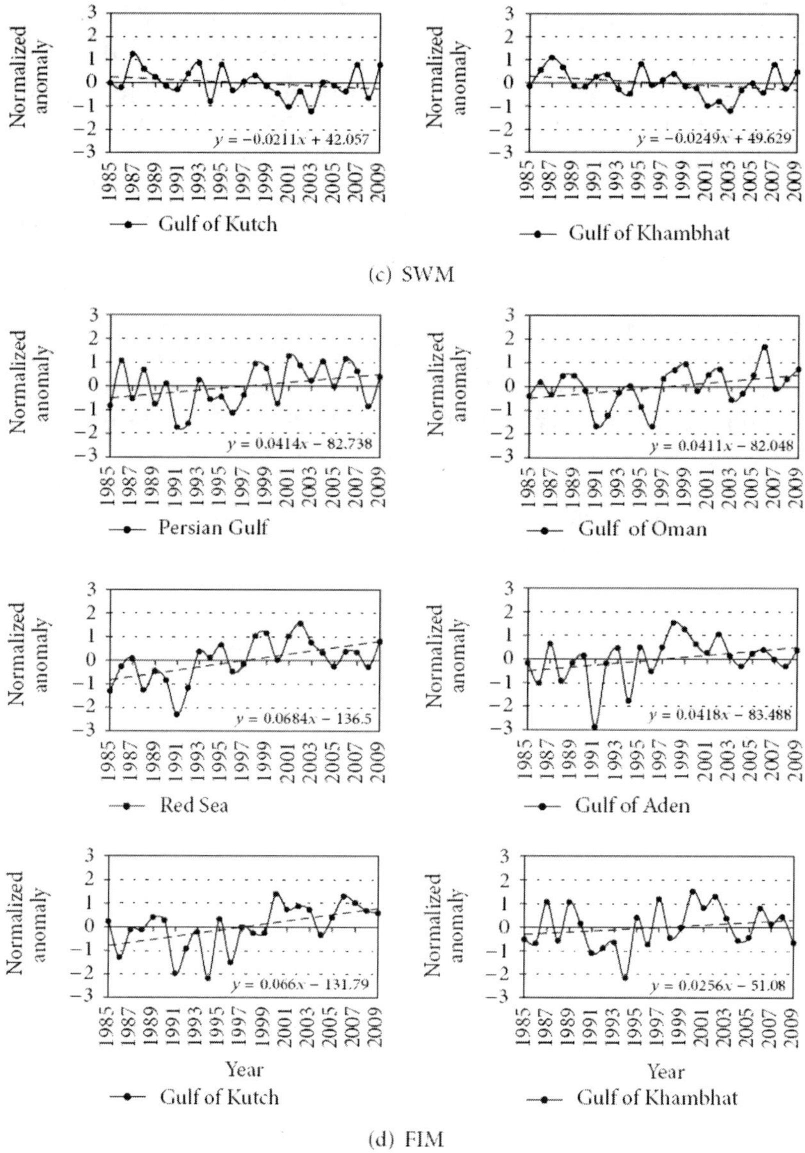

Figure 4: Comparison of the seasonal normalised SST anomaly from 1985 to 2009 of the Persian Gulf, Gulf of Oman, Red Sea, Gulf of Aden, Gulf of Kutch, and Gulf of Khambhat. (a) Northeast monsoon (NEM) (December–March), (b) spring intermonsoon (SIM) (April-May), (c) southwest monsoon (SWM) (June–September), and (d) fall intermonsoon (FIM) (October-November).

During the northeast monsoon (NEM) (December–March), (Figure 4(a)), the normalized SST anomaly increased (warming) in all eastern and western gulfs of the Arabian Sea and the Red Sea, but the increase was more in the western gulfs and the Red Sea as compared to the eastern gulfs, with the maximum rate of increase of SST in the Red Sea (0.55/decade). A similar interannual variability pattern during the NEM season was observed in the Persian Gulf, Gulf of Oman, and the Red Sea on one hand and Gulf of Aden, Gulf of Kutch, and Gulf of Khambhat on the other. From 1986 to 1992, in the Persian Gulf, Gulf of Oman, and the Red Sea, a decrease in SST anomaly was observed (cooling), which was followed by a phase of increasing SST anomaly (warming) from 1992 to 1999. From 1999 to 2005, the SST anomaly decreased substantially. However, from 2005 to 2009, again a warming phase with increasing SST anomaly was observed. In the year 1998-1999, an El Niño year, a very high positive SST anomaly was observed in the western gulfs of the Arabian Sea and the Red Sea. The warming was remarkable, as in the Persian Gulf and the Red Sea, the positive anomaly reached up to +2.05 and +1.38, respectively. Even though the Gulf of Oman showed a high positive deviation of +1.5 from the climatological mean in 1999, but its maximum positive anomaly (i.e., maximum warming) was observed during 2002 (+1.92), which was again an El Niño year. Amongst the eastern gulfs of the Arabian Sea and the Gulf of Aden, a similar pattern of interannual variation was observed, wherein from 1986 to 1990, the SST anomaly increased remarkably, followed by a phase of decreasing SST anomaly from 1990 to 1995. However, this cooling phase got extended in the Gulf of Aden up to 1997. From 1995 to 2001-2002, the SST anomaly increased in the two eastern gulfs, whereas in the Gulf of Aden, the increase was noticed up to 2003. From 2001-2002 (2003 in case of Gulf of Aden) to 2005, the SST anomaly again decreased, resulting in a cooling phase. However, from 2005 onwards, like the rest of the western gulfs of the Arabian Sea and the Red Sea, a remarkable increase in SST anomaly was observed in the Gulf of Aden and the two eastern gulfs of the Arabian Sea. In the Gulf of Aden the maximum positive SST anomaly was in 1998 (+1.98). However, in Gulf of Kutch, the maximum positive deviation from the climatological mean was noted during 2001 (+1.47), whereas in Gulf of Khambhat it was observed to be during 2009 (+1.48). During NEM, the Persian Gulf, Gulf of Oman, and the Red Sea exhibited the maximum negative deviation of SST from the climatological mean (i.e., maximum cooling) during 1992 (−1.8, −2.28, and −2.29, resp.), whereas Gulf of Kutch, and Khambhat showed the maximum negative deviation observed during 1995 (−1.15 and −1.36, resp.).

During the spring intermonsoon (SIM) (April-May) (Figure 4(b)), the normalized SST anomaly increased (warming) in all the eastern and the western gulfs of the Arabian Sea and the Red Sea, with maximum increase in the Gulf of Oman and the Red Sea (at the rate of 0.58/decade). It was observed that from 2005 onwards, during the SIM, the normalized SST anomalies increased substantially in the Persian Gulf, Gulf of Oman, Red Sea, Gulf of Kutch, and Gulf of Khambhat. The interannual variability pattern of the eastern gulfs of the Arabian Sea and the Gulf of Aden was found to be similar, beginning with a cooling phase (decreasing SST anomaly) from 1985 to 1991-1992, followed by a warming phase (increasing SST anomaly) which continued up to 1998. Whereas in the Gulf of Aden, the next phase of cooling was noticed from 1998 till 2003, in the eastern gulfs, a time lag of a year was found, with Gulf of Kutch extending the period of cooling from 1998 to 2004 and Gulf of Khambhat extending the period from 1998 to 2005. In the eastern gulfs of the Arabian Sea, the maximum positive deviation of SST from the climatological mean (i.e., maximum warming) during SIM was in 2002. Amongst the rest of the western gulfs of the Arabian Sea and the Red Sea, the pattern of interannual variability of SST was observed to be similar during the SIM. In the Persian Gulf, Gulf of Oman, and the Red Sea, from 1985 to 1991-1992, a cooling phase (with decreasing SST anomaly) was found. It was followed by a decade of warming from 1991-1992 to 2001, when the SST anomaly increased significantly. In these gulfs and the Red Sea, from 2001 onwards till 2005, another cooling phase was observed. However, in the last 5 years, that is, from 2005 to 2009 a substantial warming was noticed with significant increase in the SST anomaly during this period. Whereas the maximum positive deviation of SST from the climatological mean in the Red Sea and the Persian Gulf was during 2001 (+1.23 and +1.55, resp.), in the Gulf of Aden, it was in 1998 (+1.8) and in the Gulf of Oman in 2000 (+1.15). The maximum negative deviation of SST from the climatological mean (i.e., maximum cooling) in the Persian Gulf and the Gulf of Oman was in 1991 (−1.4 and −2.8, resp.), whereas in the Red Sea, Gulf of Aden, Gulf of Kutch, and Gulf of Khambhat, it was in 1992 (−2.1, −1.32, −0.92, and −1.41, resp.).

During the southwest monsoon season (SWM) (June–September) (Figure 4(c)), a cooling trend was observed in the eastern gulfs of the Arabian Sea, whereas in the western gulfs a warming trend was found in the last 25 years, with the maximum increase of SST anomaly in the Red Sea (0.77/decade). A similar pattern of SST anomaly was observed in the eastern gulfs of Arabian Sea on one hand and the western gulfs of Arabian Sea and the Red Sea on the other hand. However, a careful examination revealed that in the last 6 years (2003–2009) a substantial warming has been taking place in both the eastern and western gulfs of the Arabian Sea during the SWM season. In the eastern and western gulfs of Arabian Sea and the Red Sea, the SST anomaly increased from 1985 to 1987-1988, followed by a phase of decreasing anomaly, which continued until 1990-1991. During the extended El Niño period of 1991–1995, a sharp rise in SST anomaly was observed in the eastern gulfs of the Arabian Sea, after which SST decreased till 2003. On the other hand in the western gulfs of the Arabian Sea and the Red Sea, the SST anomaly increased substantially from 1991 till 1998 (with maximum positive anomaly in 1998). However, similar to the eastern gulfs, the SST anomalies again decreased from 1998 to 2003. But from 2003 onwards, a sharp rise of SST was observed till 2009, in both the eastern gulfs and the western gulfs of the Arabian Sea and the Red sea, with the maximum positive deviation of SST anomaly in the year 2007. The Persian Gulf showed positive SST anomalies in all the years from 1993 to 2009. Even in the Red Sea from 1993 onwards the anomalies have been positive, except for the years 2000 and 2004. The years with strong positive SST anomaly (more than +1.0) during SWM season were 1995, 1997, 1998, and 2002 in the Gulf of Aden; 1995, 1997, 1998, 2001, 2007, and 2009 in the Red Sea; 1998, 2002, and 2006 in the Persian Gulf; 1987 and 1993 in the Gulf of Kutch; and 1987 in the Gulf of Khambhat. None of the years showed strong positive SST anomalies (more than +1.0) in the Gulf of Oman. On the other hand, the years with strong negative SST anomaly (more than −1.0) during SWM season were 1985, 1991, and 1994 in the Gulf of Aden; 1985, 1989, 1990, and 1991 in the Red Sea; 1991 in the Persian Gulf; 1989, 1991, and 2003 in the Gulf of Oman; 2001 and 2003 in the Gulf of Kutch and the Gulf of Khambhat.

During the fall intermonsoon (FIM) (October-November), an increasing trend of the normalized anomalies for the eastern and the western gulfs of the Arabian Sea and the Red sea (Figure 4(d)) was observed. The Red Sea and the Gulf of Kutch had a much sharper increase in positive SST anomalies (0.68/decade and 0.66/decade, resp.) than the rest of the gulfs. For both the eastern gulfs of the Arabian Sea a similar interannual variability pattern was observed. In the Gulf of Kutch and Gulf of Khambhat, the SST anomaly increased from 1985-1986 to 1989-90 followed by a substantial cooling, which continued until 1994 (with maximum negative anomaly in 1994). From 1994 to 2000, a warming phase was followed by a cooling phase, which continued up to the year 2004 in the two eastern gulfs. However, in the last 4 years, that is, from 2004 till 2008, the SST anomaly again increased, with 2006 showing a high SST anomaly. Overall in the Gulf of Kutch and Gulf of Khambhat, a sharp rise of normalized anomalies (warming) was found from 1994 onwards, though the rise was more pronounced in the Gulf of Kutch, in which, from 1997 onwards, (except for 2004) none of the years showed negative departure from the climatological mean during this season. On the other hand, in the western gulfs of the Arabian Sea and the Red Sea, even though the pattern of interannual variability during the FIM season was similar, a significant difference in the magnitude of variability was noticed especially in the Persian Gulf and the Red Sea, owing to their enclosed locations. In the Persian Gulf, Gulf of Oman, Red sea, and Gulf of Aden the normalized SST anomalies decreased from 1985 to 1991 (cooling) and thereafter increased (warming) remarkably until 1998-1999. Eventually from 1998-1999 to 2003-2004, the SST anomalies again decreased significantly. However, similar to the eastern gulfs of the Arabian Sea, the western gulfs and the Red Sea also exhibited warming from 2003-2004 to 2009. During the FIM, in the Persian Gulf and the Gulf of Oman, apart from 1991, 1996 also showed intense cooling with negative SST anomaly reaching up to −1.67.

The influence of El Niño and La Niña on the interannual SST pattern during different seasons was observed. However, not all the warming and cooling events had similar effect on the SST anomaly. Besides, the geographical location of the particular gulf also played a crucial role in determining the impact of the ENSO.

3.2.3. Coefficient of Variation of Annual and Seasonal SST of the Eastern and Western Gulfs of Arabian Sea and Red Sea from 1985 to 2009

The coefficient of variation (CV) gives the magnitude of the interannual variability. CV values of the annual and seasonal SST of the eastern and the western gulfs of the Arabian Sea and the Red Sea showed large variation amongst these water bodies as seen in Table 1.

Table 1: % Coefficient of variation for annual and seasonal SST.

Gulf/Sea	Annual	NEM	SIM	SWM	FIM
Persian Gulf	18.02%	7.88%	11.77%	2.68%	7.77%
Gulf of Oman	11.1%	3.85%	8.37%	1.93%	4.61%
Red Sea	8.02%	3.56%	4.65%	2.76%	3.83%
Gulf of Aden	6.37%	2.20%	4.68%	2.63%	3.84%
Gulf of Kutch	9.35%	5.06%	3.63%	10.07%	3.14%
Gulf of Khambhat	6.5%	4%	2.13%	4.90%	2.83%

The largest variation in annual SST was found in the Persian Gulf (mean CV of 18.02%), followed by the Gulf of Oman (mean CV of 11.1%), whereas Gulf of Aden was least variable (mean CV of 6.37%). The interannual variability in seasonal mean SST was highest in the Persian Gulf in all the seasons except for the SWM season. During SWM season, the maximum variability was observed to be in the Gulf of Kutch (mean CV of 10.07%) followed by Gulf of Khambhat (mean CV of 4.9%). Gulf of Aden and Gulf of Oman were least variable during the NEM and SWM seasons, whereas Gulf of Khambhat was least variable during SIM and FIM seasons.

3.3. Comparison of Normalised SST Anomaly of the Eastern and Western Gulfs of Arabian Sea and Red Sea with Multivariate ENSO Index and Role of Indian Ocean Dipole

The El Niño/Southern Oscillation (ENSO) is a phenomenon that integrates both atmospheric and oceanic parameters and is measured by various indices like the Southern Oscillation Index (SOI), Multivariate ENSO Index (MEI), Nino 3.4, and so forth. [48–51]. El Niño and La Niña result from interaction between the ocean and the atmosphere in the tropical Pacific. The warm El Niño to cold La Niña conditions oscillate on an average of every three to five years. Of all these indices, MEI, a representative of SST, sea surface pressure, zonal and meridional components of the surface wind, air temperature and cloudiness of the sky, gives more structured information about ENSO phenomena and its correlation with surface temperature [52]. Hence, in this study, MEI was used to find the relationship between ENSO events and the change in the SST pattern in the gulfs of the Arabian Sea as seen in Figure 5.

Besides ENSO, the Indian Ocean Dipole (IOD) is also known to be affecting the interannual SST anomalies. One phase of the IOD (positive IOD or PIOD) causes a cooling of SST in the eastern tropical Indian Ocean, while the western tropical Indian Ocean (western Arabian Sea) tends to experience a warming of sea surface temperatures. The other phase (negative IOD or NIOD) in contrast involves high SST in the eastern Indian Ocean and low in the west [49, 53, 54]. Some dipole events occur without ENSO events, and some ENSO events do not accompany the dipole events. When an ENSO event occurs, the dipole tends to be positive during El Niño and negative during La Niña. However, the amplitude of the dipole varies greatly from year to year.

1991, 1994, 1997-1998 are reported to be the positive IOD years, whereas 1992, 1996, and 1998 are reported to be negative IOD years [55].

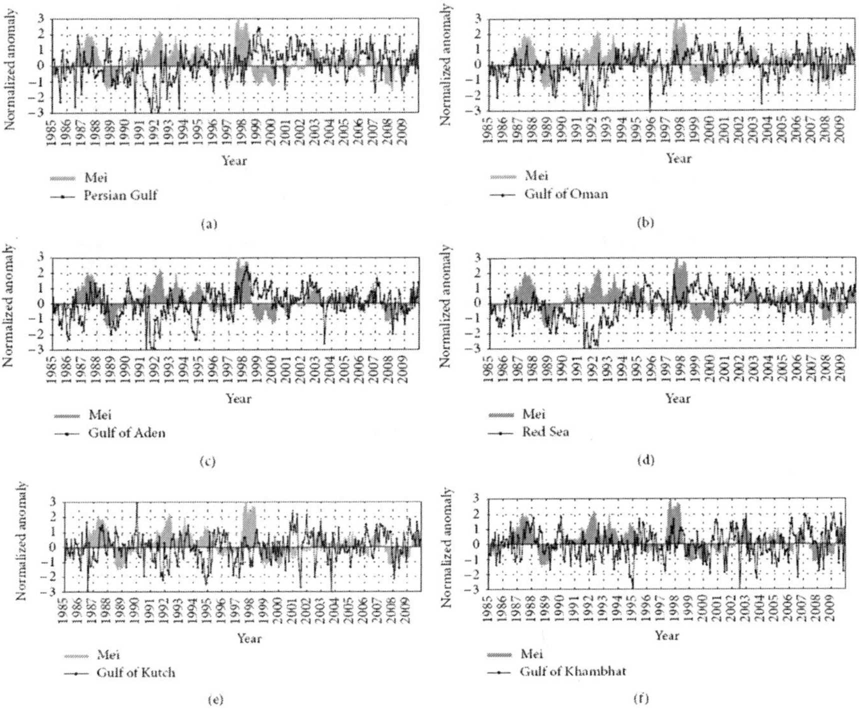

Figure 5: Comparison of normalised SST anomaly of (a) Persian Gulf, (b) Gulf of Oman, (c) Gulf of Aden, (d) Red Sea, (e) Gulf of Kutch, and (f) Gulf of Khambhat, from 1985 to 2009 along with Multivariate ENSO Index (MEI).

The major El Niño events of 1986-1987, 1991-1992, 1997-1998, 2002-2003, 2006-2007, and 2009-2010 and the La Niña events of 1988-1989, 1998-1999, 2000-2001, and 2007-2008 are evident from the peaks and troughs of the MEI values, with positive MEI values indicating a warming event (El Niño) and the negative MEI value indicating a cooling event (La Niña). The influence of the ENSO on SST of the Gulfs of the Arabian Sea can be clearly seen. However, one to one connection was not found. Besides, not all the El Niño or La Niña events had similar impact in the eastern and the western gulfs of the Arabian Sea. For example, the El Niño of 1991-1992 and 1994-1995 did not influence the eastern gulfs of the Arabian Sea as compared with the one of 1997-1998. The year 1997-1998 was also a positive IOD year. Hence, the extent of warming especially in the western gulfs of the Arabian Sea and the Red Sea was high. Besides, in 1992, a high negative anomaly in the western gulfs and the Red Sea was observed, owing to the influence of the negative IOD. Moreover, the impact of an IOD was seen more in the western domains of the Arabian Sea. During the extended El Niño of 1991–1995, the normalized positive SST anomaly increased remarkably in all the gulfs except the Gulf of Kutch and Gulf of Khambhat. Similarly, the stronger El Niño of 1997-1998 had a major impact on all gulfs except for the Gulf of Kutch. However, a delayed effect was observed in the Persian Gulf, Gulf of Oman, and Red Sea with peaks of normalised anomalies in 1998-1999. The La Niña of 1988-1989 resulted in cooling of Gulf of Oman, Red Sea, and Gulf of Aden but had no influence in the Gulfs of Kutch and Khambhat, whereas the La Niña of 2008 affected all the gulfs except for the Red Sea.

4. DISCUSSION

Time series studies of the oceanographic parameters are useful for understanding seasonal, interannual, and decadal impact of climate change. Numerous investigations have been conducted in the Arabian Sea basin to study the effect of global warming. Rupa Kumar et al. [17] reported that from 1904 to 1994 the Arabian Sea has warmed by 0.5°C. Recent studies have also confirmed about the warming of the Arabian Sea by 0.16°C/decade from 1971 to 2002 [18] and 0.10°C/decade from 1982 to 2006 [56]. Kailasam and Rao [20] using GISS Surface Temperature observed that there has been an increasing trend of SST, with strong positive SST anomalies (0.8°C) during the period of 1981–2009 over the Northern Indian Ocean.

However, as the gulfs of the Arabian Sea have unique oceanographic features and are quite distinct from the open ocean, it becomes essential to study them individually. These gulfs are single water masses, and therefore studies conducted in few of the coastal areas cannot be generalized for the whole of the gulfs. In the present study, a comparative analysis of SST in different months and seasons was carried out for the eastern and western gulfs of the Arabian Sea and the Red Sea from 1985 to 2009 to study the regional effect of the warming of the Arabian Sea. Annually the SST has increased significantly in all the western gulfs of the Arabian Sea and the Red sea, with maximum increase of normalized SST anomaly in the Red sea (0.64/decade) and the Persian Gulf (0.40/decade). However, the increase was not significant for the eastern gulfs of the Arabian Sea (Gulf of Kutch and Gulf of Khambhat).

The analysis by Kumar et al. [19] about the anthropogenic induced global warming disrupting the decadal cycle of SST of Arabian Sea after 1995 was also observed in the present study in the eastern gulfs of the Arabian Sea (Gulf of Kutch and Gulf of Khambhat), where the annual SST increased rapidly from 1995 onwards. However, in the western gulfs (Persian Gulf, Gulf of Oman, and Gulf of Aden) and the Red Sea, the disruption of the decadal cycle of SST resulting in a sharp rise of SST was observed from the year 1991-1992. Therefore, a remarkable time lag of (3 years) between the eastern and the western gulfs of the Arabian Sea and the Red sea was observed in the annual increase of SST. There was similarity in the interannual variability pattern of SST amongst the western gulfs and the Red sea on one hand and the eastern gulfs on the other.

Annually for western gulfs of the Arabian Sea, that is, the Persian Gulf, Gulf of Oman, and Gulf of Aden, cooling was observed from 1985 to 1991, and thereafter from 1992 to 2009, warming was found. Whereas in the eastern gulfs of the Arabian Sea, that is, the Gulf of Kutch and Gulf of Khambhat, a significant cooling was observed from 1985 to 1994, and thereafter from 1995 to 2009, a sharp increase in normalized anomaly was found. In the three western gulfs of the Arabian Sea, (Persian Gulf, Gulf of Oman, and Gulf of Aden), the maximum negative normalized SST anomaly was observed in 1991. In the Red Sea the maximum negative normalized SST anomaly was found in 1992 (negative IOD). On the other hand in both the eastern gulfs of the Arabian Sea, maximum negative normalized SST anomaly was observed in 1994. In the western gulfs of Arabian Sea and the Red Sea, 1997-1998, 1999, 2001, 2002, 2006, 2007, and 2009 were the years with high positive normalized SST anomalies, whereas in the eastern gulfs of the Arabian Sea, the years 2002, 2007, and 2009 showed strong positive SST anomalies. Most of these years are reported to be the El Niño years like 1997-98, 2002, 2006, and 2009 which affected the SST variability pattern.

The seasonal SST analysis for the western gulfs of the Arabian Sea and the Red Sea showed an increasing trend of SST in all the four seasons. However, in the eastern gulfs of the Arabian Sea during the SWM season, cooling was observed. During the NEM and the SIM seasons, similar interannual SST variability pattern was observed in the Persian Gulf, Gulf of Oman, and the Red Sea on one hand and Gulf of Aden, Gulf of Kutch and Gulf of Khambhat on the other. During the SWM and FIM seasons similar interannual SST variability pattern was in the eastern gulfs of Arabian Sea on one hand and the western gulfs of Arabian Sea and the Red Sea on the other.

During the NEM season, in the Persian Gulf, Gulf of Oman, and the Red Sea, an alternate cycle of cooling and warming phases was observed from 1986 to 1992, 1992 to 1999, 1999 to 2005, and 2005 to 2009, whereas in the eastern gulfs of the Arabian Sea and the Gulf of Aden, the alternate cycle of cooling and warming phases was observed from 1986 to 1990, 1990 to 1995 (1997 in Gulf of Aden), 1995 to 2001-2002 (2003 in Gulf of Aden), 2001-2002 to 2005, and 2005 to 2009. The Persian Gulf, Gulf of Oman, and the Red Sea exhibited maximum cooling during 1992, owing to the negative IOD. During the NEM season, the years 1987, 1989, and 2008 showed the influence of La Niña, with high negative SST anomalies in the western gulfs of the Arabian Sea and the Red Sea. But in the eastern gulfs of the Arabian Sea, only the influence of La Niña of 2008 was visible. Similarly of all the El Niño years, 1991–1995, 2002, and 2006 resulted in a higher positive SST anomaly in the eastern and the western gulfs of the Arabian Sea and the Red sea.

Similarly, during the SIM season, alternate warming and cooling phases were noticed in the eastern gulfs of the Arabian Sea and the Gulf of Aden from 1985 to 1991-1992, 1991-1992 to 1998, 1998 to 2004-2005 (2003 for Gulf of Aden) and 2004-2005 to 2009. In the eastern gulfs of the Arabian Sea, the maximum positive SST anomaly during SIM was in 2002 that was an El Niño year. Amongst the rest of the western gulfs of the Arabian Sea and the Red Sea, the alternate phases of warming and cooling were observed from 1985 to 1991-1992, 1991-1992 to 2001, 2001 to 2005, and 2005 to 2009. The maximum negative deviation of SST from the climatological mean in the Red Sea, Gulf of Aden, Gulf of Kutch, and Gulf of Khambhat was in 1992, owing to the negative IOD. During the SIM, the influence of El Niño of 1991–1995, 1997-1998, 2002 and 2009 and La Niña of 1987 and 1989 could be seen on the SST anomalies of the eastern and the western gulfs and the Red Sea.

In the SWM season, the periods of warming and cooling included 1985 to 1987-1988, 1987-1988 to 1990-1991, 1991 to 1995 (1998 for the western gulfs and the Red Sea), 1995 to 2003, and 2003 to 2009. The Persian Gulf and the Red Sea showed positive SST anomalies in all the years from 1993 to 2009. The years with strong positive SST anomaly were 1995, 1997, 1998, and 2002 in the Gulf of Aden; 1995, 1997, 1998,

2001, 2007, and 2009 in the Red Sea; 1998, 2002 and 2006 in the Persian Gulf; and 1987 and 1993 in the Gulf of Kutch; and 1987 in the Gulf of Khambhat, of which most of the years like 1995, 1997, 1998, 2002, 2006, and 2009 were the El Niño years. However the influence of La Niña was not seen during SWM season. The years with negative SST anomalies included 1985, 1991, and 1994 in the Gulf of Aden; 1985, 1989, 1990, and 1991 in the Red Sea; 1991 in the Persian Gulf; 1989, 1991, and 2003 in the Gulf of Oman; 2001 and 2003 in the Gulf of Kutch and the Gulf of Khambhat.

During the FIM season, an increasing trend of the normalized anomalies for the eastern and the western gulfs of the Arabian Sea and the Red sea was observed. In the eastern gulfs, the alternate phases of increasing and decreasing SST anomaly included 1985-1986 to 1989-1990, 1989-1990 to 1994, 1994 to 2000, 2000 to 2004, and 2004 to 2009. In the western gulfs of the Arabian Sea and the Red Sea the phases of warming and cooling included 1985 to 1991, 1991 to 1998-1999, 1998-99 to 2003-2004, and 2003-2004 to 2009. The years with high negative SST anomalies during the FIM were 1986, 1988, 1991, and 1994 in the Gulf of Aden, whereas in the Red Sea, these were 1985, 1988, and 1991. Of these 1985 and 1988 were the La Niña years. The years with strong positive SST anomalies during the FIM were 1986, 1998, 2001, 2004, and 2006 in the Persian Gulf; 1999 and 2006 in Gulf of Oman; 1998 and 2002 in Gulf of Aden; and 1998, 2001, and 2002 in the Red Sea. Of these 1986, 1988, 2002, and 2006 were the El Niño years.

The contrasting features between the eastern and western gulfs of the Arabian Sea could be attributed to their difference in oceanographic features, current systems, and also anthropogenic impact. Being enclosed in semiarid surroundings in the northern edges of the Indian subtropical zone, the western gulfs of the Arabian Sea come under the direct influence of multiple atmospheric pressure systems like the impact of the Siberian high pressure system, the El Niño, Southern Oscillation, the North Atlantic Oscillation, and the IOD [57]. The influence of the ENSO and the IOD affecting the interannual variability of SST was found. However, one to one connection was not found. Besides, not all the El Niño or La Niña events had similar impact in the eastern and the western gulfs of the Arabian Sea. The year 1997-1998 was an ENSO dipole year. Hence the extent of warming especially in the western gulfs of the Arabian Sea and the Red Sea was high. Besides, in 1992, a high negative anomaly in the western gulfs and the Red Sea was observed, owing to the strong influence of the negative IOD. During the extended El Niño of 1991–1995, the normalized positive SST anomaly increased drastically in all the gulfs except the Gulf of Kutch and Gulf of

Khambhat. Similarly, the stronger El Niño of 1997-1998 had a major impact on all gulfs except for the Gulf of Kutch. The atmospheric temperature over the Persian Gulf including the Gulf of Oman has been reported to be increasing rapidly and is projected to further increase up to 4°C by the end of the century [58]. There are reports of the weakening of the Siberian high pressure system resulting in a decline in the zonal component of wind and hence an increase in surface temperature [59]. Besides, the regional orography of the surrounding land masses which includes deserts and mountains also affects the physical oceanographic processes of the western gulfs of the Arabian Sea. However, the most significant contributor for the sharp rise of SST in the western gulfs of the Arabian Sea that has resulted in disruption of the decadal SST cycle much ahead of that of the Arabian Sea basin and the eastern Gulf is the increasing anthropogenic induced warming. There has been tremendous increase in human activities in the countries surrounding the western gulfs of the Arabian Sea which have contributed to the higher rate of warming along these western gulfs of the Arabian Sea as compared to the Arabian Sea basin and its eastern gulfs.

5. CONCLUSION

Sea surface Temperature (SST) is one of the key oceanographic parameters, exerting an influential role in many of the meteorological and oceanographic processes. From 1985 to 2009, the SST of the eastern and western gulfs of the Arabian Sea and the Red sea showed a clear cut signal of warming. The rate of increase of SST was highest in the Persian Gulf followed by the Red Sea. From 1991-1992 onwards, the western gulfs of the Arabian Sea and the Red sea exhibited a sharp increase in SST, whereas in the eastern gulfs of the Arabian Sea, the warming was more pronounced from 1995 onwards. Seasonal differences were also found with respect to increasing temperature. Interannual variability pattern was found to be similar in the annual and seasonal SST amongst the western gulfs and the Red Sea. Similarity in SST variability was also found in the two of the eastern gulfs of the Arabian Sea. Red Sea was found to be warming significantly across all the seasons and months. The largest variation in annual SST was found in the Persian Gulf (CV of 18.02%). The interannual variability in seasonal mean SST was highest in the Persian Gulf in all the seasons except for the SWM season. During SWM season, the maximum variability was observed in the Gulf of Kutch (CV of 10.07%) followed by

Gulf of Khambhat (CV of 4.9%). The influence of ENSO and IOD on the SST variability pattern was also found. The impact of rising temperature on other physical and biological parameters of the gulfs of the Arabian Sea and the Red Sea needs to be scrutinized further.

The difference in the oceanographic features of the eastern and western gulfs of the Arabian Sea and the Red Sea, from the open oceans, makes them unique ecosystem. Hence, caution is needed to analyze them individually and not to generalize the findings of the entire basin upon them.

ACKNOWLEDGMENTS

The authors are thankful to NASA'S Jet Propulsion Laboratory's Physical Oceanographic Centre and Climate Diagnostic Centre for providing the data.

REFERENCES

1. S. Solomon, D. Qin, M. Manning et al., Eds., Climate Change 2007: The Physical Science Basis. Contribution of Working Group I to the Fourth Assessment Report of the Intergovernmental Panel on Climate Change, Cambridge University Press, Cambridge, UK, 2007.

2. A. J. Richardson and D. S. Schoeman, "Climate impact on plankton ecosystems in the Northeast Atlantic," Science, vol. 305, no. 5690, pp. 1609–1612, 2004.

3. M. J. Behrenfeld, R. T. O'Malley, D. A. Siegel et al., "Climate-driven trends in contemporary ocean productivity," Nature, vol. 444, no. 7120, pp. 752–755, 2006.

4. R. Ji, C. S. Davis, C. Chen, D. W. Townsend, D. G. Mountain, and R. C. Beardsley, "Influence of ocean freshening on shelf phytoplankton dynamics," Geophysical Research Letters, vol. 34, no. 24, Article ID L24607, 2007.

5. B. S. Halpern, S. Walbridge, K. A. Selkoe et al., "A global map of human impact on marine ecosystems,"Science, vol. 319, no. 5865, pp. 948–952, 2008.

6. J. E. Hansen and S. Lebedeff, "Global trends of measured surface air temperature," Journal of Geophysical Research: Solid Earth, vol. 92, no. B1, pp. 345–355, 1987.

7. J. Hansen, "Public understanding of global climate change," in Carl Sagan's Universe, Y. Terzian and E. Bilson, Eds., pp. 247–253, Cambridge University Press, New York, NY, USA, 1997.

8. C. K. Folland, N. A. Rayner, S. J. Brown, et al., "Global temperature change and its uncertainties since 1861," Geophysical Research Letters, vol. 28, no. 3, pp. 2621–2624, 2001.

9. N. A. Rayner, D. E. Parker, E. B. Horton et al., "Global analyses of sea surface temperature, sea ice, and night marine air temperature since the late nineteenth century," Journal of Geophysical Research: Atmospheres, vol. 108, no. D14, p. 4407, 2003.

10. S. C. Doney, "Oceanography: plankton in a warmer world," Nature, vol. 444, no. 7120, pp. 695–696, 2006

11. F. Montaigne, "The global fish crisis," National Geographic, vol. 211, no. 4, pp. 32–99, 2007.

12. P. D. Jones and T. M. L. Wigley, "Estimation of global temperature trends: what's important and what isn't," Climate Change, vol. 100, pp. 59–69, 2007.

13. D. W. J. Thompson, J. M. Wallace, P. D. Jones, and J. J. Kennedy, "Identifying signatures of natural climate variability in time series of global mean surface temperature: methodology and insights," Journal of Climate, vol. 22, pp. 6120–6141, 2009.

14. S. A. Henson, J. L. Sarmiento, J. P. Dunne et al., "Is global warming already changing ocean productivity?" Biogeosciences Discussions, vol. 6, no. 6, pp. 10311–10354, 2009.

15. J. Hansen, R. Ruedy, M. Sato, and K. Lo, "Global surface temperature change," Reviews of Geophysics, vol. 48, no. 4, Article ID RG4004, 2010.

16. O. P. Singh, "Recent trends in summer temperatures over the North Indian Ocean," Indian Journal of Marine Sciences, vol. 29, no. 1, pp. 7–11, 2000.

17. R. Rupa Kumar, K. Krishna Kumar, R. G. Ashrit, S. K. Patwardhan, and G. B. Pant, "Climate change in India: observations and model projections," in Climate Change and India: Issues, Concerns and Opportunities, P. R. Shukla, S. K. Sharma, and P. Venkata Ramana, Eds., Tata McGraw-Hill Publishing Company Limited, New Delhi, India, 2002.

18. D. R. Kothawale, A. A. Munot, and H. P. Borgaonkar, "Temperature variability over the Indian Ocean and its relationship with Indian summer monsoon rainfall," Theoretical and Applied Climatology, vol. 92, no. 1-2, pp. 31–45, 2008.

19. S. P. Kumar, R. P. Roshin, J. Narvekar, P. K. D. Kumar, and E. Vivekanandan, "Response of the Arabian Sea to global warming and associated regional climate shift," Marine Environmental Research, vol. 68, no. 5, pp. 217–222, 2009.

20. M. K. Kailasam and R. S. Rao, "Impact of global warming on tropical cyclones and monsoons," in Global Warming, S. A. Harris, Ed., 2010, http://www.intechopen.com/books/global-warming/impact-of-global-warming-on-indian-monsoons.

21. C. R. C. Sheppard and A. L. S. Sheppard, "Corals and coral communities of Arabia," Fauna of Saudi Arabia, vol. 12, pp. 3–170, 1991.

22. C. Sheppard, A. Price, and C. Roberts, Marine Ecology of the Arabian Region: Patterns and Processes in Extreme Tropical Environments, Academic Press, London, UK, 1992.

23. R. F. G. Ormond and S. A. Banaimoon, "Ecology of intertidal macroalgal assemblages on the Hadramout coast of southern Yemen, an area of seasonal upwelling," Marine Ecology Progress Series, vol. 105, no. 1-2, pp. 105–120, 1994.

24. M. Goren and M. Dor, An Updated Checklist of the Fishes of the Red Sea CLOFRES 11, Academy of Sciences and Humanities, Jerusalem, Israel, 1994.

25. S. Al-Muzaini and P. G. Jacob, "Marine plants of the Arabian Gulf," Environment International, vol. 22, no. 3, pp. 369–376, 1996.

26. J. M. Kemp, "Zoogeography of coral reef fishes of the Gulf of Aden," Fauna of Arabia, vol. 18, pp. 293–321, 2000.

27. B. Riegl, "Climate change and coral reefs: different effects in two high-latitude areas (Arabian Gulf, South Africa)," Coral Reefs, vol. 22, no. 4, pp. 433–446, 2003.

28. S. L. Coles, Coral Species Diversity and Environmental Factors in the Arabian Gulf and the Gulf of Oman: A Comparison to the Indo-Pacific Region, vol. 507 of Atoll Research Bulletin, National Museum of Natural History Smithsonian Institution, Washington, DC, USA, 2003.

29. A. M. Dixit, P. Kumar, L. Kumar, K. D. Pathak, and M. I. Patel, Economic Valuation of Coral Reef Systems in Gulf of Kachchh. Final Report. World Bank Aided Integrated Coastal Zone Management (ICZM) Project, Gujarat Ecology Commission, 2010.

30. PERSGA/GEF, The Red Sea and Gulf of Aden Regional Network of Marine Protected Areas. Regional Master Plan, PERSGA Technical Series no. 1, PERSGA, Jeddah, Saudi Arabia, 2002.

31. C. Sheppard, M. Al-Husiani, F. Al-Jamali et al., "The Gulf: a young sea in decline," Marine Pollution Bulletin, vol. 60, no. 1, pp. 13–38, 2010.

32. W. Gladstone, B. Curley, and M. R. Shokri, "Environmental impacts of tourism in the Gulf and the Red Sea," Marine Pollution Bulletin, 2012.

33. W. Hamza and M. Munawar, "Protecting and managing the Arabian Gulf: past, present and future,"Aquatic Ecosystem Health and Management, vol. 12, no. 4, pp. 429–439, 2009.

34. C. Sheppard, "The main issues affecting coasts of the Indian and Western Pacific oceans: a meta-analysis from Seas at the Millennium," Marine Pollution Bulletin, vol. 42, no. 12, pp. 1199–1207, 2001.

35. A. N. Al-Ghadban and A. R. G. Price, "Dredging and infilling," in The Gulf Ecosystem, Health and Sustainability, N. Y. Khan, M. Munawar,

and A. Price, Eds., pp. 207–218, Bakhuys, Leiden, The Netherlands, 2002.

36. S.-Y. Chao, T. W. Kao, and K. R. Al-Hajri, "A numerical investigation of circulation in the Arabian Gulf,"Journal of Geophysical Research, vol. 97, no. 7, pp. 11219–11236, 1992.

37. R. M. Reynolds, "Physical oceanography of the Gulf, Strait of Hormuz, and the Gulf of Oman—results from the Mt Mitchell expedition," Marine Pollution Bulletin, vol. 27, pp. 35–59, 1993.

38. M. A. Al Saafani and S. S. C. Shenoi, "Water masses in the Gulf of Aden," Journal of Oceanography, vol. 63, no. 1, pp. 1–14, 2007.

39. V. S. Kumar and K. A. Kumar, "Waves and currents in tide-dominated location off Dahej, Gulf of Khambhat, India," Marine Geodesy, vol. 33, no. 2-3, pp. 218–231, 2010.

40. E. Kalnay, M. Kanamitsu, R. Kistler et al., "The NCEP/NCAR 40-year reanalysis project," Bulletin of the American Meteorological Society, vol. 77, no. 3, pp. 437–471, 1996.

41. M. Rouault, B. Pohl, and P. Penven, "Coastal oceanic climate change and variability from 1982 to 2009 around South Africa," African Journal of Marine Science, vol. 32, no. 2, pp. 237–246, 2010.

42. P. C. McClain, "Overview of the satellite data applications in the climate and earth sciences laboratory of NOAA'S National Environmental Satellite, data and information services," in Proceeding of the US-India Symposium-Cum Workshop on Remote Sensing Fundamentals and Applications, pp. 1–18, 1985.

43. S. Krishnaswami and R. Nair, "JGOFS (India)—introduction," Current Science, vol. 71, pp. 831–833, 1996.

44. J. Fang, S. Piao, Z. Tang, C. Peng, and W. Ji, "Interannual variability in net primary production and precipitation," Science, vol. 293, no. 5536, p. 1723, 2001.

45. J. G. Colborn, Thermal Structure of the Indian Ocean, IIOE Monograph no. 2, An East West Press, 1975.

46. R. R. Rao, R. L. Molinari, and J. F. Festa, "Evolution of the climatological near-surface thermal structure of the tropical Indian

Ocean. 1. Description of mean monthly mixed layer depth, and sea surface temperature, surface current, and surface meteorological fields," Journal of Geophysical Research, vol. 94, no. 8, pp. 10801–10815, 1989.

47. T. G. Prasad and M. Ikeda, "A numerical study of the seasonal variability of Arabian Sea high-salinity water," Journal of Geophysical Research C, vol. 107, no. 11, pp. 18-1–18-12, 2002.

48. G. T. Walker, "Correlation in seasonal variations of weather," Memoirs of the India Meteorological Department, vol. 24, pp. 333–345, 1924.

49. N. H. Saji, B. N. Goswami, P. N. Vinayachandran, and T. Yamagata, "A dipole mode in the tropical Indian ocean," Nature, vol. 401, no. 6751, pp. 360–363, 1999.

50. D. W. J. Thompson, J. M. Wallace, P. D. Jones, and J. J. Kennedy, "Identifying signatures of natural climate variability in time series of global-mean surface temperature: methodology and insights," Journal of Climate, vol. 22, pp. 6120–6141, 2009.

51. G. Foster, J. D. Annan, P. D. Jones et al., "Comment on "influence of the Southern Oscillation on tropospheric temperature" by J. D. McLean, C. R. de Freitas, and R. M. Carter," Journal of Geophysical Research: Atmospheres, vol. 115, no. D9, Article ID D09110, 2010.

52. K. Wolter and M. S. Timlin, "Monitoring ENSO in COADS with a seasonally adjusted principal component index," in Proceedings of the 17th Climate Diagnostics Workshop, pp. 52–57, NOAA/NMC/CAC, NSSL, Oklahoma Climatological Survey, CIMMS and the School of Meteorology, University of Oklahoma, Norman, Okla, USA, 1993.

53. P. J. Webster, A. M. Moore, J. P. Loschnigg, and R. R. Leben, "Coupled ocean-atmosphere dynamics in the Indian Ocean during 1997-98," Nature, vol. 401, no. 6751, pp. 356–360, 1999.

54. T. Yamagata, S. K. Behera, S. A. Rao, and N. H. Saji, "Comments on 'dipoles, temperature gradients, and tropical climate anomalies'," Bulletin of the American Meteorological Society, vol. 84, no. 10, pp. 1418–1422, 2003.

55. M. Nagura and M. Konda, "The seasonal development of an SST anomaly in the Indian Ocean and its relationship to ENSO," Journal of Climate, vol. 20, no. 1, pp. 38–52, 2007.

56. I. M. Belkin, "Rapid warming of large marine ecosystems," Progress in Oceanography, vol. 81, no. 1–4, pp. 207–213, 2009.

57. T. Shinoda, H. H. Hendon, and M. A. Alexander, "Surface and subsurface dipole variability in the Indian Ocean and its relation with ENSO," Deep-Sea Research Part I, vol. 51, no. 5, pp. 619–635, 2004.

58. G. A. Meehl, T. F. Stocker, and W. D. Collins, "Global climate projections," in Climate Change 2007: The Physical Science Basis. Contribution of Working Group I to the Fourth Assessment Report of the Intergovernmental Panel on Climate Change, S. Solomon, D. Qin, M. Manning, et al., Eds., pp. 80–87, Cambridge University Press, New York, NY, USA, 2007.

59. S. A. Piontkovski, M. Al-Gheilani, P. B. Jupp, R. A. Al-Azri, and A. K. Al-Hashmi, "Interannual changes in the sea of oman ecosystem," The Open Marine Biology Journal, vol. 6, pp. 38–52, 2012.

2

COMMUNITY STRUCTURE OF SHALLOW ROCKY SHORE FISH IN A TROPICAL BAY OF THE SOUTHWESTERN ATLANTIC

Camilo Moitinho Ferreira[1] , Ericka O. Cavalcanti Coni[1] , Diego V. Medeiros[2] , Cláudio L. S. Sampaio[3] , José Amorim Reis-Filho[4] , Francisco Barros[5] , Miguel Loiola[6] , José de Anchieta Cintra da Costa Nunes[4] *

[1]Programa de Pós Graduação em Ecologia e Conservação, Universidade Estadual da Paraíba. Instituto de Gestão, Estudos e Monitoramento Ambiental (IGEMA). Ecology & Environmental Science, School of Biological Sciences, University of Adelaide, South Australia, SA 5005, Australia. (Av. das Baraúnas, 351 - Campus Universitário, CEP: 58109-753, Campina Grande, PB, Brazil. Rua Eduardo Pinto, nº 49, CEP: 42835-000, Arembepe, Camaçari, Bahia, Brazil. Southern Seas Ecology Laboratories, room 128b, Darling Building, DX 650 418).

[2]Programa de Pós Graduação em Ecologia e Monitoramento Ambiental, Universidade Federal da Paraíba. (Rua da Mangueira, CEP: 58297-000 Rio Tinto, PB, Brazil).

[3]Universidade Federal de Alagoas, Polo Penedo, Departamento de Engenharia de Pesca. Laboratório de Ictiologia e Conservação (LIC). (Av. Beira Rio s/nº, Centro Histórico, CEP: 57200-000, Penedo, Alagoas, Brazil).

[4]Laboratório de Ecologia Bentônica (LEB), Universidade Federal da Bahia. Programa de Pós Graduação em Ecologia e Biomonitoramento, Universidade Federal da Bahia. Instituto de Educação, Ciência e Utilidade Sócio-Ambiental (ECUS). (Rua Barão de Geremoabo, s/n Ondina, CEP: 40170-115, Salvador, BA, Brazil. Rio Vermelho, CEP: 41940-250, Salvador, Bahia, Brazil).

[5]Universidade Federal da Bahia, Laboratório de Ecologia Bentônica. (Rua Barão de Jeremoabo, s/n Ondina, CEP: 40170-115, Salvador, BA, Brazil. Rio Vermelho, CEP: 41940-250, Salvador, Bahia, Brazil).

[6]Laboratório de Estudos de Recifes de Corais e Mudanças Globais (RECOR), Universidade Federal da Bahia. (Rua Barão de Jeremoabo, CEP: 40170-115, Salvador, BA, Brazil. Rio Vermelho, CEP: 41940-250, Salvador, Bahia, Brazil).

ABSTRACT

Southwestern Atlantic Ocean rocky shores sustain important reef fish communities. However, those communities in tropical regions are not well understood, especially in Brazil. In this present article we assess community parameters of reef fishes such as composition, trophic organization and their relationships with physical and biological factors on four tropical rocky shores in Todos os Santos Bay, southwestern Atlantic. During six months, a total of 80 visual censuses were performed, in which 3,582 fish belonging to 76 species were recorded. Territorial herbivorous fish and turf algae were dominant at all the sites. The spatial variability of fish community structure was related to the benthic cover composition and depth. The high abundance of territorial herbivores and mobile invertebrate feeders could be associated with high levels of turf cover, low wave exposure and shallow waters. Moreover, this fact could be a consequence of the low density of roving herbivores and large carnivores probably due to the pressure of intense fishing activity. Thus complementary studies are needed to evaluate the actual conservation status of these rocky shore reefs, singularly located habitats connecting inner and outer reefs in Todos os Santos Bay.

KEYWORDS

Todos os Santos Bay; Rocky shores; Fish communities; Habitat structure; Trophic groups; Visual census

1. INTRODUCTION

Coastal environments, such as coral reefs and rocky shores are amongst the most diverse and productive ecosystems in the world (PAULY; CHRISTENSEN, 1995; REAKA-KUDLA, 1997). Besides their ecological relevance, reef systems are also economically important for thousands of people living in tropical coastal areas (RUSS, 1991; COSTANZA et al., 1997; MOBERG; FOLKE, 1999; RUSS; ALCALA, 1999). However, all these environments are in a critical state of degradation, mainly due to natural and anthropogenic factors such as global climate change, marine pollution, overfishing and unplanned coastal development (HUGHES, 1994; DUFFY, 2003; PANDOLFI et al., 2003;BELLWOOD et al., 2004; WILKINSON, 2008; BRUCE et al., 2012; FRANCINI-FILHO et al., 2008).

The Brazilian reef environments encompass oceanic reef systems, such as the archipelagoes of Fernando de Noronha, St. Peter and St. Paul, Trindade/Martim Vaz (rocky islands) and the Rocas Atoll (a biogenic island), as well as coastal reef systems (FLOETER; GASPARINI, 2000). The country's eastern and northeastern coastlines primarily present coral reef structures, while the southern and southeastern regions are predominated by rocky shore structures (FLOETER et al., 2006). Rocky shores are considered coastal environments formed by rock outcrops located in the transitional zone between terrestrial and aquatic environments, in which the submerged part harbors complex marine life, including animals and plants of various phyla (FERREIRA et al., 2001). Compared to coral reefs, Brazilian rocky shores show relatively low structural complexity. However, their high species richness and reef fish diversity are well recognized (FERREIRA et al., 2001; HOSTIM-SILVA et al., 2006; RANGEL et al., 2007; CHAVES; MONTEIRO-NETO, 2009).

The Brazilian northeastern reef systems (e.g. rocky and coral reefs) are considered "hotspots" of biodiversity due to the high occurrence of endemic and threatened species, particularly coral and fish fauna (LEÃO et al., 2003;MOURA, 2003). However, the majority of reef fish community structure studies have been performed on coral reef systems (FERREIRA et al., 2001; NUNES et al., 2013). Knowledge about rocky shores in tropical regions of Brazil is insufficient, especially considering their fauna and flora in the northern and northeastern regions. Such data deficiency hampers the creation and implementation of effective conservation and management policies.

Several studies have shown habitat complexity and reef fish community associations (e.g. LUCKHURST; LUCKHURST, 1978; FERREIRA et al.,

2001; WILSON et al., 2007; FRANCINI-FILHO; MOURA, 2008; MEDEIROS et al., 2011;PEREIRA-FILHO et al., 2011; CONI et al., 2013). Topographic complexity is an important factor that is generally positively associated with reef fish species richness and diversity, particularly as it provides shelter from predators (HIXON; BEETS, 1993). Additionally, it may also modify the competition, interaction and survival dynamics within the community (JONES, 1988; GUNST, 2002; FRANCINI-FILHO et al., 2010). In southeastern Brazil, the habitat complexity of rocky shores is primarily determined by the presence or absence of crevices, a consequence of the superposition of rocks (FERREIRA et al., 2001). To date, there are no studies that evaluate the relationship between reef fish communities and the biotic and physical variables of rocky shores in northeastern Brazil.

Here we present the reef fish species associated with the rocky shores at Todos os Santos Bay's northern entrance (near the municipality of Salvador), as well as describe the reef fish community's spatial patterns (including trophic organization, density and species richness) and the relationship between fish density and environmental variables (physical structure and benthic cover). The Todos os Santos Bay region harbors a wide variety of marine organisms and encompasses complex environments. However, studies, particularly about fish communities, are still insufficient. Given this, this study is a first effort to characterize the fish community structure off tropical rocky shores in the northeastern region of Brazil. It is believed that this will provide essential knowledge to support future decisions for conservation and management strategies in the Todos os Santos Bay, and all tropical rocky shores in the northeast of Brazil. A succinct history of conservation in these areas is also presented.

2. MATERIAL AND METHODS

2.1 Study Area

Todos os Santos Bay is the second largest semi- enclosed bay in Brazil covering approximately 1,000 km2(LESSA et al., 2000). This area harbors a high level of biodiversity, comprising distinct habitats such as coral reefs, mangrove swamps and seagrass beds (CASTRO; PIRES, 2001; LEÃO et al., 2003; REIS-FILHO et al., 2010; REIS-FILHO et al., 2011). The bay has a stable annual water temperature, with low temperature fluctuation varying generally between 23º and 27º C (CIRANO; LESSA, 2007).

The present study was carried out at the mouth of Todos os Santos bay, located off the city of Salvador. The four rocky shores (CV - Corredor da Vitória, PB - Porto da Barra, FB - Farol da Barra, and MP - Marco Polo) (Figure 1) studied are distributed along approximately four kilometers of shore line, with an area of ~0.04km2. These are characterized as being relatively shallow (max. 10 m depth) with the benthic substrates composed predominantly of turf algae, macroalgae and the zoanthids *Palythoa caribaeorum* and *Zoanthus sociatus*. However, other benthic organisms, such as the black sea urchin (*Echinometra lucunter*), ascidians and coral colonies of *Favia gravida, Agaricia agaricites, Montastraea cavernosa, Mussismilia hispida, Siderastrea* spp., and the two species of fire corals *Millepora alcicornis* and *Millepora nitida*, are also found within the area. These rocky shores are heavily exploited for tourism activities, artisanal fisheries, ornamental marine organism trade and spearfishing (SAMPAIO; NOTTIGHAM, 2008; NUNES et al., 2012). Despite the water's typically remaining clear throughout the year (occasionally reaching over 40 meters of visibility), during our study horizontal visibility ranged from 5 to 20 m.

2.2 Fish community structure

Free diving was conducted on all four rocky shores between August 2007 and January 2008 to assess the structure of fish communities and benthic cover. To assess the reef fish species richness and density, 20 visual censuses were randomly performed at each site using strip transects of 30 x 2 m (adapted from BROCK, 1954), to a total of 80 censuses. All the fishes observed were identified to species level in accordance with the specific literature (CARVALHO-FILHO, 1999; HOSTIM-SILVA et al., 2006). Additionally, recent changes in the classification of the Serranidae and Scaridae families were taken into consideration and adopted as proposed by CRAIG and HASTINGS (2007) and WESTNEAT and ALFARO (2005).

Then species were organized in nine trophic categories, following RANDALL (1967), FERREIRA et al. (2004), FROESE and PAULY (2014), as follows: CAR = Carnivores (those feeding on a variety of invertebrates and/or fish), PIS = Piscivores (particularly fish feeders), SIF = Sessile invertebrate feeders (feeding primarily on sessile benthic invertebrates associated with the hard or nearby soft-substrate), MIF = Mobile invertebrate feeders (feeding on a variety of mobile benthic species), OMN = Omnivores (feeding on a variety of organisms, both animal and plant materials), PLANK = Planktivores (feeding primarily on macro- and micro-zooplankton), ECTFE = Ectoparasite feeders (represented by the species *Elacatinus figaro*), RH = Roving herbivorous (large and mobile species of fish which feed primarily on a variety of algae and detritus) and TH = Territorial herbivorous (herbivorous fish which defend their territories against other fishes).

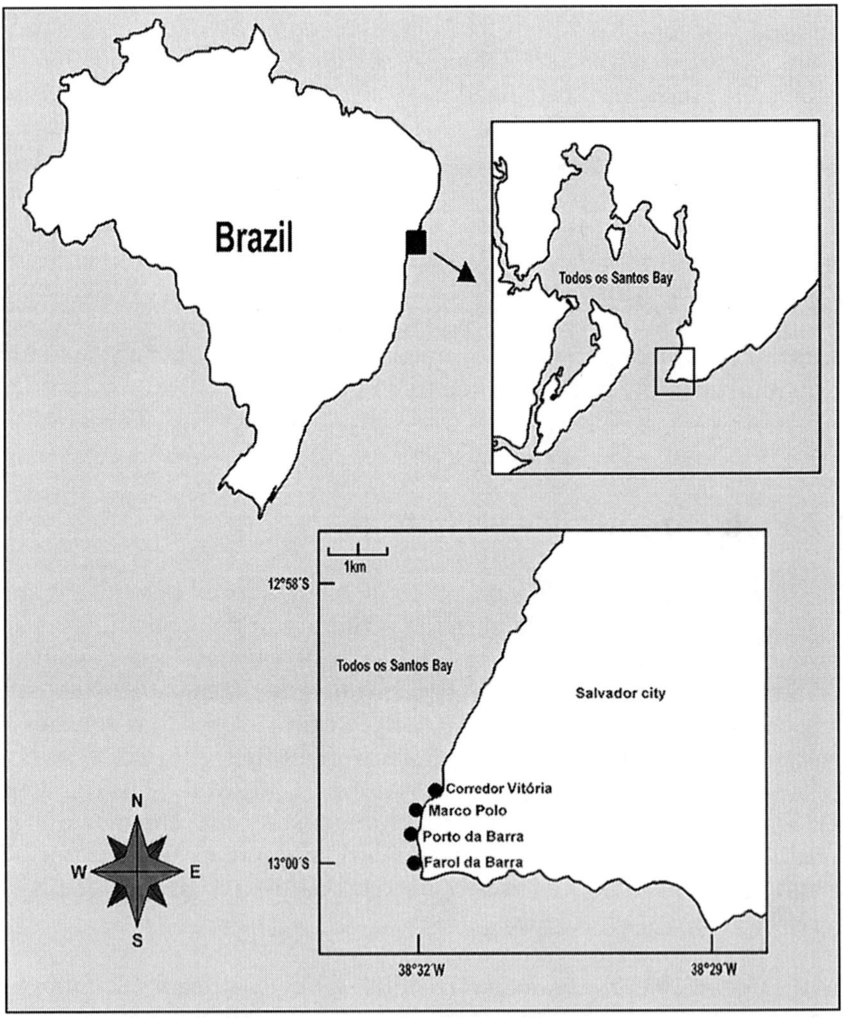

Figure 1 Map of the Todos os Santos Bay, Northeastern Brazil, showing the Rocky shores studied.

2.3 Habitat complexity

Rugosity (topographic complexity) was measured for each site as an index, using the link-chain method proposed by LUCKHURST and LUCKHURST (1978). In each transect the rugosity and depths were calculated from three measurements (taken from the beginning, middle and end of the line). To evaluate the relative benthic cover, the line-point intercept transect method was used (OHLHORST et al., 1988; ROGERS et al., 1994). The number of observations was divided by the total number of points observed, to estimate the relative coverage of benthic groups (CROSBY; REESE, 1996). A total of 480 observations were measured at each location. Only major functional groups of benthic organisms were considered in this study: macroalgae, turf algae (epilithic algae and macroalgae recruits smaller than 5mm), crustose coralline algae, scleractinian corals, fire corals, octocorals, zoanthids, sponges, ascidians, echinoderms, echionoderms and barnacles.

2.4 Data analysis

One-Way Analysis of Variance (ANOVA) was used to evaluate spatial differences on density, trophic structure (through the density of nine trophic groups) and habitat complexity (rugosity, depth and benthic cover). To compare the fish species richness between sites, random boot-strapping permutations of the data (1000 times) were performed to calculate the mean species accumulation curves and their standard deviation (GOTELLI; COLWEL, 2001). Accumulation curves were calculated using the *specaccum* tool, in the *vegan* package (OKSANEN et al., 2013) of the R software. When significant differences were found, Student-Newman-Keuls (SNK) multiple comparisons were used as a *post hoc* test. Densities data were log-transformed [log (x + 1)] and the relative cover was transformed to arcsin ($\sqrt{x+1}$) in order to meet normality and homoscedasticity assumptions (ZAR, 1999).

Multidimensional scaling (MDS) was performed based on an integrating matrix of fish fauna assemblage and benthic cover for all sampling sites. This assisted in summarizing potential similarities on the structure of fish and benthic assemblages over the space (Bray-Curtis). This method is generally the most efficient ordination method for ecological communities (MCCUNE; GRACE, 2002). Additionally, an analysis of similarity (ANOSIM), an effective test towards statistical metrics, was performed to evaluate significant differences between groups of samples corresponding to the four rocky shores sampled. A similarity perceptual test (pairwise ANOSIM) was added in order to clarify the similarity levels amongst sites. Finally, a Canonical Correspondence Analysis (CCA) was used to analyze the relationship between the fish community structure and explanatory habitat variables (rugosity, depth and benthic composition). This ordination represents a powerful technique used to understand how multiple species simultaneously respond to environmental factors, extracting significant gradients from ecological matrices (TER BRAAK; VERDONSCHOT, 1995). To increase the model power, a forward selection procedure was used to include only the most important variables. Finally, the Monte Carlo test (999 permutations) was incorporated in the final model to check if the axes were significant. Species present in less than 5% of the samples were excluded from MDS, ANOSIM, and CCA analysis.

3. RESULTS

3.1 Composition patterns of rocky shore fishes

In the present study, 3,582 fishes belonging to 76 species distributed in 35 families were registered (Table 1), averaging 0.75 individuals per m2. The species accumulation curves demonstrated that the species richness was different between the sites. Marco Polo and Porto da Barra rocky shores showed the highest species richness, followed by Corredor da Vitória and Farol da Barra (MP > PB = CV > FB) (Figure 2).

Regarding fish density, there was a significant difference amongst sampled sites ($F = 2.76$; $p = 0.04$; SNK test: PB = FB > CV > MP) with higher densities at Corredor da Vitória (1.18 ± 0.22 fish.m-2), followed by Porto da Barra (1.01 ± 0.11 fish.m-2), Farol da Barra (0.88 ± 0.09 fish.m-2) and Marco Polo (0.65 ± 0.05 fish.m-2). The ten most abundant species in decreasing order were: *Stegastes fuscus, Ophioblennius trinitatis, Acanthurus bahianus, Haemulon aurolineatum, Abudefduf saxatilis, Acanthurus chirurgus, Pseudupeneus maculatus, Sparisoma axillare, Pempheris schomburgki* and *Labrisomus nuchipinnis*. The Brazilian damselfish *S. fuscus* was clearly the dominating species at all the sites with its higher density found at Corredor da Vitória and Farol da Barra. Following *S. fuscus, O. trinitatis* was the second most dominant fish species at Porto da Barra and Farol da Barra, *H. aurolineatum* at Marco Polo and *A. saxatilis* at Corredor da Vitória (Figure 3). The abundance of the other six species varied moderately amongst sites. The species which were not considered in the graph displayed densities lower than 0.016 fish.m-2.

Table 1 Species list (Mean density (ind.m^{-2}) ± Standard error) observed in the dives, organized in phylogenetic order, according to[Nelson (2006)]

Family	Species	Vitória	Marco Polo	Porto da Barra	Farol da Barra
MURAENIDAE	*Gymnothorax vicinus* (Castelnau. 1855)	0.001 ± 0.001	0.001 ± 0.001	0.001 ± 0.001	-
	Gymnothorax funebris Ranzani. 1840	0.001 ± 0.001	-	0.001 ± 0.001	0.001 ± 0.001
	Gymnothorax moringa (Cuvier.1829)	0.001 ± 0.001	0.001 ± 0.001	0.002 ± 0.001	-
OPHICHTHIDAE	*Ahlia* sp.	-	-	0.001 ± 0.001	-
	Myrichthys breviceps (Richardson. 1848)	0.002 ± 0.001	-	0.002 ± 0.001	-
	Myrichthys ocellatus (Lesueur. 1825)	0.001 ± 0.001	-	0.002 ± 0.001	0.001 ± 0.001
	Ophictis ophis (Linnaeus. 1758)	0.004 ± 0.004	0.001 ± 0.001	-	-
SYNODONTIDAE	*Synodus* sp.	0.003 ± 0.002	0.001 ± 0.001	0.006 ± 0.003	-
OGCOCEPHALIDAE	*Ogcocephalus vespertilio* (Linnaeus. 1758)	0.001 ± 0.001	0.004 ± 0.002	-	-
HOLOCENTRIDAE	*Holocentrus adscensionis* (Osbeck. 1771)	0.003 ± 0.003	0.020 ± 0.009	0.011 ± 0.005	0.002 ± 0.002
	Myripristis jacobus Cuvier. 1829	0.001 ± 0.001	-	0.001 ± 0.001	-
FISTULARIIDAE	*Fistularia tabacaria* Linnaeus. 1758	0.001 ± 0.001	-	0.001 ± 0.001	-
SCORPAENIDAE	*Scorpaena plumieri* Bloch. 1789	-	0.001 ± 0.001	0.001 ± 0.001	0.001 ± 0.001
	Scorpaenodes caribbaeus Meek and Hildebrand.1928	0.003 ± 0.002	0.001 ± 0.001	-	0.001 ± 0.001
SERRANIDAE	*Rypticus saponaceus* (Bloch and Schneider. 1801)	-	0.001 ± 0.001	-	-
	Rypticus subbifrenatus (Gill. 1861)	-	-	0.001 ± 0.001	-
	Rypticus bistrispinus (Mitchill. 1818)	0.002 ± 0.001	0.001 ± 0.001	0.005 ± 0.003	-
	Serranus baldwini (Evermann and Marsch. 1900)	-	0.001 ± 0.001	-	-

EPINEPHELIDAE	*Cephalopholis fulva* (Linnaeus. 1758)	0.016 ± 0.006	0.008 ± 0.004	0.007 ± 0.003	0.003 ± 0.002
	Epinephelus adscensionis (Osbeck. 1765)	-	0.001 ± 0.001	0.002 ± 0.001	0.004 ± 0.002
APOGONIDAE	*Apogon* sp.	-	-	0.002 ± 0.002	-
CARANGIDAE	*Caranx bartholomaei* (Cuvier. 1833)	0.003 ± 0.003	-	-	0.001 ± 0.001
GERREIDAE	*Eucinostomus* sp.	0.002 ± 0.001	0.002 ± 0.001	0.004 ± 0.003	0.008 ± 0.008
HAEMULIDAE	*Anisotremus surinamensis* (Bloch. 1791)	-	0.001 ± 0.001	-	-
	Anisotremus virginicus (Linnaeus. 1758)	-	0.001 ± 0.001	-	0.001 ± 0.001
	Haemulon aurolineatum Cuvier. 1830	0.073 ± 0.068	0.046 ± 0.037	0.048 ± 0.047	-
	Haemulon parra (Desmarest. 1823)	0.006 ± 0.003	0.008 ± 0.007	0.038 ± 0.021	0.006±0.005
	Haemulon steindachneri (Jordan and Gilbert.1882)	0.024 ± 0.006	0.009 ± 0.006	0.003 ± 0.002	-
	Haemulon squamipinna Rocha and Rosa. 1999 *	0.001 ± 0.001	-	-	0.001 ± 0.001

SCIAENIDAE	*Odontoscion dentex* (Cuvier. 1830)	-	-	0.008 ± 0.008	-
	Pareques acuminatus (Bloch and Schneider.1801)	0.034 ± 0.031	0.002 ± 0.002	0.001 ± 0.001	0.002 ± 0.002
MULLIDAE	*Mulloidichthys martinicus* (Cuvier. 1829)	0.001 ± 0.001	-	-	-
	Pseudupeneus maculatus (Bloch. 1793)	0.013 ± 0.006	0.029 ± 0.015	0.029 ± 0.013	0.042 ± 0.024
PEMPHERIDAE	*Pempheris schomburgki* Müller and Troschel.1848	0.016 ± 0.016	-	0.063 ± 0.063	-
CHAETODONTIDAE	*Chaetodon striatus* Linnaeus. 1758	0.011 ± 0.004	0.009 ± 0.003	0.020 ± 0.006	0.010 ± 0.006
POMACANTHIDAE	*Holocanthus tricolor* (Bloch. 1795)	-	0.001 ± 0.001	-	-
	Pomacanthus paru (Bloch. 1787)	-	-	-	0.001 ± 0.001
KYPHOSIDAE	*Kyphosus* sp.	0.004 ± 0.002	0.001 ± 0.001	0.006 ± 0.006	-
CIRRHITIDAE	*Amblycirrhitus pinos* (Mowbray. 1927)	0.002 ± 0.001	0.001 ± 0.001	-	-
POMACENTRIDAE	*Abudefduf saxatilis* (Linnaeus. 1758)	0.019 ± 0.017	0.011 ± 0.008	0.019 ± 0.017	0.013 ± 0.010
	Chromis multilineata (Guichenot. 1853)	-	0.002 ± 0.001	0.036 ± 0.022	-
	Stegastes fuscus (Cuvier. 1830) *	0.572 ± 0.077	0.285 ± 0.070	0.271 ± 0.022	0.393 ± 0.040
	Stegastes pictus (Castelnau. 1855) *	0.003 ± 0.003	-	0.001 ± 0.001	0.003 ± 0.002
	Stegastes variabilis (Castelnau. 1855)	-	0.002 ± 0.001	0.007 ± 0.003	0.016 ± 0.006
LABRIDAE	*Halichoeres brasiliensis* (Bloch. 1791) *	0.004 ± 0.002	0.001 ± 0.001	0.001 ± 0.001	0.003 ± 0.002
	Halichoeres penrosei (Starks. 1913) *	-	0.004 ± 0.002	0.023 ± 0.001	-
	Halichoeres poeyi (Steindachner. 1867)	0.020 ± 0.006	0.014 ± 0.006	0.014 ± 0.006	0.009 ± 0.004
	Thalassoma noronhanum (Boulenger. 1890) *	-	-	0.001 ± 0.001	-
	Xyricthis novacula (Linnaeus. 1758)	-	0.001 ± 0.001	-	-
	Sparisoma amplum (Ranzani. 1842) *	-	0.001 ± 0.001	-	0.002 ± 0.001
	Sparisoma axillare (Steindachner. 1878) *	0.029 ± 0.014	0.027 ± 0.012	0.019 ± 0.008	0.018 ± 0.007
	Sparisoma frondosum (Agassiz. 1831) *	0.003 ± 0.002	-	-	-

	Sparisoma sp.	-	-	0.007 ± 0.006	0.024 ± 0.018
LABRISOMIDAE	*Labrisomus cricota Sazima*. Gasparini and Moura. 2002*	-	-	-	0.001 ± 0.001
	Labrisomus nuchipinnis (Quoy and Gaimard. 1824)	0.013 ± 0.004	0.007 ± 0.005	0.014 ± 0.004	0.038 ± 0.007
	Malacoctenus sp. *	0.016 ± 0.005	0.017 ± 0.005	0.028 ± 0.011	0.043 ± 0.013
CHAENOPSIDAE	*Emblemariopsis signifera* (Ginsburg. 1942)	0.006 ± 0.003	-	-	-
BLENNIIDAE	*Entomacrodus* sp.	-	-	0.002 ± 0.002	0.001 ± 0.001
	Ophioblennius trinitatis Miranda-Ribeiro. 1919 *	0.004 ± 0.003	0.043 ± 0.012	0.167 ± 0.023	0.159 ± 0.028
	Scartella cristata (Linnaeus. 1758)	0.006 ± 0.004	0.005 ± 0.003	0.008 ± 0.004	0.004 ± 0.002
	Parablennius sp.	0.004 ± 0.002	0.001 ± 0.001	0.003 ± 0.002	-
GOBIIDAE	*Bathygobius soporator* (Valenciennes. 1837)	-	-	-	0.001 ± 0.001
	Coryphopterus sp.	-	0.001 ± 0.001	-	-
	Elacatinus figaro Sazima. Moura and Rosa. 1996 *#	-	0.006 ± 0.006	-	-
ACANTHURIDAE	*Acanthurus bahianus* Castelnau. 1855	0.066 ± 0.037	0.023 ± 0.009	0.082 ± 0.019	0.019 ± 0.008
	Acanthurus chirurgus (Bloch. 1787)	0.083 ± 0.083	0.011 ± 0.010	0.005 ± 0.004	0.016 ± 0.011
	Acanthurus coeruleus Bloch and Schneider. 1801	0.001 ± 0.001	0.004 ± 0.003	0.003 ± 0.002	-
SPHYRAENIDAE	*Sphyraena barracuda* (Walbaum. 1792)	-	-	0.002 ± 0.002	-
BOTHIDAE	*Bothus ocellatus* (Agassiz. 1831)	0.003 ± 0.002	0.001 ± 0.001	0.002 ± 0.001	-
MONACANTHIDAE	*Cantherhines pullus* (Ranzani. 1842)	0.002 ± 0.001	0.005 ± 0.003	0.005 ± 0.002	0.002 ± 0.001
OSTRACIIDAE	*Acanthostracion polygonius* Poey. 1876	0.001 ± 0.001	0.001 ± 0.001	0.001 ± 0.001	-
TETRAODONTIDAE	*Canthigaster figueiredoi* Moura and Castro. 2002 *	-	0.003 ± 0.002	-	-
	Sphoeroides greeleyi Gilbert. 1900	0.004 ± 0.003	-	-	0.007 ± 0.006
	Sphoeroides spengleri (Bloch. 1785)	0.011 ± 0.005	0.025 ± 0.012	0.017 ± 0.005	0.023 ± 0.006
DIODONTIDAE	*Chilomycterus spinosus* (Linnaeus. 1758)	-	0.001 ± 0.001	-	-

* species endemic to Brazil; # endangered species.

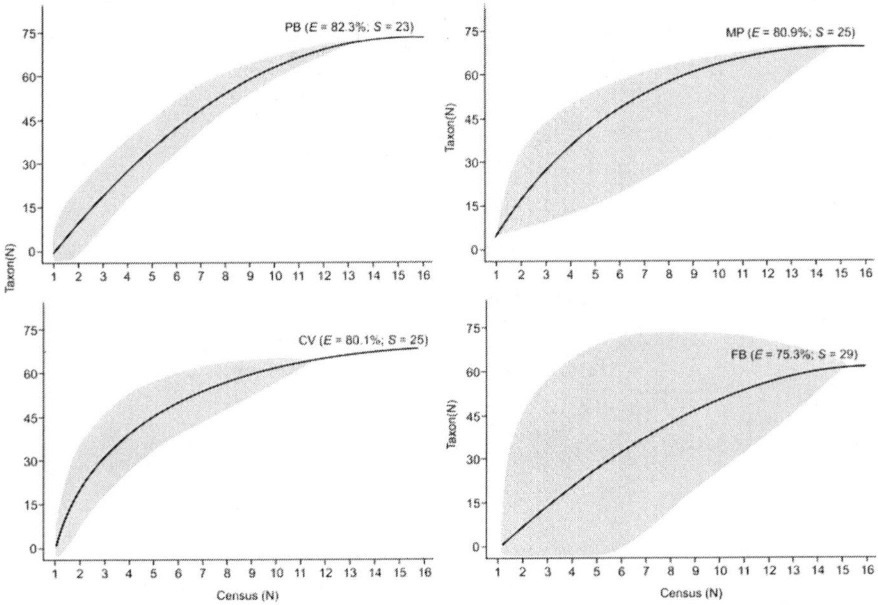

Figure 2 Species accumulation curves obtained by the sixteen underwater visual census in four sampling areas, namely, PB: Porto da Barra; MP: Marco Polo; CV: Corredor da Vitória; FB: Farol da Barra. The shaded area shows the estimated standard deviation. The efficiency of the inventories (*E*), as the percentage of the censused taxa that potentially occur in the area, and the number of samples (*S*) necessary to achieve a complete inventory, are also shown.

The rocky shore fish structure differed significantly between sites (ANOSIM global test: R = 0.196; $p = 0.0001$) and significant pairwise differences between sites were recorded (Table 2). The MDS ordination diagrams showed the formation of at least three different groups in the rocky shore fish communities' structure (Figure 4a). Corredor da Vitória had relatively greater dissimilarity than the other sites, although Marco Polo also presented a quite dissimilar pattern.

All sites were dominated by territorial herbivorous fishes (~52% of fishes), followed by mobile invertebrate feeders (~18%) and roving herbivores (~13%). The other categories were not significantly representative in this study (Figure 5). Only two categories showed significant differences between sites: the territorial herbivores (p = 0.01; F = 3.907) and the piscivores ($p < 0.05$; F = 3.398), with the latter being one of the less representative groups (0.87%). Territorial herbivores, mainly represented by the species *S. fuscus* and *O. trinitatis*, showed the highest density values at the Farol da Barra (65.2%) and Marco Polo (50.7%) reefs. It is important to highlight that only small piscivores were observed in the census, with the lizard fish, *Synodus* sp. being the most dominant species.

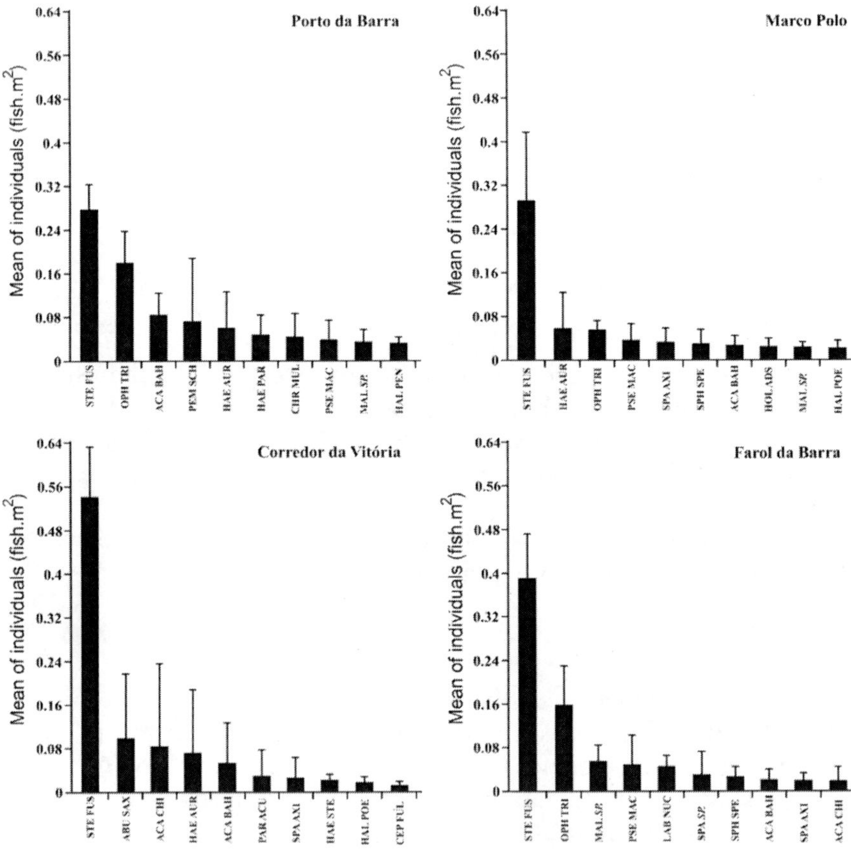

Figure 3 Mean (± Standard deviation - SE) of density of the 10 most abundant species recorded. *Stegastes fuscus, Ophioblennius trinitatis, Acanthurus bahianus, Haemulon aurolineatum, Abudefduf saxatilis, Acanthurus chirurgus, Pseudupeneus maculatus, Sparisoma axillare, Pempheris schomburgki* and *Labrisomus nuchipinnis*.

Table 2 Fish assemblage and environmental variables results of ANOSIM (Global R) and Pairwise test (Significance level)

	Global R	Significance level
Fish assemblage		
PB, MP	0.13	0.002
PB, CV	0.472	0.0001
PB, FB	0.158	0.0004
MP, CV	0.133	0.008
MP, FB	0.109	0.009
CV, FB	0.246	0.0001
Benthic cover		
PB, MP	0.144	0.004
PB, CV	0.845	0.0001
PB, FB	0.187	0.005
MP, CV	0.76	0.0001
MP, FB	0.251	0.0004
CV, FB	0.847	0.0001

PB: Porto da Barra; MP: Marco Pólo; CV: Corredor Vitória; FB: Farol da Barra.

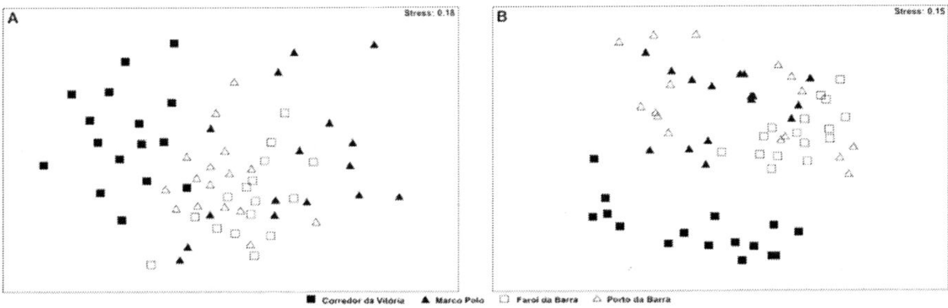

Figure 4 Multidimensional scaling (MDS) based on Bray-Curtis similarity index for: (a) fish communities and; (b) benthic cover compositions.

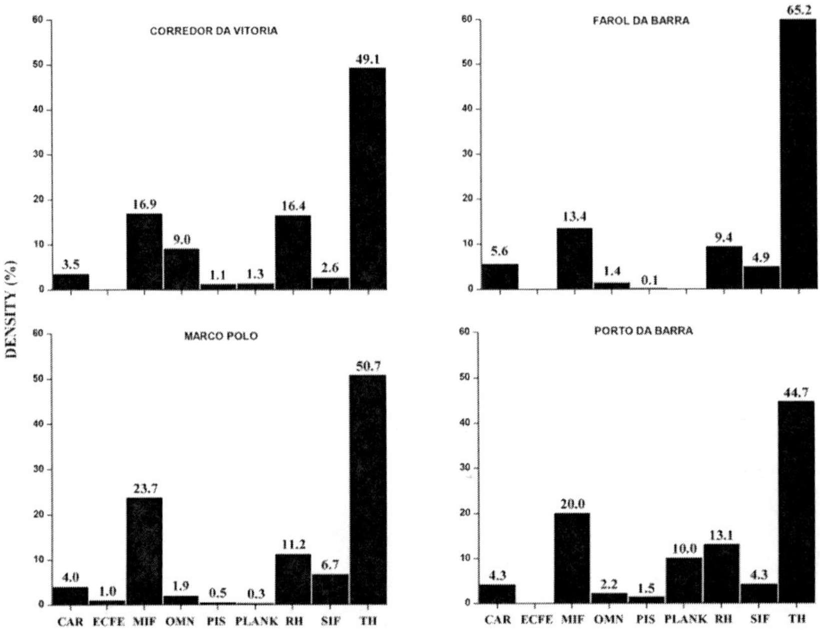

Figure 5 Density (%) of the trophic composition at each sampled site. CAR: Carnivores; ECTFE: Ectoparasite feeders; MIF: Mobile invertebrate feeders; OMN: Omnivores; PIS: Piscivores; PLK: Planktivores; RH: Roving herbivorous; SIF: Sessile invertebrate feeders; TH: Territorial herbivorous.

3.2 Relationships between fish communities and environmental variables

The ANOVA results showed that rocky shores were homogeneous for the variables: roughness, sand, soft coral, echinoderm cover and rocks; and heterogeneous for the variables: depth, turf algae cover, ascidians, crustose algae, zoanthids, sponges, macroalgae, coral and barnacles (Table 3). Regarding the benthic cover, by using a multivariate approach, significant differences were observed between the four rocky shores studied (ANOSIM global test: $R = 0.47$; $p = 0.001$). The comparison of the pairwise ANOSIM test exhibited an analogous pattern to those found for fish structure, with Corredor da Vitória being the site with the greatest dissimilarity (Table 2). The MDS diagram showed the formation of different groups with Corredor da Vitória being less similar to the other sites, confirming the pairwise ANOSIM results (Figure 4b).

Turf, macroalgae, roughness, crustose coralline algae and sponge cover were, in decreasing order, the five most important variables influencing fish community distribution, representing 42.5% of data variability explained on the first axis. The Monte Carlo test showed that the first two axes of the CCA ordination diagram were significant ($p < 0.001$) and accounted for 42.5% and 26.1% respectively, of the variance amongst samples, species and variables (Figure 6). The first axis was positively related to roughness and substrate that was dominated by sponges and macroalgae. The second axis was positively associated with roughness, sponges and turf, and negatively associated with macroalgae and crustose coralline algae. There was a clear distinction in species distribution regarding habitat variables. Turf cover clearly influenced most of the species. *Stegastes fuscus, P. maculatus* and *Eucinostomus* sp. showed a noticeable positive association with turf cover. On the other hand, the distribution and abundance of *A. chirurgus* and *Cephalopholis fulva* were highly influenced by roughness, and *O. trinitatis* distribution was influenced by crustose coralline algae cover.

3.3 Composition and structure patterns of rocky shore fish communities

The present study is a pioneer research in evaluating the structure of rocky reef fish in northeastern Brazil. Other studies within the same region, which have focused on deep reefs, tidal pools, insular and coastal coral reefs (FEITOZA et al., 2005; GODINHO & LOTUFO, 2010; MEDEIROS et al., 2010; CHAVES et al., 2010). The MDS diagram in this study presented a slight overlapping of the fish's composition between sites. It is important to note that the sites studied share high densities of some species, such as *S. fuscus, O. trinitatis* and *A. bahianus*, which likely explains the MDS result. Yet despite this, the sites are remarkably different regarding their abiotic and biotic characteristics, which may be the principal factor responsible for the differences found in the rocky reef fish community structure. Food availability, directly associated with benthic cover variation, is one of the major factors that may influence these differences. However, other factors should also be taken into account, given that all sites show dissimilarities concerning wave exposure, number of visitors (tourists and fishermen and boat traffic.

Table 3 Environmental variables obtained at sample sites. Mean (± S. E.) and results of ANOVA and SNK tests

	PB	MP	CV	FB	F statistic	*p*	SNK
Roughness	1.34 ± 0.22	1.34 ± 0.28	1.29 ± 0.16	1.21 ± 0.12	1.49	0.22	-
Depth	2.2 ± 0.79	3.34 ± 2	2.57 ± 0.94	1.7 ± 0.36	5.41	0.002	MP > CV ≈ PB > FB
Turf algae	34.75 ± 15.68	41.62 ± 23.75	-	37.21 ± 14.84	22.7	0.000	MP > PB ≈ FB > CV
Ascidians	4.15 ± 4.63	4.56 ± 6.30	2.06 ± 2.93	-	4.06	0.011	PB ≈ MP > CV ≈ FB
Crustose algae	7.46 ± 8.46	3.73 ± 5.69	5.81 ± 7.13	12.26 ± 6.96	4.15	0.01	FB > PB ≈ MP ≈ CV
Sand	18.71 ± 11.47	8.53 ± 23.08	4.58 ± 11.92	8.94 ± 13.43	2.34	0.082	-
Soft coral	8.1 ± 12.15	5.4 ± 23.08	5.20 ± 11.92	-	1.98	0.12	-
Equinoderms	6.43 ± 11.01	4.56 ± 11.19	5.83 ± 11.19	7.88 ± 6.64	0.6	0.61	-
Zoanthids	9.13 ± 11.88	6.02 ± 10.03	10.82 ± 13.84	23.71 ± 17.38	5.27	0.003	FB > PB ≈ MP ≈ CV
Sponges	1.03 ± 1.58	1.03 ± 2.32	4.57 ± 6.43	9.98 ± 15.31	4.3	0.008	CV > PB ≈ MP ≈ FB
Macro algae	12.47 ± 13.12	10.82 ± 4.53	3.74 ± 8.60	0.61 ± 1.30	4.3	0.008	CV > PB ≈ MP ≈ FB
Coral	3.31 ± 6.08	7.47 ± 7.82	3.73 ± 4.01	1.03 ± 2.32	3.81	0.014	MP > PB ≈ CV ≈ FB
Rocks	2.07 ± 4.18	7.69 ± 15.28	9.35 ± 12.94	1.24 ± 3.41	2.42	0.74	-
Barnacles	-	5.82 ± 12.78	0.20 ± 0.83	-	3.23	0.028	MP > PB ≈ CV ≈ FB

PB: Porto da Barra; MP: Marco Pólo; CV: Corredor Vitória; FB: Farol da Barra.

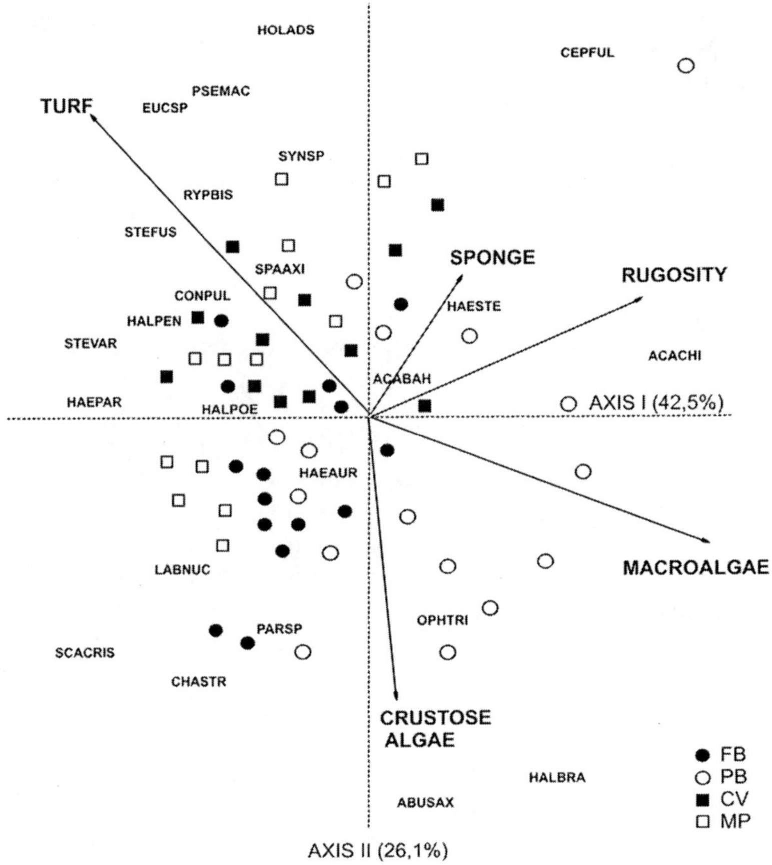

Figure 6 Canonical Correspondence Analysis (CCA) ordination. Species are represented by six letter codes. FB: Farol da Barra; PB: Porto da Barra; CV: Corredor da Vitória; MP: Marco Polo.

Compared to other rocky shores along the Brazilian coast, in either tropical or subtropical regions, the area in which the present study was carried out is much shallower and narrower (FERREIRA et al., 2001; FLOETER et al., 2007; RANGEL et al., 2007; CHAVES; MONTEIRO, 2009). Despite these differences, this study presented a higher total density of fish (0.75 ind.m-2) than did the rocky shores studied by Ferreira et al. (2001) (0.37 ind.m-2) and CHAVES; MONTEIRO (2009) (0.74 ind.m-2), yet a lower density than was observed by RANGEL et al. (2007) (1.83 ind.m-2) andFLOETER et al. (2007) (1.92 ind.m-2). The presence of a clear variation related to synergic features such as coastal proximity, habitat complexity, anthropogenic activities and fish densities might be strongly related to latitudinal gradients (MORA, 2008; CHAVES; MONTEIRO-NETO, 2009; NUNES et al., 2012). Although subtropical areas may have higher species richness than tropical regions in the South Atlantic, tropical rocky shores can potentially sustain a higher variety of reef fish species than was previously thought, and could exceed the number of species found in the southwestern Atlantic's subtropical waters (LUIZ-JR et al., 2008). However, caution should be exercised when drawing comparisons between other studies and the results obtained here, since the previous studies on rocky shore reefs were performed with different scope, methodologies, sampling effort and temporal and spatial standardization scales.

The slight spatial variation in species richness found here might be related to the interaction of several factors. Operating both on micro and macro scales, these factors are responsible for differences in species richness, even in small geographical regions (KARLSON; CORNELL, 1998). An important factor already well documented around the world is the intensity and frequency of the effects of natural disturbances on the local richness (CONNEL, 1978; SOUSA, 1979). In general, environments affected by intermediate disturbances potentially hold a higher richness than those affected more severely. This is due to the propensity of intermediate disturbances to slow down the competitive exclusion, occasionally opening new spaces for colonization (CONNEL, 1978). In addition, natural disturbances have induced the fragmentation and presence of different successional stages, increasing the diversity and availability of microhabitats and consequently the number of species (HUSTON, 1985;CHESSON, 1997; JONES; SYMS, 1998). While microhabitat diversity has not been assessed in this study it is believed that the four rocky shores studied present differences in the frequency not only of natural disturbance (e.g. wave exposure) but also of anthropogenic disturbance (NUNES et al., 2013). Additionally, the differences in the number of species present could also be attributed to this. However, to confirm this hypothesis the intensity and frequency of disturbance should be investigated, seeing that the wave exposure in the study area varies seasonally.

Stegastes fuscus (Pomacentridae) was the most abundant species at all the sites, possibly because this species has a preference for sites with low exposure zones, sheltered places (FLOETER et al., 2007), low abundance of competitors (such as large roving herbivorous fish), and high food availability (turf). The Brazilian endemic and cryptic species *O. trinitatis* (Bleniidae) was the second most abundant fish found at two sites, probably for the same reasons as *S. fuscus*. It is possible to think that herbivorous blennies such as *O. trinitatis* evolved to occupy a different niche to damselfishes. Thus they need not compete for food supplies, instead accessing turf algae areas opportunistically. In this way, these two species would be able to share the same area and increase both their densities. In the coral reefs of Todos os Santos Bay, *Coryphopterus* spp. (Gobiidae) are the dominating ones among the cryptic species, a pattern which can also be observed on other southeastern Brazilian rocky shores (FERREIRA et al., 2001; FLOETER et al., 2007; CHAVES; MONTEIRO-NETO, 2009). In addition, the reason why *H. aurolineatum, A. chirurgus, A. bahianus* and *A. saxatiles* showed high abundance at some study sites could possibly be because these species are recognized by their characteristic behavior of forming large schools on reef structures, seeking protection, food and reproduction areas (FERREIRA et al., 2004; PEREIRA et al., 2011; FRANCINI-FILHO et al., 2012). The geographical position of the site (the Todos os Santos Bay's mouth), could thus contribute to housing a large quantity of reef fish larvae from reefs in the inner portion. This would provide settlement areas for larvae with rapid development and those with very slow growth rates, unable to reach coral reef areas inside Todos os Santos Bay.

Our results demonstrate that territorial fishes are extremely abundant at all sites and the most representative species in this group by far was *Stegastes fuscus* (~78%). It is well known that territorial fish, especially *S. fuscus*, are common inhabitants on tropical and shallow reefs where their main food resource (turf) grows better (FERREIRA et al., 1998; CECCARELLI et al., 2001; FLOETER et al., 2007). Turf algae were the most abundant benthic organisms in all the areas studied, which could explain the dominance of territorial fishes. The second most representative trophic group (MIF) includes fishes adapted to a variety of conditions. First of all, the Haemulidae genus showed the highest abundance (40%), possibly due to their characteristic behavior of forming large and typical schools of up to hundreds of individuals when they are recruits and juveniles (PEREIRA et al., 2011; PEREIRA; FERREIRA, 2013). In addition, some of the sites have an elevated water flow which favors fish species with good swimming skills, such as the members of the Labridae family, which was the second in their contribution to the observed patterns of MIF (FULTON et al., 2001; NUNES et al., 2013). Furthermore, the density of macroinvertebrates inside the *S. fuscus* territory is higher than that outside it, suggesting that the density of MIFs and *S. fuscus* might be directly associated (FERREIRA et al., 1998).

The trophic structure results on the Barra rocky shores are different from those reported by RANGEL et al. (2007), where omnivores obtained the highest densities, a pattern that has been observed in other subtropical Brazilian rocky shore studies (FERREIRA et al., 2004; CHAVES; MONTEIRO-NETO, 2009). FERREIRA et al. (2004) state that roving/territorial herbivores and invertebrate feeders are dominant at lower latitudes. However, MOURA and FRANCINI-FILHO (2006) argue that carnivores (fish and/ or invertebrate feeders) were the most abundant group, followed by omnivore, herbivore and planktivore fish on the southern coast of the state of Bahia. The majority of studies cited in this paragraph, as well as our own, were carried out in areas of intense anthropogenic pressure, with the exception of MOURA and FRANCINI-FILHO (2006), whose study was performed in a Marine Protected Area (the Abrolhos bank). Anthropogenic disturbances are known to have a major impact upon the food web, interfering throughout the whole community structure. Regarding trophic structure, it is likely that our results were influenced by abiotic and biotic characteristics, as the sites are relatively more sheltered and have warmer waters than other rocky shore regions. To illustrate this, the low abundance of the planktivorous fish (*Chromis* spp.) is likely due to the decrease of wave exposure. It is commonly understood that planktivorous fish prefer turbulent waters with a higher availability of plankton (FLOETER et al., 2007).

3.4 Relationship between fish communities and environmental variables

Subtidal benthic communities on rocky shores are likely to have a less diverse algae community in the Todos os Santos bay (MARINS et al., 2008). In the present study, turf algae cover was the dominant benthic organism by far. This same benthic structure pattern had been found by former studies on Brazilian coral reefs (FRANCINI-FILHO; MOURA, 2008; FRANCINI-FILHO et al., 2013) and rocky shores (FERREIRA et al., 2001; FLOETER et al., 2007). Turf algae is recognized as a conspicuous organism in shallow areas due to its dependence upon sunlight (FERREIRA et al., 1998). However, many studies have shown that they can be detrimental to the vitality of coral reefs (VERMEIJ et al., 2010; BAROTT et al., 2012). Nevertheless, these effects remain controversial as they may vary depending on turf and corals species composition (BIRRELL et al., 2005), with some coral species appearing to be unaffected by them (JOMPA; MCCOOK, 2003). Furthermore, turf algae can have positive consequences upon communities, serving as a food resource for many organisms (FERREIRA; GONÇALVES, 2006; FRANCINI-FILHO et al., 2010).

In the CCA results the majority of the observed species were positively associated with turf algae. This association with some species such as *S. fuscus* and *Sparisoma axillare* is to be expected, given that turf algae is their preferred food source (FLOETER et al., 2007; FRANCINI-FILHO et al., 2010). Yet surprisingly, the same pattern also occurred with other species, such as *P. maculatus* and *Eucinostomus* sp. These species are typical fossorial feeders, foraging mostly over sand (CARVALHO-FILHO, 1999). A potential explanation for this pattern is the fact that large areas with an interface of sand and turf substrate are easily seen across the study sites. Another possible explanation is that in association with turf algae, there are many other components, such as invertebrates and microorganisms composition, detritus and sediment (CHOAT, 1991), which could also explain their association with invertebrate feeders (e.g. *Haemulon parrae, Halichoeres poey* and *H. penrosei*).

Another important positive relationship observed was between *Acanthurus* spp. and roughness. *Acanthurus bahianus* and *A. chirurgus* are common inhabitants of complex habitats, such as those with fire corals (CONI et al., 2013). Even though the Barra rocky shore did not have a great amount of fire coral, its roughness can provide sufficient protection for a variety of fishes, such as the carnivorous *C. fulva*. The higher habitat complexity can increase the rates of predation success, due to the amount of prey availability (HIXON, 1993).*Ophioblennius trinitatis* showed a high association with crustose coralline cover mainly at Porto da Barra and Farol da Barra, sites with the highest rates of crustose algae cover (see Table 3). It is likely that this fish scrapes algae microfilm that overgrows crustose coralline algae.

The present results show the high variability between fish community structure and habitat complexity relationship found in reef systems. SALE (1991) observed that structuring habitat factors can differ between locations and that fish communities are not predictable. MEDEIROS et al. (2010), studying Brazilian coastal reefs, found that depth and substrate composition, particularly macroalgae cover, were positive determinants for the fish community structure. According to these authors, the presence of macroalgae is one of the components that enhance habitat complexity, and thus provide shelter for small and/or juvenile individuals of many reef fish species. However, for the four rock shores sampled, food availability (e.g., turf algae) might be the principle factor responsible for the fish community structure.

3.5 Conservation remarks

The rocky shores studied are under intense human pressure, primarily due to pollution, fishing and/or tourism activities (authors, per. obs.). Clear waters, shallow depths and free access encourage intense fishing practices in the region, potentially making it the area most frequented by spear fishermen on the coast of the state of Bahia (NUNES et al., 2012). Currently, reports of marine ornamental fish in these areas are becoming rare, mainly due to the intense capture of these species since the 1970s (SAMPAIO; NOTTHINHAM, 2008). To illustrate this, during the study period, there were only a few reports of the neon goby *Elacatinus figaro*. Due to its bright color, this species is extremely attractive to the ornamental industry, and it is currently classified as endangered by the Brazilian Red List (IUCN) (MACHADO et al., 2008; ROSA; LIMA, 2008; SAMPAIO; NOTTHINHAM, 2008).

Until the mid-1980s, the dog snappers *Lutjanus jocu*, the great barracuda *Sphyraena barracuda*, the stingrays*Dasyatis* spp. and the Atlantic goliath grouper *Ephinephelus itajara* were observed regularly (authors, per. obs.). Currently, just few individuals and/or only juveniles can be sporadically spotted in the area. Another evident reduction is in the biomass of roving herbivorous fishes, such as *Scarus trispinosus*, which is known to be an important functional group keeping the algal and coral populations under control (BELLWOOD et al., 2004; HUGHES, 1994, 2007, 2010; MCCOOK, 1999; SZMANT, 2002; NORSTRÖM et al., 2009). The extirpation of these fish from the reef systems can lead to severe negative consequences (DUFFY, 2003). It is widely accepted that the predator-prey relationship is a key to the ecological balance of the communities, and therefore that the removal of predators from reef environments could affect the entire reef community structure (FLOETER et al., 2007; FRANCINI-FILHO et al., 2010). Undoubtedly, the intense fishing exploitation on the Barra rocky shores has reduced the abundance of competitors and predators (such as roving herbivorous and large carnivorous fish, both fisheries target species), and consequently, increased the abundance of territorial herbivorous and invertebrate feeders (non-fisheries species). The degree of impact this has had upon the rocky shores remains unknown, given that this region has been exploited for approximately five centuries.

Although information regarding the community structure of Brazilian reef fishes is relatively widespread, the majority of the studies undertaken were performed on tropical coral reefs and rocky shores in temperate zones (EBELING; HIXON, 1991). In comparison, there are few studies involving the associations among fishes on tropical rocky shores

(CONI et al., 2007, 2010; NUNES et al., 2007a, 2007b; MEDEIROS et al., 2009a; 2009b; MAIA-NOGUEIRA et al., 2010; MEDEIROS et al., 2011). Thus this region still remains largely unknown. The purpose of this study was to take the initial step to contribute to broadening future studies on shallow coral reefs, shipwrecks and deep reefs near the rocky shores studied here. Ultimately, the northeastern rocky shores deserve special attention, since they are located at the mouth of the Todos os Santos Bay, an area accorded top priority for biological conservation (PASSOS et al., 2002). In addition, these rocky shores lie in a strategic position connecting the internal and external reef environments of the bay and contributing directly to the continuous interchange of species.

ACKNOWLEDGEMENTS

We thank Laís Chaves and Rodrigo Maia-Nogueira for their valuable help during the fieldwork and for kindly reviewing this manuscript. This study received no specific grant from any funding agency, commercial or non-profit sectors. FB was supported by a fellowship PQ-CNPq (Nº 302642/2008-0).

REFERENCES

1. BAROTT, K. L.; WILLIAMS, G. J.; VERMEIJ, M. J.; HARRIS, J.; SMITH, J. E.; ROHWER, F. L.; SANDIN, S. A. Natural history of coral-algae competition across a gradient of human activity in the Line Islands. Mar. Ecol. Prog. Ser., v. 460, p. 1-12, 2012.

2. BELLWOOD, D. R.; HUGHES, T. P.; FOLKE, C.; NYSTRÖM, M. Confronting the coral reef crisis. Nature, v. 429, p. 827-833, 2004.

3. BIRRELL, C. L.; MCCOOK, L. J.; WILLIS, B. L. Effects of algal turfs and sediment on coral settlement. Mar. Pollut. Bull., v. 51, n. 1/4, p. 408-414, 2005.

4. BROCK, V. E. A preliminary report on a method of estimating reef fish populations. J. Wild. Manag., v. 18, n. 3, p. 297-308, 1954.

5. BRUCE, T.; MEIRELLES, P. M.; GARCIA, G.; PARANHOS, R.; REZENDE, C. E.; DE MOURA, R. L.; FRANCINI FILHO, R. B.; CONI, E. O. C.; VASCONCELOS, A. T.; AMADO FILHO, G.; HATAY, M.; SCHMIEDER, R.; EDWARDS, R.; DINSDALE, E.; THOMPSON, F. L. Abrolhos bank reef health evaluated by means of water quality, microbial diversity, benthic cover, and fish biomass data. PLoS ONE, v. 7, n. 6, p. e36687, 2012.

6. CARVALHO-FILHO, A. Peixes: costa brasileira. 3.ed. São Paulo: Melro, 1999. 320 p.

7. CASTRO, C. B.; PIRES, D. O. Brazilian coral reefs: what we already know and what is still missing. Bull. Mar. Sci., v. 69, n. 2, p. 357-371, 2001.

8. CECCARELLI, D. M.; JONES G. P.; MCCOOK, L. J. Territorial damselfishes as determinants of the structure of benthic communities on coral reefs. Oceanogr. Mar. Biol. Ann. Rev., v. 39, p. 355-389, 2001.

9. CHAVES, L. C. T.; MONTEIRO-NETO, C. Comparative analysis of rocky reef fish community structure in coastal islands of south-eastern Brazil. J. Mar. Biol. Assoc. U. K., v. 89, n. 3, p. 609-619, 2009.

10. CHAVES, L. C. T.; NUNES, J. A. C. C.; SAMPAIO, C. L. S. Shallow reef fish communities of south Bahia coast, Brazil. Braz. J. Oceanogr., v. 58, n. spe 4, p. 33-46, 2010.

11. CHESSON, P. L. Diversity maintenance by integration of mechanisms over various scales. Proc. 8th Int. Coral Reef Symp., v. 1, p. 405-410, 1997.

12. CHOAT, J. H. The biology of herbivorous fishes on coral reefs. In: SALE, P. F. (Ed.). The Ecology of Fishes on Coral Reefs. San Diego: Academic press, 1991. p. 120-155.

13. CIRANO, M.; LESSA, G. C. Oceanographic characteristics of Baía de Todos os Santos Bay, Brasil. Braz. J. Geophys., v. 25, n. 4, p. 363-387, 2007.

14. CONI, E. O.; NUNES, J. A. C. C.; SAMPAIO, C. L. S. *Halichoeres penrosei* (Labridae), a sporadic cleaner wrasse. J. Mar. Biol. Assoc. U. K., v. 1, p. 1-2, 2007.

15. CONI, E. O. C.; NUNES, J. A. C. C.; FERREIRA, C. M.; MAIA-NOGUEIRA, R.; MEDEIROS, D. V.; SAMPAIO, C. L. S. The Spanish hogfish *Bodianus rufus* (Labridae) acting as cleaner of nocturnal fish in the north-east of Brazil. Mar. Biod. Rec., v. 3, n. e23, p. 2, 2010.

16. CONI, E. O. C.; FERREIRA, C. M.; MOURA, R. L.; MEIRELLES, P. M.; KAUFMAN, L.; FRANCINI-FILHO, R. B. An evaluation of the use of branching fire-corals (*Millepora* spp.) as refuge by reef fish in the Abrolhos Bank, eastern Brazil. Environ. Biol. Fish., v. 96, n. 1, p. 45-55, 2013.

17. CRAIG, M. T.; HASTINGS, P. A. A molecular phylogeny of the groupers of the subfamily Epinephelinae (Serranidae) with a revised classification of the Epinephelini. Ichthyol. Res., v. 54, n. 1, p. 1-17, 2007.

18. CROSBY, M. P.; REESE, E. S. A Manual for monitoring coral reefs with indicator species: Butterflyfishes as Indicators of Change on Indo-Pacific Reefs. Maryland: Silver Spring, 1996. 47 p.

19. DE RAEDEMAECKER, F.; MILIOU, A.; PERKINS, R. Fish community structure on littoral rocky shores in the Eastern Aegean Sea: Effects of exposure and substratum. Estuar. Coast. Shelf Sci., v. 90, n. 1, p. 35-44, 2010.

20. DUFFY, J. E. Biodiversity loss, trophic skew and ecosystem functioning. Ecol. Lett., v. 6, n. 8, p. 680-687, 2003.

21. EBELING, A. W.; HIXON, M. A. Tropical and Temperate reef fishes: Comparison of Community Structures. In: SALE, P. F. S. (Ed.). The Ecology of Fishes on Coral Reefs. San Diego: Academic press, 1991. p. 509-563.

22. FEITOZA, B. M.; ROSA, E. S.; ROCHA, L. A. Ecology and zoogeography of deep reef fishes in northeastern Brazil. Bull. Mar. Sci., v. 76, n. 3, p. 725-742, 2005.

23. FERREIRA, C. E. L.; GONÇALVES, J. E. A.; COUTINHO, R.; PERET, A. C. Herbivory by the Dusky Damselfish,*Stegastes fuscus* (Cuvier, 1830) in a tropical rocky shore: effects on the benthic community. J. Exp. Mar. Biol. Ecol., v. 229, n. 2, p. 241-264, 1998.

24. FERREIRA, C. E. L.; GONÇALVES, J. E. A.; COUTINHO, R. Fish community structure and habitat complexity in a tropical rocky shore. Environ. Biol. Fish., v. 61, n. 4, p. 353-369, 2001.

25. FERREIRA, C. E. L.; GONÇALVES, J. E. A. Community structure and diet of roving herbivorous fishes in the Abrolhos Archipelago, south-western Atlantic. J. Fish Biol., v. 69, n. 4, p. 1533-1551, 2006.

26. FERREIRA, C. E. L.; FLOETER, S. R.; GASPARINI, J. L.; JOYEUX, J. C.; FERREIRA, B. P. Trophic structure patterns of Brazilian reef fishes: a latitudinal comparison. J. Biogeogr., v. 31, n. 7, p. 1093-1106, 2004.

27. FLOETER, S. R. F.; GASPARINI, J. L. G. The southwestern Atlantic reef fish fauna: composition and zoogeographic patterns. J. Fish Biol., v. 56, n. 5, p. 1099-1114, 2000.

28. FLOETER, S. R.; HALPERN, B. S.; FERREIRA, C. E. L. Effects of fishing and protection on Brazilian reef fishes. Biol. Conserv., v. 128, n. 3, p. 391-402, 2006.

29. FLOETER, S. R.; KROHLING, W.; GASPARINI, J. L.; FERREIRA, C. E. L.; ZALMON, I. Reef fish community structure on coastal islands of the southeastern Brazil: the influence of exposure and cover. Environ. Biol. Fish., v. 78, n. 2, p. 147-160, 2007.

30. FRANCINI-FILHO, R. B.; MOURA, R. L.; THOMPSON, F. L.; REIS, R. M.; KAUFMAN, L.; KIKUCHI, R. K. P.; LEÃO, Z. M. A. N. Diseases leading to accelerated decline of reef corals in the largest South Atlantic reef complex (Abrolhos Bank, eastern Brazil). Mar. Pollut. Bull., v. 56, n. 5, p. 1008-1014, 2008.

31. FRANCINI-FILHO, R. B.; MOURA, R. L. Dynamics of fish assemblages on coral reefs subjected to different management regimes in the Abrolhos Bank, eastern Brazil. Aquat. Conserv., v. 18, n. 7, p. 1166-1179, 2008.

32. FRANCINI-FILHO, R. B.; FERREIRA, C. M.; CONI, E. O. C.; MOURA, R. L. KAUFMAN, L. Foraging activity of roving herbivorous reef fish (Acanthuridae and Scaridae) in eastern Brazil: influence of resource availability and interference competition. J. Mar. Biol. Assoc. U. K., v. 90, n. 3, p. 481-492, 2010.

33. FRANCINI-FILHO, R. B.; CONI, E. O. C.; FERREIRA, C. M.; ALVES, A. C.; AMADO-FILHO, G. Group nest clearing behavior by the sergeant

major Abudefduf saxatilis (Pisces: Pomacentridae). Bull. Mar. Sci., v. 88, n. 2, p. 195-196, 2012.

34. FRANCINI-FILHO, R. B.; CONI, E. O. C.; MEIRELLES, P. M.; AMADO-FILHO, G. M.; THOMPSON, F. L.; PEREIRA-FILHO, G. H.; BASTOS, A. C.; ABRANTES, D. P.; FERREIRA, C. M.; GIBRAN, F. Z.; GÜTH, A. Z.; SUMIDA, P. Y. G.; OLIVEIRA, N. L.; KAUFMAN, L.; MINTE-VERA, C. V.; MOURA, R. L. Dynamics of coral reef benthic assemblages of the Abrolhos Bank, eastern Brazil: inferences on natural and anthropogenic drivers. PLoS ONE, v. 8, n. 1, p. e54260, 2013.

35. FROESE, R.; PAULY, P. FishBase. World Wide Web electronic publication. Available at: < www.fishbase.org >. Accessed: 02/2014.

36. FULTON, C. J., BELLWOOD, D. R.; WAINWRIGHT, P. C. The relationship between swimming ability and habitat use in wrasses (Labridae). Mar. Biol., v. 139, n. 1, p. 25-33, 2001.

37. GODINHO, W. O.; LOTUFO, T. M. C. Local *v.* microhabitat influences on the fish fauna of tidal pools in northeast Brazil. J. Fish Biol., v. 76, n. 3, p. 487-501, 2010.

38. GOTELLI, N.; COLWELL, R. K. Quantifying biodiversity: procedures and pitfalls in the measurement and comparison of species richness. Ecol. Lett., v. 4, n. 4, p. 379-391, 2001.

39. GUST, N. Scarid biomass on the northern Great Barrier Reef: the influence of exposure, depth and substrata. Environ. Biol. Fish., v. 64, n. 4, p. 353-366, 2002.

40. HIXON, M. Predation, prey refuges, and the structure of coral-reef fish assemblages. Ecol. Monogr., v. 63, n. 1, p. 77-101, 1993.

41. HIXON, M. A.; BEETS, J. P. Predation, prey refuges and the structure of coral-reef fish assemblages. Ecol. Monogr., v. 63, n. 1, p. 77-101, 1993.

42. HOSTIM-SILVA, M.; ANDRADE, A. B.; MACHADO, L. F.; GERHARDINGER, L. C.; DAROS, F. A.; BARREIROS, J. P.; GODOY, E. A. S. Peixes de costão rochoso de Santa Catarina: Ilha do Arvoredo. Itajaí: Universidade do Vale do Itajaí, 2006. 134 p.

43. HUGHES, T. P. Catastrophes, phase shifts, and large-scale degradation of a Caribbean coral reef. Science, v. 265, n. 5178, p. 1547-1551, 1994.

44. HUGHES, T. P.; RODRIGUES, M. J.; BELLWOOD, D. R.; CECCARELLI, D.; HOEGH-GULDBERG, O.; MCCOOK, L.; MOLTSCHANIWSKYJ, N.; PRATCHETT, M. S.; STENECK, R. S.; WILLIS, B. Phase shifts, herbivory and the resilience of coral reefs to climate change. Curr. Biol., v. 17, n. 4, p. 360-365, 2007.

45. HUGHES, T. P.; NICHOLAS, A. J.; GRAHAM, J. B. C.; JACKSON, P. J. M.; ROBERT, S. S. Rising to the challenge of sustaining coral reef resilience. Trends Ecol. Evol., v. 25, n. 11, p. 633-641, 2010.

46. HUSTON, M. A. Patterns of species diversity on coral reefs. Annu. Rev. Ecol. Syst., v. 16, p. 149-177, 1985.

47. JOMPA, J.; MCCOOK, L. J. Contrasting effects of turf algae on corals: massive *Porites* spp. are unaffected by mixed-species turfs, but are killed by the red alga Anotrichium tenue. Mar. Ecol. Prog. Ser., v. 258, p. 79-86, 2003.

48. JONES, G. P. Ecology of rocky reef fish of north-eastern New Zealand: a review. N. Z. J. Mar. Freshwater Res., v 22, n. 3, p. 445-462, 1988.

49. JONES, G. P.; SYMS, C. Disturbance, habitat structure and the ecology of fishes on coral reefs. Aust. J. Ecol., v. 23, n. 3, p. 287-297, 1998.

50. KARLSON, R. H.; CORNELL, H. V. Scale-dependent variation in local vs. regional effects on coral species richness. Ecol. Monogr., v. 68, n. 2, p. 259-274, 1998.

51. LEÃO, Z. M. A. N.; KIKUCHI, R. K. P.; TESTA, V. Corals and Coral Reefs of Brazil. In: CORTÉS, J. (Ed.). Latin American Coral Reefs. Amsterdam: Elsevier, 2003. p. 9-52.

52. LESSA, G. C.; BITTENCOURT, A. C. S. P.; BRICHTA, A.; DOMINGUEZ, J. M. L. A reevaluation of the late quaternary sedimentation in Todos os Santos Bay (BA), Brazil. An. Acad. Bras. Cienc., v. 72, n. 4, p. 573-590, 2000.

53. LUCKHURST, B. E.; LUCKHURST, K. Analysis of influence of substrate variables on coral reef fish communities. Mar. Biol., v. 49, p. 317-324, 1978.

54. LUIZ-JR, O. J; CARVALHO-FILHO, A.; FERREIRA, C. E. L.; FLOETER, S. R.; GASPARINI, J. L.; SAZIMA, I. The reef fish assemblage of the Lage dos Santos Marine State Park, Southwestern Atlantic: annotated checklist with comments on abundance, distribution, trophic structure, symbiotic association, and conservation. Zootaxa, v. 1807, p. 1-25, 2008.

55. MACHADO, A. B. M.; DRUMMOND, G. M.; PAGLIA, A. P. (Eds.). Livro Vermelho da Fauna Brasileira Ameaçada de Extinção. Brasília: Ministério do Meio Ambiente, 2008. p. 188-189.

56. MAIA-NOGUEIRA, R.; MEDEIROS, D. V.; NUNES, J. A. C. C.; SAMPAIO, C. L. S. Banded butterflyfish *Chaetodon striatus* (Chaetodontidae) cleaning the green turtle, *Chelonia mydas* (Cheloniidae). Mar. Biod. Rec., v. 3, p. e116, 2010.

57. MARINS, B. V.; BRASILEIRO, P. S.; BARRETO, M. B. B.; NUNES, J. M. C.; YONESHIGUE-VALENTIM, Y.; AMADO FILHO, G. M. Subtidal benthic marine algae of the Todos os Santos Bay, Bahia State, Brazil. Oecol. Bras., v. 12, n. 2, p. 229-242, 2008

58. MCCOOK, L. J. Macroalgae, nutrients and phase shifts on coral reefs: scientific issues and management consequences for the Great Barrier Reef. Coral Reefs, v. 18, n. 4, p. 357-367, 1999.

59. MCCUNE, B.; GRACE, J. B. Analysis of ecological communities. Gleneden Beach: MjM Software Design. 2002. p. 190-191.

60. MEDEIROS, D. V.; NUNES, J. A. C. C.; SAMPAIO, C. L. S. A mutton hamlet *Alphestes afer* (Bloch, 1793) reproductive event in northeast Brazil. Panam. J. Aquat. Sci., v. 4, n. 2, p. 212-215, 2009a.

61. MEDEIROS, D. V.; NUNES, J. A. C. C.; REIS-FILHO, J. A.; SAMPAIO, C. L. S. Yellowline arrow crab*Stenorhynchus seticornis* (Brachyura: Majidae) acting as a cleaner of reef fish, eastern Brazil. Mar. Biodivers. Rec., v. 4, n. e68, 2009b.

62. MEDEIROS, P. R.; GREMPEL, R. G.; SOUZA, A. T.; ILARRI, M. I.; ROSA, R. S. Non-random reef use by fishes at two dominant zones in a tropical, algal-dominated coastal reef. Environ. Biol. Fish., v. 87, n. 3, p. 237-246, 2010.

63. MEDEIROS, P. R.; ROSA, R. S.; FRANCINI-FILHO, R. B. Dynamics of fish assemblages on a continuous rocky reef and adjacent

unconsolidated habitats at Fernando de Noronha Archipelago, tropical western Atlantic. Neotrop. Ichtyol., v. 9, n. 4, p. 869-879, 2011.

64. MOBERG, F.; FOLKE, C. Ecological goods and services of coral reef ecosystems. Ecol. Econ., v. 29, n. 2, p. 215-233, 1999.

65. MORA, C. A clear human footprint in the coral reefs of the Caribbean. Proc. Roy. Soc. Lond. B., v. 275, n. 1636, p. 767-773, 2008.

66. MOURA, R. L. Brazilian reefs as priority areas for biodiversity conservation in the Atlantic Ocean. Proc. 9th Int. Coral Reef Symp., v. 9, n. 2, p. 917-920, 2003.

67. MOURA, R. L.; FRANCINI-FILHO, R. B. Reef and shore fishes of the Abrolhos Region, Brazil. In: DUTRA, G. F.; ALLEN, G. R.; WERNER, T.; MCKENNA, A. S. (Eds.). A Rapid Marine Biodiversity Assessment of the Abrolhos Bank, Bahia, Brazil. Washington: Conservation International, 2006. p. 40-55.

68. NELSON, J. S. Fishes of the World. 4th ed. New York: Wiley, 2006. 601 p.

69. NORSTRÖM, A.V.; NYSTRÖM, M.; LOKRANTZ, J.; FOLKE, C. Alternative states on coral reefs: beyond coral - macroalgal phase shifts. Mar. Ecol. Prog. Ser., v. 376, p. 295-306, 2009.

70. NUNES, J. A. C. C.; CHAVES, L. T.; MAIA-NOGUEIRA, R.; SAMPAIO, C. L. S. Association between juvenile reef fish and the Caribbean reef squid *Sepioteuthis sepioidea* on north-eastern Brazilian coastal reefs. J. Mar. Biol. Assoc. U. K., v. 87, n. 3, p. 761-762, 2007a.

71. NUNES, J. A. C. C.; MAIA-NOGUEIRA, R.; SAMPAIO, C. L. S. "Friendly behavior" between two species of Myrichthys in Brazilian waters. Coral Reefs, v. 26, p. 199, 2007b.

72. NUNES, J. A. C. C.; MEDEIROS, D. V.; REIS-FILHO, J. A.; SAMPAIO, C. L. S.; BARROS, F. Reef fishes captured by recreational spearfishing on reefs of Bahia state, northeast Brazil. Biota. Neotrop., v. 12, n. 1, p. 179-185, 2012.

73. NUNES, J. A. C. C.; SAMPAIO, C. L. S.; BARROS, F. How wave exposure, group size and habitat complexity influence foraging and population

densities in fishes of the genus *Halichoeres* (Perciformes: Labridae) on tropical rocky shores. Mar. Biol., v. 160, n. 9, p. 2383-2394, 2013.

74. OHLHORST, S. L.; LIDDELL, W. D.; TAYLOR, R. J.; TAYLOR, J. M. Evaluation of reef census techniques. Proc. 6th Int. Coral Reef Symp., v. 2, p. 319-324. 1988.

75. OKSANEN, J.; BLANCHET, F. G.; KINDT, R.; LEGENDRE, R.; MINCHIN, P. R.; O'HARA R. B.; SIMPSON, G. L.; SOLYMOS, P.; STEVENS, M. H. H.; WAGNER, H. Vegan: community ecology package, R Package Version 2.0-10, 2013.

76. PANDOLFI, J. M.; BRADBURY, R. H.; SALA, E.; HUGHES, T. P.; BJORNDAL, K. A.; COOKE, R. G.; MCARDLE, D.; MCCLENACHAN, L.; NEWMAN, M. J. H.; PAREDES, G.; WARNER, R. R.; JACKSON, J. B. C. Global trajectories of the long-term decline of coral reef ecosystems. Science, v. 301, n. 5635, p. 955-958, 2003.

77. PASSOS, C. M. O.; PEREIRA, R. M.; MONTEIRO, M. M. Avaliação e ações prioritárias para a conservação da biodiversidade das zonas costeiras marinhas. Brasília: Ministério do Meio Ambiente, 2002.

78. PAULY, D.; CHRISTENSEN, V. Primary production required to sustain global fisheries. Nature, v. 374, p. 255-257, 1995.

79. PEREIRA, P. H. C.; FEITOSA, J. L. L.; FERREIRA, B. P. Mixed-species schooling behavior and protective mimicry involving coral reef fish from the genus *Haemulon* (Haemulidae). Neotrop. Ichthyol., v. 9, n. 4, p. 741-746, 2011.

80. PEREIRA P. H. C.; FERREIRA, B. P. Effects of life phase and schooling patterns on the foraging behaviour of coral-reef fishes from the genus Haemulon. J. Fish. Biol., v. 82, n. 4, p. 1226-1238, 2013.

81. PEREIRA-FILHO, G. H.; AMADO-FILHO, G. M.; GUIMARÃES, S. M. P. B.; MOURA, R. L.; SUMIDA, P. Y. G.; ABRANTES, D. P.; BAHIA, R. G.; GÜTH, A. Z.; JORGE, R. R.; FRANCINI FILHO, R. B. Reef fish and benthic assemblages of the Trindade and Martin Vaz Island group, southwestern Atlantic. Braz. J. Oceanogr., v. 59, n. 3, p. 201-212, 2011.

82. RANDALL, J. E. Food habits of reef fishes of the West Indies. Stud. Trop. Oceanogr., v. 5, p. 665-847, 1967.

83. RANGEL, C. A.; CHAVES, L. C. T.; MONTEIRO-NETO, C. Baseline assessment of the reef fish assemblage from Cagarras archipelago, Rio de Janeiro, southeastern Brazil. Braz. J. Oceanogr., v. 55, n. 1, p. 7-17, 2007.

84. REAKA-KUDLA, M. L. The global biodiversity of coral reefs: a comparison with rain forests. In: REAKA-KUDLA, M. L.; WILSON, D. E.; WILSON, E. O. (Eds.). Biodiversity II. Washington: Joseph Henry Press, 1997. pp 83-108.

85. REIS-FILHO, J. A.; BARROS, F.; NUNES, J. A. C. C.; FERREIRA, A. Estuarine ichthyofauna of the Paraguaçu River, Todos os Santos Bay, Bahia, Brazil. Biota Neotrop., v. 10, n. 4, p. 301-311, 2010.

86. REIS-FILHO, J. A.; BARROS, F; NUNES, J. A. C. C.; SAMPAIO, C. L. S.; SOUZA, G. B. G. Moon and tide effects on fish capture in a tropical tidal flat. J. Mar. Biol. Assoc. U. K., v. 1, p. 1-9, 2011.

87. ROGERS, C. S.; GARRISON, G.; GROBER, R.; HILLIS, Z. M.; FRANKE, M. A. Coral Reef Monitoring Manual for the Caribbean and Western Atlantic. St John: National Park Service, 1994. 114 p.

88. ROSA, R. S.; LIMA, F. C. T. Peixes. In: BARBOSA, A.; MACHADO, M.; DRUMMOND, G. M.; PAGLIA, A. P. (Eds.). Livro vermelho da fauna brasileira ameaçada de extinção. Belo Horizonte: Fundação Biodiversitas, 2008. p. 8-285.

89. RUSS, G. R. Coral reef fisheries: effects and yields. In: SALE, P. F. (Ed.). The Ecology of Fishes on Coral Reefs. San Diego: Academic press, 1991. p. 601-635.

90. RUSS, G. R.; ALCALA, A. C. Management histories of Sumilon and Apo Marine Reserves, Philippines, and their influence on national marine resource policy. Coral Reefs, v. 18, n. 4, p. 307-319, 1999.

91. SALE, P. F. The ecology of fishes on coral reefs. London: Academic Press, 1991. 754 p.

92. SAMPAIO, C. L. S.; NOTTINGHAM, M. C. Guia para identificação de peixes ornamentais. Brasília: Ibama, 2008. 205 p.

93. SOUSA, W. P. Disturbance in marine intertidal boulder fields: the noneequilibrium maintenance of species diversity. Ecology, v. 60, n. 6, p. 1225-1239, 1979.

94. SZMANT, A. M. Nutrient enrichment on coral reefs: Is it a major cause of coral reef decline? Estuaries, v. 25, n. 4, p. 743-766, 2002.

95. TER BRAAK, C. J. F.; VERDONSCHOT, P. F. M. Canonical correspondence analysis and related multivariate methods in aquatic ecology. Aquat. Sci., v. 57, n. 3, p. 255-289, 1995.

96. VERMEIJ, M. J. A.; VAN MOORSELAAR, I.; ENGELHARD, S.; HÖRNLEIN, C.; VONK, S. M.; VISSER, P. M. The effects of nutrient enrichment and herbivore abundance on the ability of turf algae to overgrow coral in the Caribbean. PLoS ONE, v. 5, n. 12, p. e14312, 2010.

97. WESTNEAT, M. W.; ALFARO, M. E. Phylogenetic relationships and evolutionary of the reef fish family Labridae. Mol. Phylogenet. Evol., v. 36, n. 2, p. 370-390, 2005.

98. WILKINSON, C. Status of coral reefs of the world: 2008. Global Coral Reef Monitoring Network. Townsville: Australian Institute of Marine Science, 2008. 304 p.

99. WILSON, S. K.; GRAHAM, N. A. J.; POLUNIN, N. V. C. Appraisal of visual assessments of habitat complexity and benthic composition on coral reefs. Mar. Biol., v. 51, n. 3, p. 1069-1076, 2007.

100. ZAR, J. H. Biostatistical analysis. 4th ed. New Jersey: Prentice-Hall, 1999. 663 p.

EOCENE (LUTETIAN) SHARK-RICH COASTAL PALEOENVIRONMENTS OF THE SOUTHERN NORTH SEA BASIN IN EUROPE: BIODIVERSITY OF THE MARINE FÜRSTENAU FORMATION INCLUDING EARLY WHITE AND MEGATOOTH SHARKS

C. G. Diedrich

PaleoLogic, Nansenstraße 8, D-33790 Halle, Germany

ABSTRACT

The Fürstenau Formation (Lutetian, Paleogene, Eocene) is based on type sections near Fürstenau in Germany (central Europe) and is built of 22 meter thick marine glauconitic and strongly bioturbated sands, clays, and a vertebrate-rich conglomerate bed. The conglomerate layer from the Early Lutetian transgression reworked Lower Cretaceous, and Paleogene marine sediments. It is dominated by pebbles from the locally mountains which must have been transported by an ancient river in a delta fan. Marine reworked Lower Cretaceous and Paleogen pebbles/fossils, were derived from the underlying deposits of northern Germany (= southern pre North Sea basin). The benthic macrofauna is cold upwelling water influenced and non-tropical, and medium divers. The vertebrate fish fauna is extremely rich in shark teeth, with about 5,000 teeth per cubic meter of gravel. The most dominant forms are teeth from sand shark ancestors Striatolamia macrota, followed by white shark ancestors Carcharodon auriculatus. Even teeth from the magatooth shark ancestor Carcharocles sokoloviare present in a moderately diverse and condensed Paleogene fish fauna that also includes rays, chimaeras, and more then 80 different bony fish. Fragmentary turtle remains are present, and few terrestrial vertebrates and even marine mammals with phocids, sirenians and possibly whales.

1. INTRODUCTION

Early research in the Dalum fossil locality in the Fürstenau region of north-west Germany (Figure 1(a)) concentrated largely on the stratigraphy and the fossil shark tooth content [1]. A preliminary analysis of the conglomerate components has also been published by Bartholomäus [2]. The site has only been excavated by private collectors, who have been active there since 1971 [2]. Whereas aragonite and calcite fossils at Dalum have all been dissolved away, they are still preserved in a sand pit at Osteroden (Figure 1(b)) from which 11,000 fish otoliths have previously been analysed by Schwarzhans [3]. The extremely rare teeth of terrestrial mammals from the Dalum and Osteroden sites were the focus of a paper by Franzen and Mörs [4].

The Paleogene (Middle Eocene) marine localities at Dalum and Osteroden (Figures 1(a) and 1(b)), in the southern pre-North Sea basin, represent an important environmental bridge between marine, coastal, deltaic swamp, and terrestrial faunas (Figure 1(c)). The marine-dominated fossils at Dalum also offer valuable information on the paleobiodiversity of the central European tropical terrestrial faunas during the Eocene.

Other important marine Eocene sites in Europe are at Helmsleben in Germany, and at Oosterzele and Balegem in Belgium (Figure 1(c)), which are also well known for their enormous quantities of shark teeth [5] and which also contain rare remains of terrestrial mammals [4]. The marine Paris Basin to the south of the Artois submarine swell (Figure 1(c)) is also one of the classic fossil regions of Europe in which many of the shark species presented herein were first found, as described in the monograph by Lériche [6], as well as by other authors. Finally, the Lower Eocene London Clay Formation in England, for example, on the Isle of Sheppey (Figure 1(c)), represents near-shore, fresh-water-influenced, coastal paleoenvironments that contain overlapping terrestrial and marine faunal remains [7].

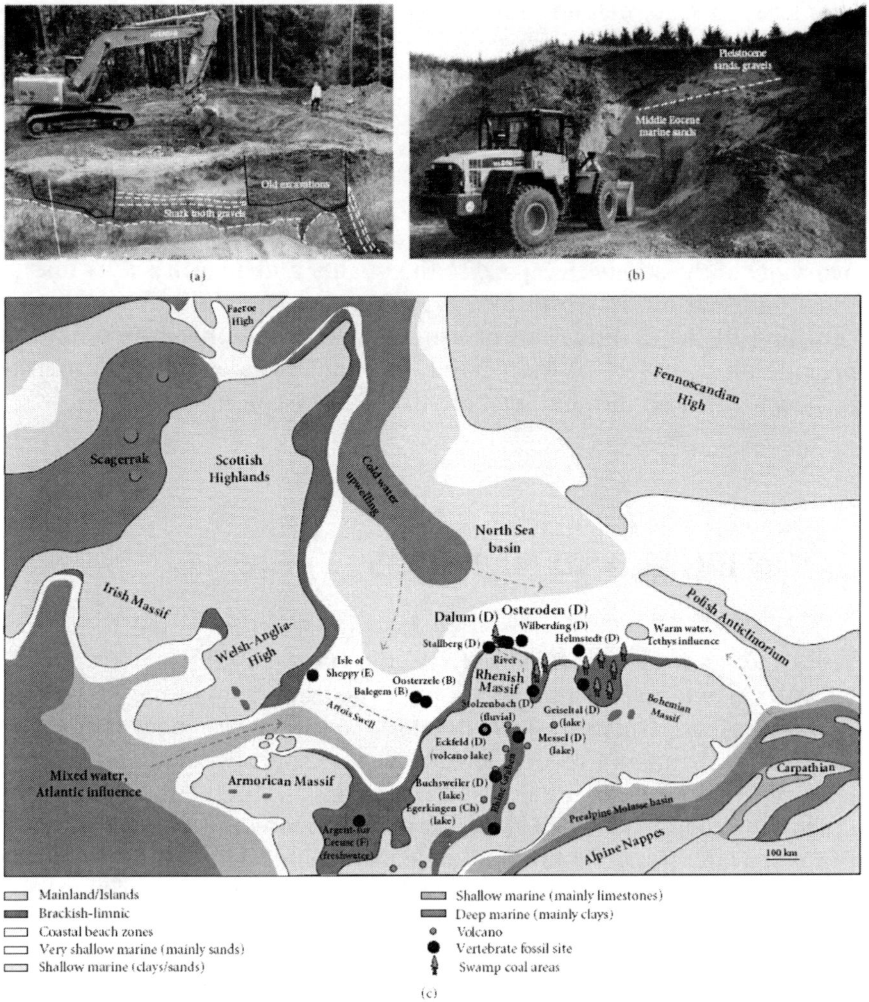

Figure 1: (a) The Dalum site near Fürstenau and (b) the Osteroden sand pit near Bippen, in north-western Germany. (c) Paleogeography of important Eocene marine and terrestrial vertebrate fossil sites of Europe (compiled from [4, 5, 9–14]).

The Eocene terrestrial fauna (Figure 1(c)) was made famous by the abundance of early mammal, bird, reptile, anuran, and fish skeletons, as well as botanical specimens and insects, from the German at UNESCO World Heritage Messel sites at Geiseltal (an Early to Middle Eocene lake site) [8] and at Messel (Middle Eocene) [9]. The Eckfeld Maar volcano lake site has also yielded important fossil material of terrestrial fauna and flora from the Middle Eocene [10], as have the Eocene coal mine at Stolzenbach [11] and the Swiss mammal site at Egerkingen, which is of Upper Eocene of age [12].

This study presents preliminary stratigraphic and sedimentological analyses focussed on the marine, terrestrially influenced, vertebrate-rich gravels at Dalum, and includes a preliminary statistically based overview of the site with regard to the quality and quantity of its fossils. This information is used for a paleoenvironmental reconstruction resulting in the compilation of a new paleogeographical map, and to provide an overview of the (mainly marine) biodiversity, and marine palaeocurrents on the southern pre North Sea basin.

2. MATERIAL AND METHODS

In May and June of 2011 the UNESCO-supported "Geo and Naturpark TERRA.Vita" near Osnabrück (NW Germany) started a systematic excavation and research project (under the leadership of the PaleoLogic private research institute) at the Dalum forest site near Fürstenau (Figure 1(a)) and at a second locality in the Osteroden sand pit, Bippen (Figure 1(b)). The eventual objective, following the completion of the research program, was to develop and promote the Shark Center at Bippen, which is a visitor and research center within the UNESCO-supported "Geo and Naturpark TERRA.Vita" of north-west Germany.

In May 2011 a 4-day field program using heavy machinery was completed at the Dalum and Osteroden sites (Figures 1(a) and 1(b)). Stratigraphic cross-sections were prepared at both of these sites (making use of 5-meter drillholes at Osteroden) indicating a combined 22-meter thick marine section (Figure 3).

A total of 180 cubic meters of conglomeratic material was excavated from the Dalum site for sieving, of which only 250 buckets (10 litres each) were sieved during the following two weeks (i.e., only 0.1% of the recovered material). It was only possible to excavate 10 cubic meters of material at the Osteroden site, which has not yet been sieved (other than for a preliminary sample) because of the intense weathering and damage shown by all of the fossils, and the large degree of root dissolution in the teeth.

Attention has therefore been focussed on the Dalum gravel, which was sieved into two different fractions of +4 mm and −4 to +1 mm, in order to obtain a good variety of gravel types (Figures 4–5), invertebrate remains (Figures 6–8), and vertebrate remains (Figures 9–15), and to facilitate the gravel pebble analysis. Only material greater then 4 mm is presented herein, which is adequate for qualitative and quantitative gravel analyses and for analysis of the main vertebrate fauna. In total, 14,437 fossils were recovered and studied, of which 95% are shark teeth and only 5% (747 specimens) are teeth from other types of animals (Figure 5).

More than half of the 13,683 shark teeth recovered were identifiable, but lower proportions of the ray teeth and fin spines and the bone fish teeth and bones. The finer material (less than 4 mm) appears (after sieving a preliminary sample) to contain many additional species, which will be recovered and examined during the coming years from the retained fine fraction of the gravel. A separate, more detailed fish analysis (covering all material >1 mm) will also be the subject of future research, after much more material has been recovered to allow a more complete picture of the fish and invertebrate paleobiodiversity in the pre-North Sea basin during the Eocene.

All the material described and illustrated herein from the 2011 excavations (14,440 fossils in total) is now housed in the Shark Center, Bippen (SCB).

3. GEOLOGY AND SEDIMENTOLOGY

3.1. Glacial Tectonics

The Tertiary sediments in north-west Germany were deformed and compressed as a result of Middle Pleistocene glacial tectonics [13]. The Ankumer range in particular is the result of proglacial folding due to a Saalian glacier, as indicated by a U-shaped moraine that extends as far as the north-western part of the Wiehengebirge mountain range near Bramsche (Figure 2()) [15]. The Eocene sediments at the Dalum (Figure2()) and Osteroden sites are therefore strongly folded. The fold axis at Dalum is oriented east to west (parallel to the main glacial fold axis, cf. Figure 2()), with a “saddle” that dips at about 10–15° towards the east. In a road cutting about 50 meters to the east of the new sampling trench and parallel to it (Figure 2()) the top of the saddle structure is missing in the road section [1] having been eroded during the Saalian and further the Weichselian Ice Age. At the new section in the forest instead the top of the saddle is preserved and the convex deformed sediment layers that would normally be expected are more or less horizontal (Figure 2()), indicating that the saddle has been compressed from above. At Osteroden the deformation is more complex, and in the recent pit the gravel layers dip at about 30–60° to the north-east. This is also parallel to the main axis of glacial deformation (Figure 2).

3.2. Paleogene Stratigraphy, Sedimentology, and Dating in North-West Germany

No firmly dated early Tertiary (Paleocene) deposits are known in north-west Germany [13].

Early Eocene (Ypresian) marine sediments have, however, been dated by microfossils such as diatoms, radiolarians, and foraminifers [16, 17], and are known from a clay pit at Wilberding, near Damme in north-west Germany [18], and also from another clay pit at Stallberg near Emsbüren (Figure 1(c)) which contains shark remains and reworked iron-carbonate concretions from the Lower Cretaceous [19]. These gray-green, glauconitic marine clays of Wilberding and Stallberg containing phosphorite nodules (phosphorites), that appear to be of Lower Eocene age, represent the earliest known marine Tertiary sediments in north-west Germany. The Stalberg section also contained many mixed layers 1 mm to 40 mm thick of volcanic ash (tuff) and marine sediments, which are explained as being the result of volcanic ash clouds from the marine strait Skagerrak (possibly also from the Eifel volcanos, e.g., the Eckfelder Maar, cf. Figure 1(c)) [20]. This was a tectonically active period that was also responsible for the Rhine Graben structure of central Europe (Figure1(c)) [20], and it was also the time when the Messel and Geiseltal fossil sites were formed in Germany, together with the Stolzenbach site and the coastal swamps of the Rhenish Massif (Figure 1(c)).

Middle Eocene (Lutetian) marine deposits for the Dalum and Osteroden sites are presented herein, and a new formation name introduced. A compilation of the two sections from Dalum and Osteroden (Figure 3) has allowed the dating of these sediments, and also enabled sedimentological and paleoenvironmental interpretations to be made.

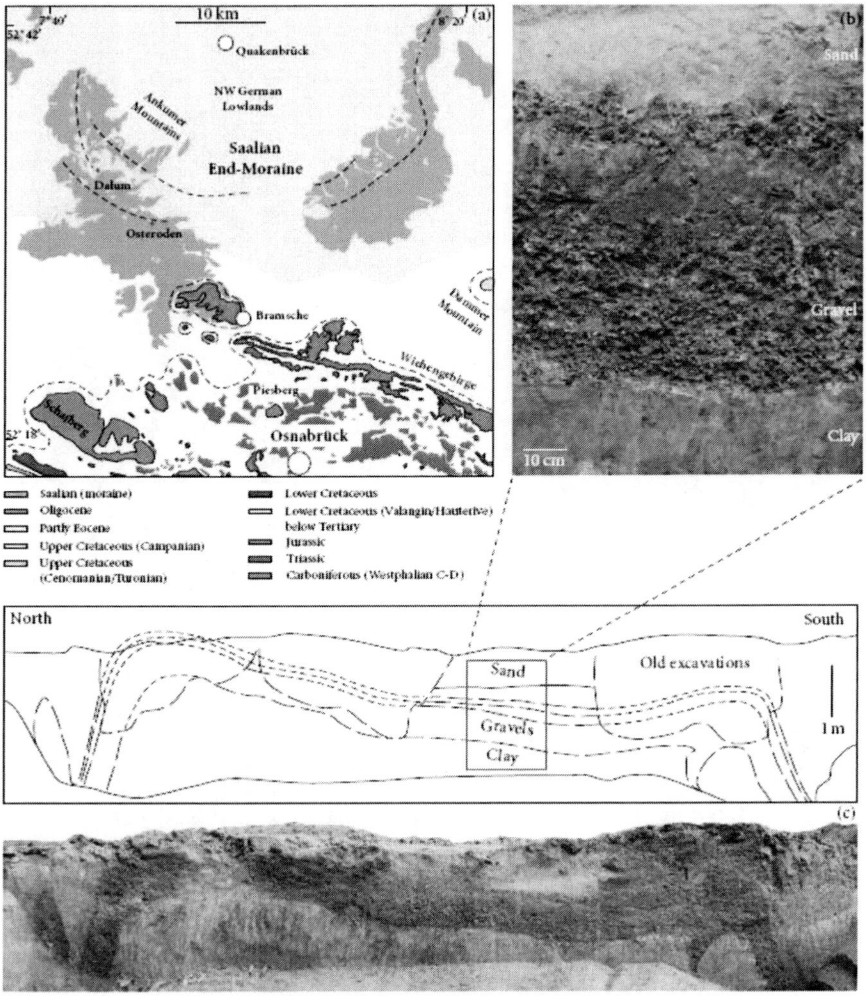

Figure 2: Glacial tectonics and distribution of Tertiary deposits as result of (a) the Saalian Ice Age end-moraine forming the Ankumer Mountains (geology redrawn and simplified from [15]), including the Eocene sites at Dalum and Osteroden; (b) north-south section through the gravels at Dalum; (c) photo of the same section at Dalum, showing the Eocene marine deposits, which where deformed into a saddle-like structure (well visible in the Eocene gravel layer) by glacial compression.

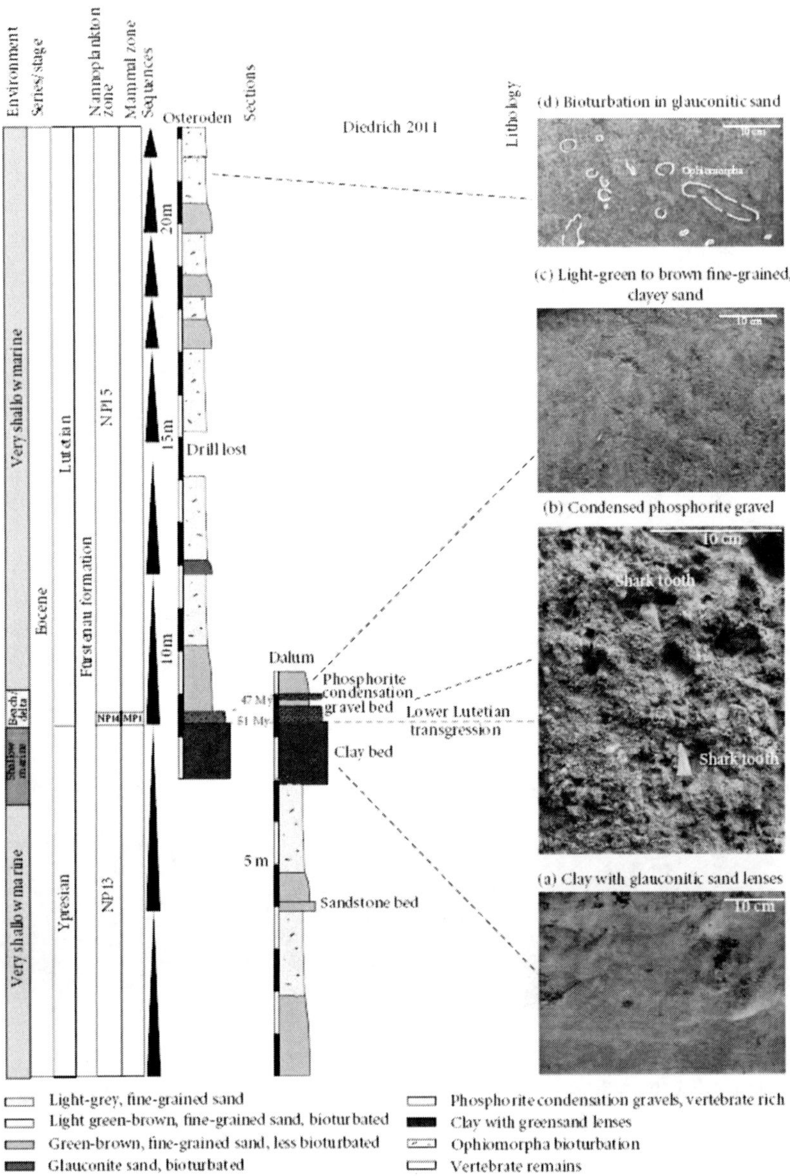

Figure 3: Stratigraphy and sedimentology of the Eocene Fürstenau Formation vertebrate sites at Osteroden and Dalum (NW Germany). (a) Clay with glauconitic sand lenses. (b) Condensed phosphorite gravel bed with abundant vertebrate remains, especially shark teeth. (c) Light-green to brown fine-grained, clayey sand, at the base of the sequence. (d) Bioturbated glauconitic sands, with Ophiomorpha and other burrows (top of sequence).

Figure 4: Middle Eocene (basal Lutetian) gravel types. (1–3(?4)) Carboniferous (Westphalian C-D, originating from the Schafberg and Piesberg mountains) quartzite variants which were already well rounded in the Paleozoic, and have only been transported a short distance since then (SCB no. Sedim-2). (5) Carboniferous conglomerate (Westphalian C-D), originating from the Schafberg and Piesberg mountains (SCB no. Sedim-3). (6–8) Upper Cretaceous (Campanian-Maastrichtian) flints (SCB no. Sedim-6), consisting of Campanian "Rijckholt" flint (6-7) from the Aachen region and northern Netherlands (SCB no. Sedim-6a) and "Maas-Egg," black flint (8) from the Maastrichtian of the northern Netherlands (SCB no. Sedim-6b). (9) Paleozoic metamorphite of uncertain origin (SCB no. Sedim-10), (10–12) red Paleozoic volcanic rocks of uncertain origin, possibly from Scandinavia (SCB no. Sedim-13), (13) Paleogene white fine-grained

sandstone of uncertain origin (SCB no. Sedim-9), (14-15) Paleogen glauconitic fine-grained sandstones (SCB no. Sedim-11), and (16) Jurassic silica/iron-carbonate concretion from the Osnabrücker Bergland (SCB no. Sedim-17). (18) Jurassic or Lower Cretaceous serpulid (SCB no. Sedim-4). (19) Iron-carbonate concretion fragment with orthicon (Bochianites) ammonite negatives from the Valanginian/Hauterivian, Lower Cretaceous (SCB no. Sedim-18). (20) Middle Eocene phosphorite nodule (SCB no. Sedim-7). (21-22) Eocene phosphatic shark coprolites (SCB no. Sel-890, 898). (23) Upper Cretaceous (Campanian) platy flint (SCB no. Sedim-8). (24) Shark tooth from Squalicorax kaupi, reworked from the Campanian, Upper Cretaceous (SCB Dal-Sel-901), labial. (25) Shark tooth from Cretolamna appendiculata, reworked from the Santonian-Campanian, Upper Cretaceous (SCB Dal-Sel-882), labial. (26) Shark tooth from Cretolamna appendiculata,reworked from the Campanian, Upper Cretaceous (SCB Dal-Sel-881), labial. (27) Lower Cretaceous (Upper Valanginian) orthicon ammonite fragment of (Bochianites sp. (SCB no. Sedim-18). (28) Lower Cretaceous (Upper Valanginian) orthicon ammonite fragment, a so-called “cat paw” (SCB no. Sedim-17). (29) Lower Cretaceous (Lower Hauteriveian) ammonite fragment of Endemoceras sp. (SCB no. Sedim-22). (30) Lower Cretaceous (Upper Hauteriveian) ammonite fragment of Aegocrioceras sp. (SCB no. Sedim-22). (31) Lower Cretaceous (Upper Valanginian) ammonite fragment of Oostrella sp. (SCB no. Sedim-18). (32) Lower Cretaceous elongated burrow preserved in iron-carbonate concretion (SCB no. Sedim-43).

3.3. The New “Fürstenau Formation” (Ypresian to Lutetian, Early to Middle Eocene)

3.3.1. Type Stratum

Fürstenau Formation, Early to Middle Eocene, after the nearby city of Fürstenau.

3.3.2. Type Localities

The Dalum site near Fürstenau and the Osteroden sand pit near Bippen, north-west Germany.

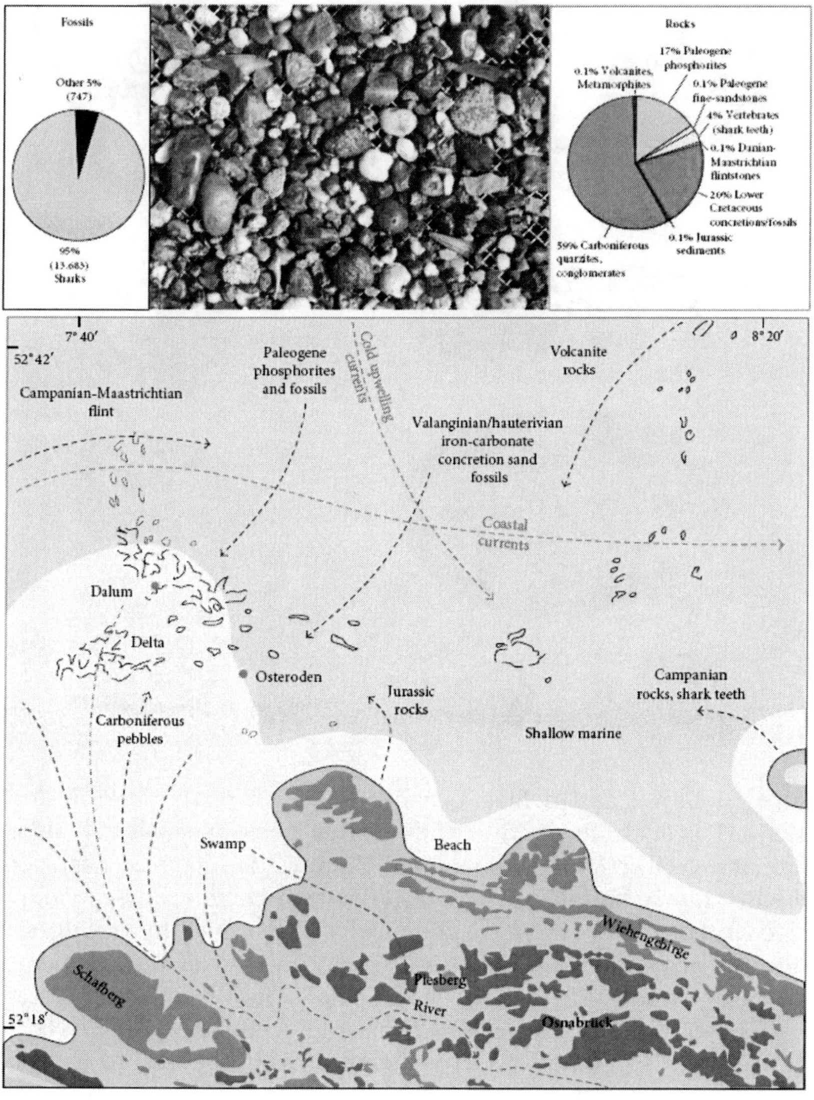

Figure 5. Basal Middle Eocene (Early Lutetian) gravel: quantitative analysis and transportation direction reconstruction. The main Paleozoic components (Carboniferous quartz) must have been transported by an ancient river running between today's Wiehengebirge and Teutoburger Wald mountain ranges, ending in a delta. Other components (Jurassic material) are from the Wiehengebirge range, whereas Lower Cretaceous pebbles and ammonite fossils are from north Germany. Flints were accumulated from different directions (Maastricht/Aachen area, and Dammer Berge; Material no. SCB Dal-Sedim-1–16).

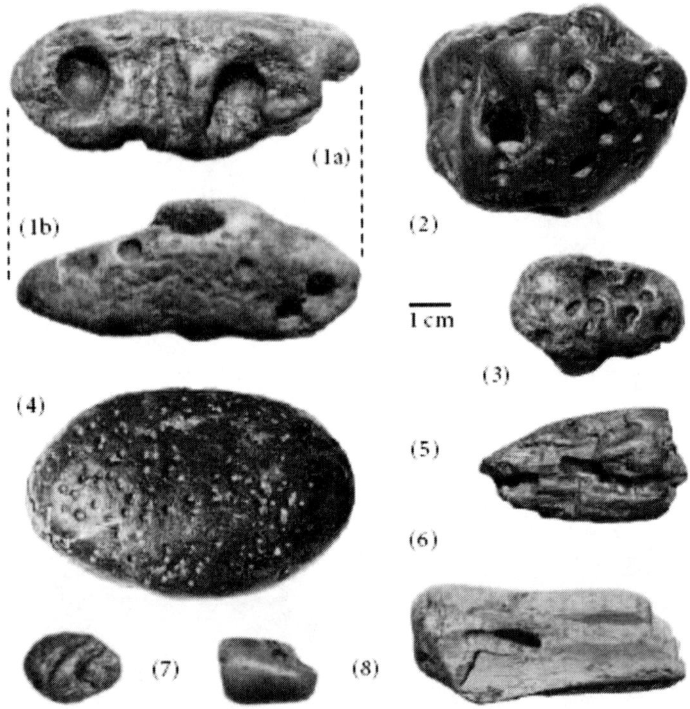

Figure 6: Bored and drilled beach pebbles from the Middle Eocene (Lutetian) gravels at Dalum. Only softer rock types, such as iron-carbonate concretions (Valanginian-Hauterivian, Lower Cretaceous) and phosphorite nodules, or wood (Paleocene-Eocene, Paleogene) were drilled by bivalves or were chemically dissoluted by sponges. (1) LargeGastrochaenolites boreholes in reworked Lower Cretaceous iron-carbonate concretion pebble (SCB no. Dal-Invert-2). (2) Small Gastrochaenolithes boreholes in reworked Eocene phosphorite (SCB no. Dal-Invert-1). (3) Small Gastrochaenolithes boreholes in reworked Lower Cretaceous iron-carbonate concretion pebble (SCB no. Dal-Invert-3). (4) Small boreholes made by Clione-type sponges in reworked Eocene phosphorite (SCB no. Dal-Invert-4). (5) Boreholes of the bivalve Teredo sp. in Paleogene (Eocene) silicified driftwood (SCB no. Dal-Invert-6). (6) Boreholes of the bivalve Teredo sp. in silicificated Eocene driftwood pieces (SCB no. Dal-Invert-7). (7) Boreholes in reworked Lower Cretaceous ammonite fragment, iron-carbonate concretion (SCB no. Dal-Invert-5a). (8) SmallGastrochaenolithes boreholes in reworked Lower Cretaceous orthicon ammonite fragment (Bochianites), iron-carbonate concretion (SCB no. Dal-Invert-5b).

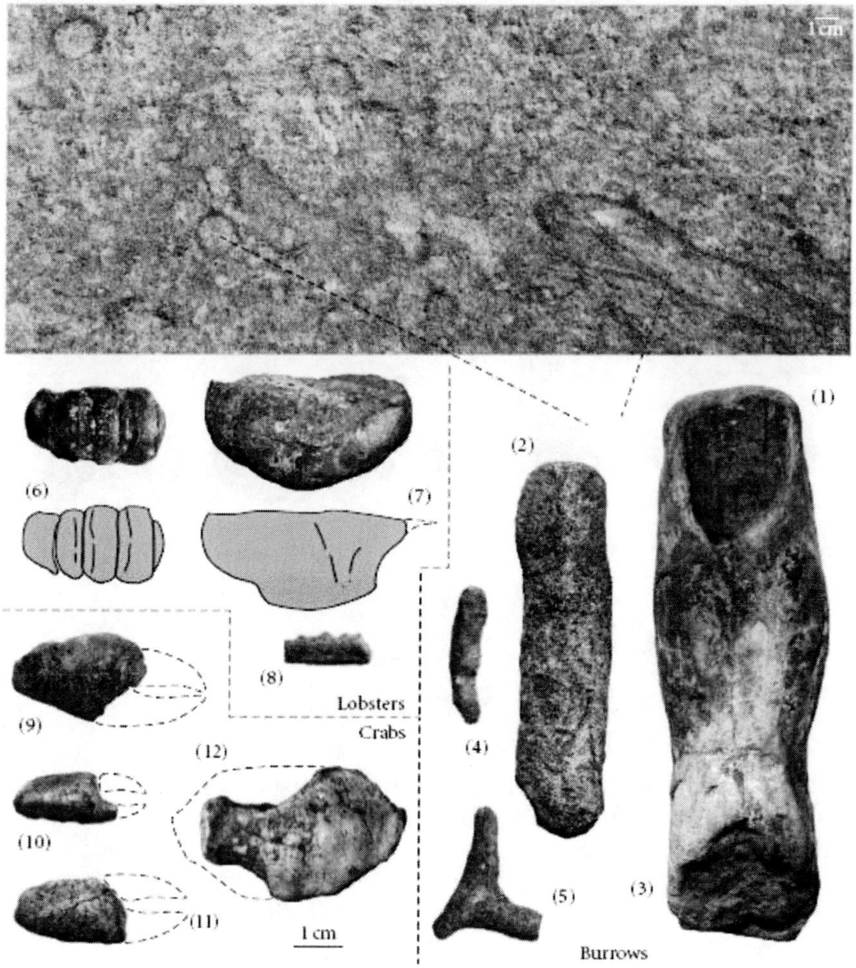

Figure 7: Crustacean remains and burrows from the Middle Eocene (Lutetian) gravels at Dalum. (1) Burrows by crustaceans, with large burrows of Ophiomorpha isp. and other smaller burrows in bioturbated glauconitic sands of the Fürstenau Fm. at the Osteroden sand pit (see also burrows in Figure. 3(d)). (2) Refilled Ophiomorpha isp. burrow fragment, phosphorite concretion (SCB no. Invert-41). (3) Encrusted Ophiomorpha isp. burrow fragment, phosphorite concretion (SCB no. Invert-38). (4) Refilled burrow fragment, phosphorite concretion (SCB no. Invert-40). (5) Triaxonic burrow fragment, phosphorite concretion (SCB no. Invert-39). (6) Abdomen fragment from the lobster Hoploparia sp., phosphorite concretion (SCB no. Invert-14), dorsal. (7) Carapace of the lobsterHoploparia sp., phosphorite concretion (SCB no. Invert-16), lateral. (8) Lower pincer of the lobster Hoploparia sp.,

phosphorite concretion (SCB no. Invert-13), lateral. (9) CrabZanthopsis large right pincer, phosphorite concretion (SCB no. Invert-18), anterior. (10) Crab Litoricola/Portunites right pincer, phosphorite concretion (SCB no. Invert-17), anterior. (11) Crab Xanthilites sp. large right pincer, phosphorite concretion (SCB no. Invert-20), anterior. (12) Crab Xanthilites sp. carapace in phosphorite concretion (SCB no. Invert-19), dorsal.

Figure 8: Mollusc and anthozoa remains from the Middle Eocene (Lutetian) gravels at Dalum. (1) Gastropod Natica sp., phosphorite steinkern (SCB no. Invert-25), dorsal and lateral. (2) Gastropod Sconsia sp., phosphorite steinkern (SCB no. Invert-27), dorsal and lateral. (3) Gastropod Cyprea sp. phosphorite steinkern (SCB no. Invert-29), dorsal and lateral. (4) Gastropod Tectonatica sp., phosphorite steinkern (SCB no. Invert-26), lateral. (5) Gastropod Volutospina sp. phosphorite steinkern (SCB no. Invert-30), lateral. (6) Gastropod Littoriniscala sp., phosphorite steinkern (SCB no. Invert-32), lateral. (7) Gastropod Xenophora sp. phosphorite steinkern (SCB no. Invert-28), dorsal and lateral. (8) Gastropod Falsifusus sp. phosphorite steinkern-negative (SCB no. Invert-31), lateral. (9) Bivalve Striarca sp. phosphorite steinkern (SCB no. Invert-34), lateral. (10) Bivalve Pteriasp., phosphorite steinkern (SCB no. Invert-35), lateral. (11) Oyster Pycnodonte sp., shell fragment, metasomatically replaced, silicified (SCB no. Invert-36), lateral. (12) AnthozoaParacyathus caryophyllus, metasomatically replaced, silicified (SCB no. Invert-37), lateral and dorsal.

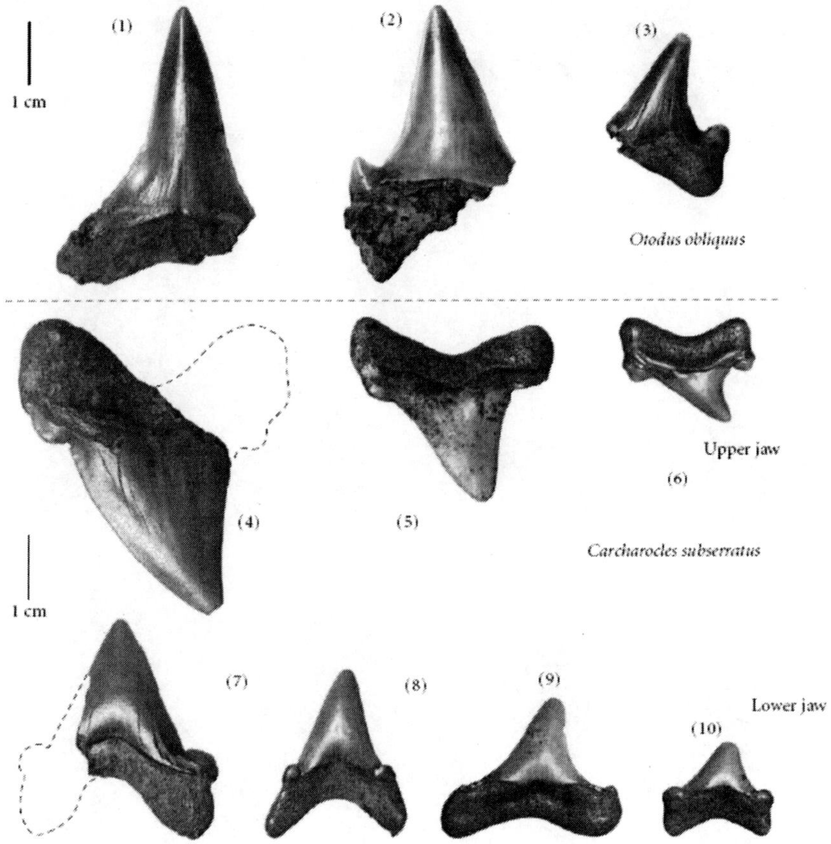

Figure 9: Shark teeth from lamniform (Odontidae) sharks (from different jaw positions) from the Middle Eocene (Lutetian) gravels at Dalum. (1–3). Otodus obliquus (Agassiz, 1843): (1) lower jaw anterior tooth (SCB no. Sel-856), (2) lower jaw anterolateral tooth (SCB no. Sel-857), and (3) lateral tooth (SCB no. Sel-858). (4–10). Carcharocles subserratus(Agassiz, 1843): (4) upper jaw anterior tooth (SCB no. Sel-869), (5) upper jaw lateral tooth (SCB no. Sel-870), (6) upper jaw posterolateral tooth (SCB no. Sel-872), (7) lower jaw anterior tooth (SCB no. Sel-873), and (8) Lower jaw anterolateral tooth (SCB no. Sel-874), (9) lower jaw lateral tooth (SCB no. Sel-876), (10) lower jaw posteriolateral tooth (SCB no. Sel-875; all in labial view).

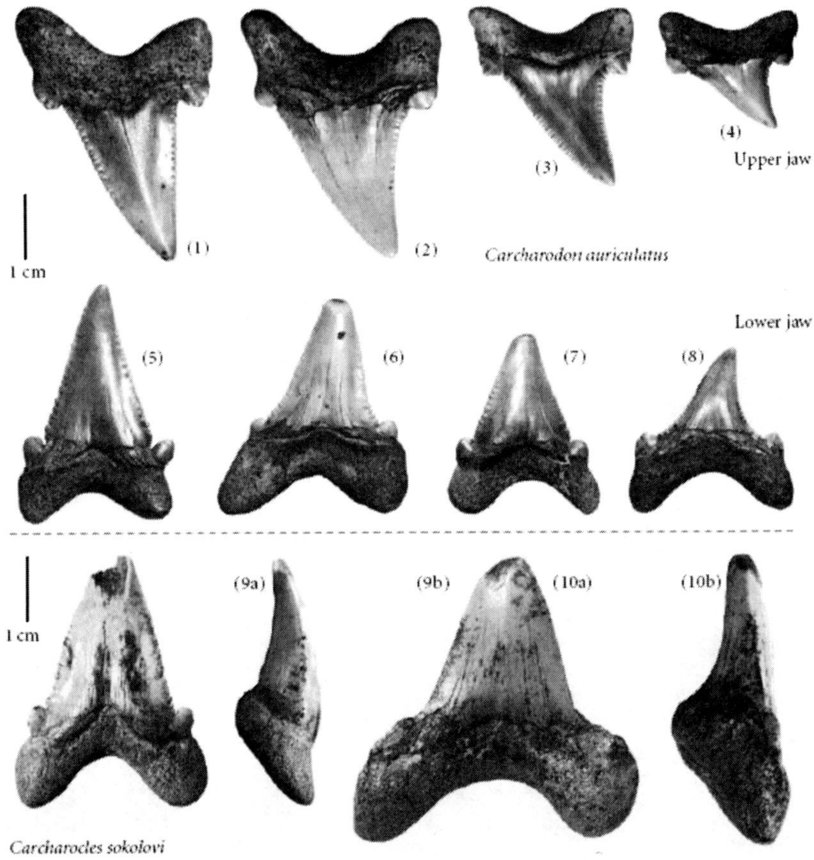

Figure 10: Shark teeth from lamniform (Carcharidae) sharks (from different jaw positions) from the Middle Eocene (Lutetian) gravels at Dalum. (1–8) Carcharodon auriculatus(Blainville, 1818): (1) upper jaw anterior tooth (SCB no. Sel-26), mirrored, (2) upper jaw anterolateral tooth (SCB no. Sel-25), (3) upper jaw anterior lateral tooth (SCB no. Sel-27), mirrored, (4) upper jaw posterior lateral tooth (SCB no. Sel-65), mirrored, (5) lower jaw anterior tooth (SCB no. Sel-4), mirrored, (6) lower jaw anterolateral tooth (SCB no. Sel-5), mirrored, (7) lower jaw anterior lateral tooth (SCB no. Sel-7), (8) lower jaw posterior lateral tooth (SCB no. Sel-66; (1-2(a)) labial and (b) lateral, (3–10), labial views). (9-10)Carcharocles sokolovi (Jaeckel, 1895): (1) lower anterior tooth (SCB no. Sel-85) and (2) lower anteriolateral tooth (SCB no. Sel-88).

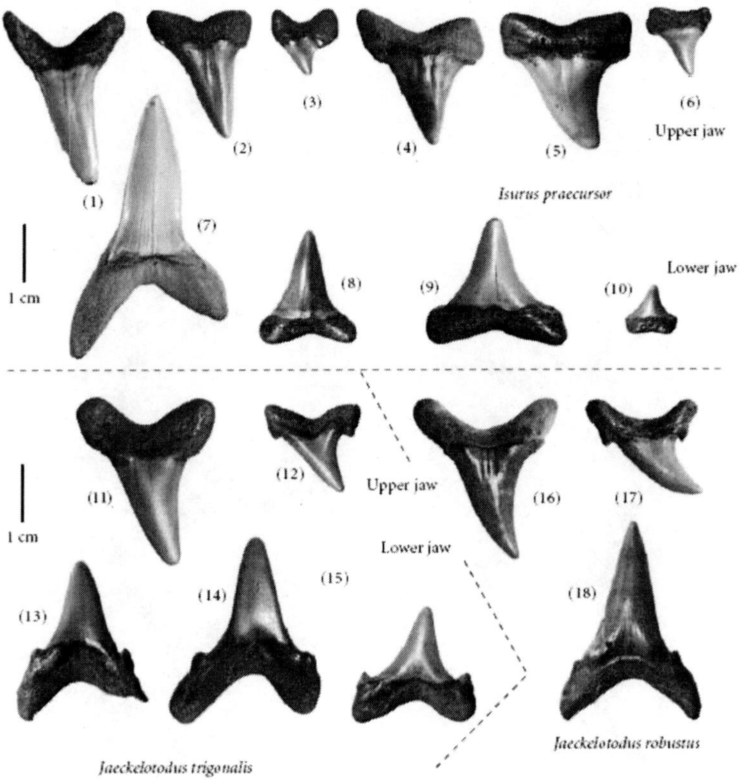

Figure 11: Shark teeth from lamniform (Lamnidae, Jaeckelotodontidae) sharks (from different jaw positions) from the Middle Eocene (Lutetian) gravels at Dalum. (1–10) Isurus praecursor (Lériche, 1905): (1) first upper anterior tooth (SCB no. Sel-101), 2. upper jaw anterior tooth (SCB no. Sel-117), (3) second upper jaw tooth (SCB no. Sel-112),(4) upper jaw anterolateral tooth (SCB no. Sel-124), (5) upper jaw anterior lateral tooth (SCB no. Sel-123), (6) upper jaw posterior lateral tooth (SCB no. Sel-125), (7) first lower jaw anterior tooth (SCB no. Sel-97), (8) second lower jaw anterolateral tooth (SCB no. Sel-109), (9) lower jaw middle lateral tooth (SCB no. Sel-104), (10) lower jaw distal lateral tooth (SCB no. Sel-108). (11–15) Jaeckelotodus trigonalis (Jaeckel, 1895): (11) upper jaw anterior tooth (SCB no. Sel-157), (12) upper jaw lateral tooth (SCB no. Sel-156), (13) lower jaw first anterior tooth (SCB no. Sel-160), (14) lower jaw anterolateral tooth (SCB no. Sel-158), (15) lower jaw lateral tooth (SCB no. Sel-159). (16–18) Jaeckelotodus robustus (Lériche, 1921): (16) upper jaw anterolateral tooth (SCB no. Sel-131), (17) upper jaw lateral tooth (SCB no. Sel-137), and (18) lower jaw anterior tooth (SCB no. Sel-129; all in labial view).

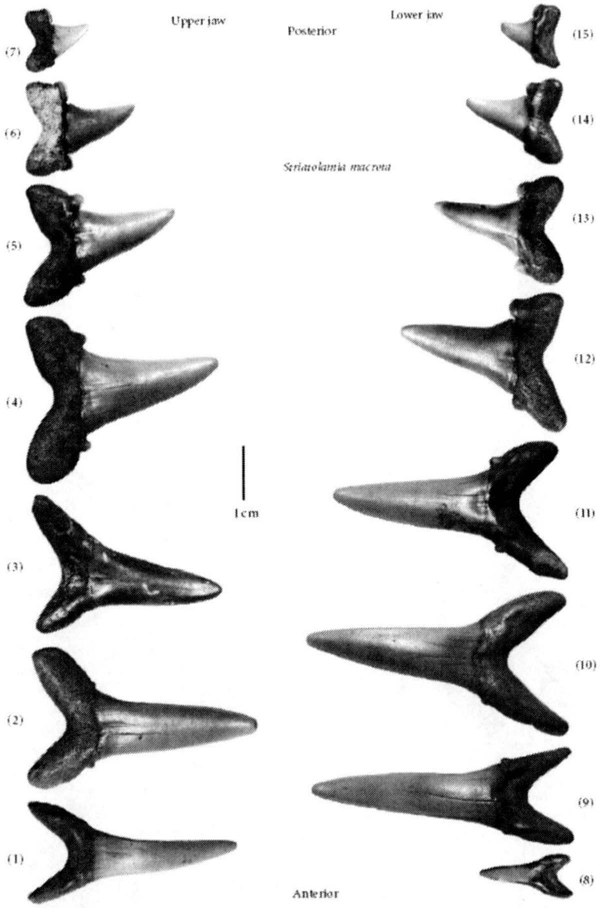

Figure 12: Shark teeth from lamnid (Odontaspididae) Striatolamia macrota (Agassiz, 1843) sharks (from different jaw positions) from the Middle Eocene (Lutetian) gravels at Dalum. (1) upper jaw first tooth (SCB no. Sel-195), (2) upper jaw second tooth (SCB no. Sel-179), (3) upper jaw third tooth (SCB no. Sel-212), (4) upper jaw fifth tooth (SCB no. Sel-195), (5) upper jaw middle lateral tooth (SCB no. Sel-172), (6) upper jaw middle lateral tooth (SCB no. Sel-171), (7) upper jaw posteriolateral tooth (SCB no. Sel-220), (8) lower jaw first (symphyseal) tooth (SCB no. Sel-167), (9) lower jaw second tooth (SCB no. Sel-168), (10) lower jaw third tooth (SCB no. Sel-169), (11) lower jaw fourth tooth (SCB no. Sel-170), (12) lower jaw anterior lateral tooth (SCB no. Sel-217), (13) lower jaw middle lateral tooth (SCB no. Sel-218), (14) lower jaw distal lateral tooth (SCB no. Sel-219), and (15) lower jaw posterolateral tooth (SCB no. Sel-214; and all in labial view).

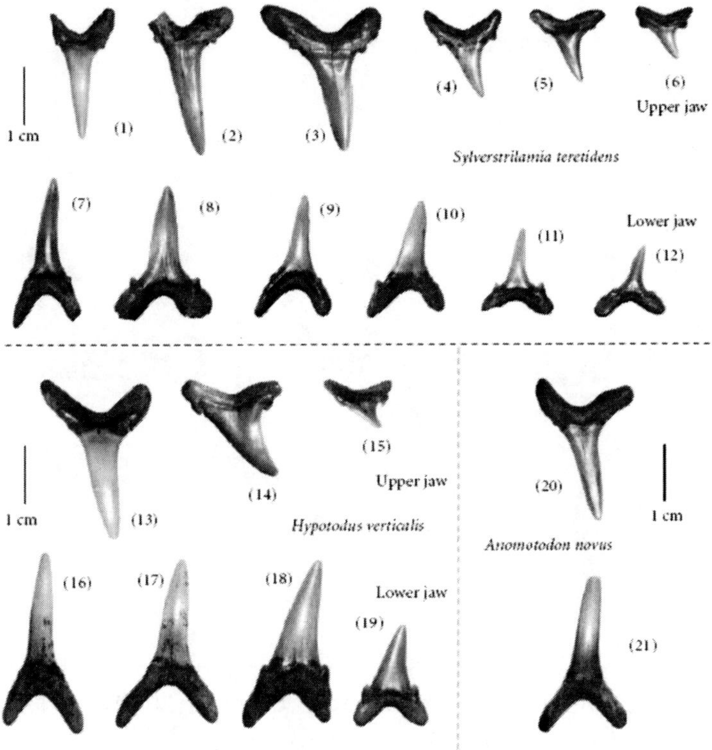

Figure 13: Shark teeth from lamniform (Odontaspididae, Mitsukurinidae) sharks (from different jaw positions) from the Middle Eocene (Lutetian) gravels at Dalum. (1–12).Sylvestrilamia teretidens (White, 1931): (1) upper jaw first anterior tooth (SCB no. Sel-832), (2) upper jaw second anterior tooth (SCB no. Sel-833), (3) upper jaw third anterior tooth (SCB no. Sel-834), (4) upper jaw middle lateral tooth (SCB no. Sel-835), mirrored, (5) upper jaw middle lateral tooth (SCB no. Sel-836), mirrored, (6) upper jaw posteriolateral tooth (SCB no. Sel-837), (7) lower jaw second anterior tooth (SCB no. Sel-838), (8) lower jaw third anterior tooth (SCB no. Sel-839), (9) lower jaw anterior tooth (SCB no. Sel-840), (10) lower jaw anterolateral tooth (SCB no. Sel-841), (11) lower jaw middle lateral tooth (SCB no. Sel-843),and (12) lower jaw posterior lateral tooth (SCB no. Sel-845), mirrored. (13–19). Hypotodus verticalis (Agassiz, 1843): (13) upper jaw anterior tooth (SCB no. Sel-823), (14) upper jaw middle lateral tooth (SCB no. Sel-824), (15) upper jaw posteriolateral tooth (SCB no. Sel-825), (16) lower jaw second anterior tooth (SCB no. Sel-827), (17) lower jaw third anterior tooth (SCB no. Sel-828), (18) lower jaw fourth anterior tooth (SCB no. Sel-829), (19) lower jaw middle lateral tooth (SCB no. Sel-826). (20–22) Anomotodon novus (Winkler, 1874): (20) upper jaw anterior tooth (SCB no. Sel-818), and (21) Lower jaw anterior tooth (SCB no. Sel-819), (22) lower jaw anterior tooth (SCB no. Sel-; 820) all in labial view).

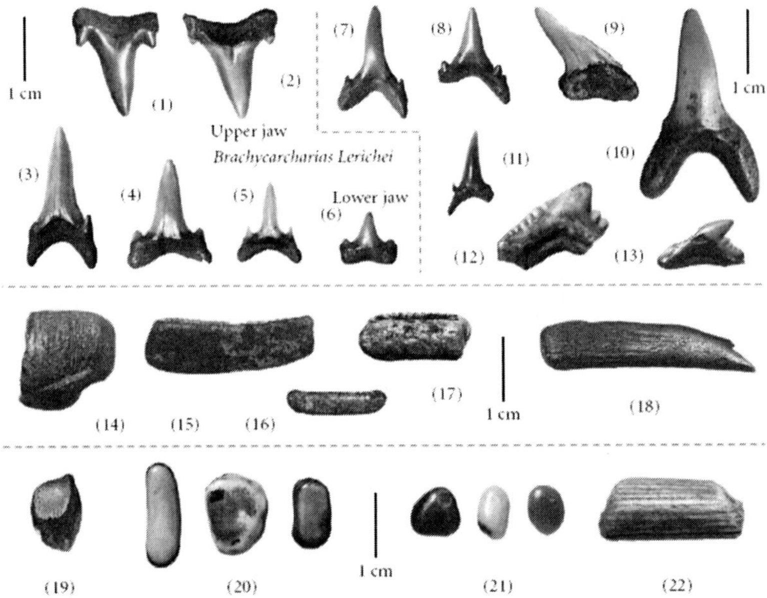

Figure 14: Shark teeth from rare shark species, rays, chimaera, and bony fish remaining from the Middle Eocene (Lutetian) gravels at Dalum. 1–6. Brachycarcharias lerichei(Casier, 1946): (1) upper jaw second anterior tooth (SCB no. Sel-849), (2) upper jaw third anterior tooth (SCB no. Sel-850), (3) lower jaw first anterior tooth (SCB no. Sel-851), (4) lower jaw third anterior tooth (SCB no. Sel-848), (5) lower jaw lateral tooth (SCB no. Sel-852), (6) lower jaw posterolateral tooth (SCB no. Sel-853). 7-8. Carcharias sp.: (7) lower jaw anterolateral tooth (SCB no. Sel-878), (8) lower jaw lateral tooth (SCB no. Sel-879). (9) Xiphodolamia ensis Leidy, 1877 upper jaw second tooth (SCB no. Sel-859). (10)Isurolamna affinis (Casier, 1946) lower jaw first anterior tooth (SCB no. Sel-865). (11)Otodus winkleri Lériche, 1905 upper jaw anterolateral tooth (SCB no. Sel-883). (12)Galeocerdo eaglesomei (White, 1955) tooth (SCB no. Sel-2). (13) Pachygaleus lefevrei(Dammeries, 1891) tooth (SCB no. Sel-1). (14) Tooth from a large unidentified ray (SCB no. Bat-7). (15) Aetobatis irregularis (Agassiz, 1843) tooth fragment (SCB no. Bat-4). (16)Leidybatis jugosus (Leidy, 1876) tooth (SCB no. Sel-902). (17) Myliobatis dixoni (Agassiz, 1843) tooth fragment (SCB no. Bat-6). (18) Pristis lathami Galeotti, 1837 rostral tooth (SCB no. Bat-9). (19) Edaphodon bucklandi Agassiz, 1843 (SCB no. Chim-1). (20) Phylodussp. teeth (SCB no. Tel-1). (21) Pycnodus sp. three lateral row teeth (SCB no. Tel-2). (22)Cylindracanthus sp. rostrum fragment (SCB no. Tel-8). (1–13 labial view).

Figure 15: Reptile and mammal remains form the Middle Eocene (Lutetian) gravels at Dalum. (1) Crocodile tooth from Diplocynodon (SCB no. Rept-1), labial and lateral. (2) Crocodile osteoderm fragment from Diplocynodon (SCB no. Rept-2), dorsal. (3) TurtleTrionyx osteoderm fragment (SCB no. Rept-3), dorsal. (4) Distal joint of a humerus from an unidentified medium-sized mammal (SCB no. Mam-1), (a) dorsal, (b) lateral, and (c) ventral. (5) Femur fragment from an unidentified large mammal (SCB no. Mam-2), dorsal and lateral.

3.3.3. Description

The Fürstenau Formation is exposed over a thickness of about 22 meters (Figure 3), the lower part of which (Early Eocene, Ypresian) consists of slightly glauconitic marine sands, with thinly bedded sandstone layers above (Dalum site). These are followed by 1 to 2 meters of thick gray-green clays with lenses of glauconitic sand (possibly still Early Eocene), the top of which have been eroded by a conglomerate bed 40 cm thick at Osteroden and 80 cm thick at Dalum (Figure 2(b)). These conglomerates can be dated herein into the basal Lutetian (basal Middle Eocene). They are fully decalcified at Dalum but are carbonatic at Osteroden where they contain calcareous shells of marine invertebrates, as well as aragonitic fish otoliths in places. The middle Middle Eocene marine glauconitic sands (NP 15 Zone) are composed of at least seven parasequence sets, starting with green, clay-rich, fine-grained sands and ending with green-gray, glauconitic, highly bioturbated (mainlyOphiomorpha), medium-grained sands (Figures 3, 7(1–3)). The gravels and the macrofauna from the Fürstenau Formation are described below.

4. ANALYSIS OF THE CONDENSED LUTETIAN PHOSPHORITE CONGLOMERATE BED

4.1. Pebble Types and Origin

The degree of rounding in the pebbles varies (Figure 4), but this property is not very useful for the Carboniferous quartzite pebbles as they are rederived Paleozoic fluvial clasts that were already rounded. Further analysis of the degree of rounding and estimates of transport distances are therefore unlikely to succeed, at least for that particular fraction of the pebbles.

Other pebble groups such as most of the flint pebbles, however, do indicate long-distance transport over a long period of time within a marine environment, while those comprised of local sediments, especially carbonates, have clearly only travelled relatively short distances.

The qualitative rock analysis reveals more then 30 different types of rock within the conglomerates from Dalum and Osteroden (Figure 4), for which provenances cannot always be determined. It has, however, been possible to identify the main rock types and their original ages (Table 1) which are important for an understanding of terrestrial or limnic influences, marine paleocurrents, and the erosional history of the region, and have allowed a new paleogeographical map to be drawn up (Figure 5).

4.1.1. Paleozoic Sedimentary Clasts

Clear to white and pale red or milky, well-rounded quartz pebbles (Figures 4(1) and 4(4)) are from the Westphalian C-D sandstone series of the Upper Carboniferous, which outcrops mainly within the Schafberg horst structure, but also at the lesser extent Piesberg and Hüggel horsts near Osnabrück [21]. These quartzite pebbles were already well rounded by previous transport during the Paleozoic and similar well-rounded quartzite pebbles can be seen in the conglomerates of the Piesberg and Schafberg horsts. These pebbles have therefore been redeposited after travelling only short distances, and do not provide any evidence of long-distance transport during the Paleogene, to their secondary deposition site.

Table 1: Dalum gravel analysis based on 3,927 pebbles (>4 mm).

Age	Rock type	Number of clasts
Paleogene (Early/Middle Eocene)	Phosphorite concretions	600
	Phosphorite burrows	1
	Phosphorite fossils	1
	Fine-grained white sandstone	10
	Vertebrate remains (mainly shark teeth)	150
Upper Cretaceous (Campanian-Maastrichtian)	Flint	10
Lower Cretaceous (Valanginian-Hauterivian)	Iron-carbonate concretion: rock fragments	980
	Iron-carbonate concretion: burrows	50
	Iron-carbonate concretion: ammonite fragments	10
Jurassic (Bathonian-Kimmeridgian)	Oolitic iron ores	1
	Oolitic carbonates	1
	Silica-iron-carbonate concretions	1
Carboniferous (Westphalian C-D)	Quartzite	2,100
	Quartzite conglomerates	10
Devonian	Black lydite	1
Volcanic rocks (age unclear)	Red volcanic rocks	1

4.1.2. Paleozoic Volcanic Clasts

Clasts of red volcanic rocks (Figures 4(11–12)) are only present as small pebbles (<3 cm) and their origin remains unclear. They were probably carried southwards from the Fennoscandian High by upwelling currents, because no such volcanic rock types are known to the south in the northern Rhenish Massif.

4.1.3. Lower Cretaceous Reworked Fossils and Clasts

The most abundant reworked fossils are ammonite fragments (Figures 4(27) and 4(31)), especially single chambered fragments (see the "cat paw" specimen in Figure 4(28)), which are all found in carbonate-iron concretions that originated from the Lower Cretaceous claystones of northern Germany. Most abundant are fragments of orthicon ammonites (Figures 4(27) and 4(28)).

The ammonite fragments, which are also found in the Lower Cretaceous claystones of north-west Germany [22], range in age from Valanginian to Hauterivian (Lower Cretaceous) and include Bochianites sp. andOosterella sp. (Upper Valanginian), Endemoceras sp. and Leopoldia sp. (Lower Hauterivian), and Aegocriocerassp. (Upper Hauterivian). All ammonites must have already been previously reworked during the Lower Eocene [19], and again during the basal Middle Eocene transgression. Rock types with similar (medium-brown) preservation to these reworked fossils contain abundant fossilized burrow casts (Figure 4(32)) and must have originated from the Lower Cretaceous marine sediments of north-west Germany. These can be distinguished from Eocene burrow casts since the latter are different and occur in phosphatic concretions (cf. Figure 7(2–5)).

4.1.4. Upper Cretaceous Clasts

The Upper Cretaceous clasts are carbonates from the Teutoburger Wald (Cenomanian to Santonian) and Dammer Berg (Campanian) regions of Germany (Figure 2(a)) that weather easily (except for the Campanian flint nodules, but these are absent from the Cenomanian to Santonian). The platy flints, which show little rounding, (Figure 4(23)) possibly originated from the Dammer Berge region. The Maastrichtian marine carbonate deposits in the studied region are not represented but typical black flint from further west in Belgium is common, including the "Maas-Eggs" (Figure 4(8)) that can sometimes be found in the Dalum gravels. Other red-to-brown flint variants resemble the "Rijckholt" flint (Figures 4(6) and 4(7)), typical of the Campanian limestones in northern Holland and on the Dutch-German border. All of these flint pebbles in the Dalum gravels (except for the platy flints) are well rounded and seem to have been transported to north-west Germany by marine paleocurrents from the west (Figure 5).

4.1.5. Jurassic Clasts

The black silica-ironstone concretions found in these gravels (Figure 4(16)), which resemble black Paleozoic lydite, are from the Middle Jurassic Soninnien beds and are well known in the Osnabrücker Bergland, as well as also being found in reworked Quaternary deposits of the region [23]. Rare, typically red, oolites (iron ores, Figure 4(17)) originate from the Middle Jurassic Macrocephaly oolith beds (also know as "Portha-Oolith or Wittekindsflöz") of the central part of the Wiehengebirge mountain range [24] (Figure 5) and, as with the carbonates, do not survive over long transport distances. In total, the sedimentary clasts from the Jurassic of the Osnabrücker Bergland, and in particular the Wiehengebirge mountain range, are poorly represented and comprise less then 1% of the total number of clasts (Figure 5).

4.1.6. Lower to Middle Eocene Clasts

The main hard rock types represented from the Lower and Middle Eocene are the phosphorite nodules (Figure4(20)) that have been reworked from the soft Eocene marine sands (including most of the marine vertebrate remains that accumulated in the Dalum and Osteroden condensed phosphorite nodule gravels), and possibly some white fine-grained sandstones (Ypresian, Figure 4(13)). The more polished pebbles appear to have come from the older Lower Eocene (Paleocene) horizons that have been reported from other Paleogene sites in north-west Germany [13] and have later been reworked and bored by organisms during the Middle Eocene (Figures 6(2) and 6(4)), while the unpolished phosphorites (Figure 4(20)) may be younger (Middle Eocene) or else have not travelled as far. Also less-transported are sporadic phosphatic shark coprolites, some of which have been reworked from earlier Eocene horizons (Figures 4(21) and 4(22)).

The quantitative rock-type analysis (Figure 5) is based on 3,927 pebbles greater than 4 mm (Table 2, Figure 5). The main rock types identified were abundant Carboniferous (Westphalian C-D) quartzites and (less common) quartzite conglomerate clasts (59%), Lower Cretaceous (Valagin/Hauterive) iron-carbonate concretions and fossils (20%), and Paleogene (Paleocene) Eocene phosphorite nodules (15%). These three main rock types derived from different directions and have been reworked from sediments of different ages (Figure 5). While the Carboniferous material must have been washed in from the south east by a river passing the Schafberg and reaching into a fluvial delta system, the Lower Cretaceous pebbles and fossils were washed opposite south and south east towards the coast, together with the Paleogene phosphorite and fossil material (Figure 5). All other rock types and components are less well represented and not important to the reconstruction of paleocurrents and erosional history.

Table 2: Fossil types at the Dalum site, based on 14,440 specimens in material >4 mm.

Fossil type	Amount
Plants	173
Burrows/boreholes in pebbles	184
Burrows (refilled, phosphatic)	43
Bivalves (fragments)	61
Gastropods	25
Crustaceans	29
Anthozoans	4
Chimaera	15
Rays	206
Sharks	13,690
Turtles	5
Crocodiles	3
Mammals	2

4.2. Fossil Content

In total, 14,437 fossils were recovered from the Dalum site excavation; these are listed by fossil type in Table 2. Nearly all of the Eocene invertebrate remains are only preserved in the reworked phosphorites; the only fossils preserved by silica replacement are the shells of larger oysters (with silica rings) and the solitary corals. The phosphatic coprolites, the silicified wood, all the bones, and in particular the teeth, were the most resistant specimens.

4.2.1. Invertebrates

The invertebrate fauna is very incomplete and largely represented by phosphorite steinkerns of shells and burrow casts that survived the enrichment processes in the Middle Eocene gravels of Dalum.

Drills and Boreholes Drills and boreholes in pebbles [25] are quite common in pebble-rich beach Eocene paleoenvironments also found in the Oligocene pebble-rich beaches of north-west Germany [26]. The mechanical bivalve drills and chemical sponge boreholes in the Eocene pebbles from the Dalum site have only been made in soft to moderately hard rock types, and in particular in 85% of the reworked Lower Cretaceous iron-carbonate nodules (as in the Oligocene beach pebbles from the Osnabrück Bergland region) [26] and ammonite fragments; they are less common in Eocene phosphorites. The two typical ichnites of Gastrochaenolites isp. produced by bivalves (Figures 6(1–3), 6(7) and 6(8)), and Clione sponge burrows (Figure 6(4)), are common in Lower Cretaceous iron-carbonate concretion pebbles and lesser in Eocene phosphorite pebbles. Teredo bivalve boreholes are also present, but only abundant in driftwood (Figures 6(5) and 6(6)).

Anthozoans: Corals without reef structures are known throughout the southern North Sea basin, represented by the solitary coral Paracyathus caryophyllus. The few specimens of this coral that have been recovered have been preserved by metasomatic silica replacement (Figure 8(12)). This solitary, nonhermatypic coral is also found in Oligocene shallow marine, beach pebble paleoenvironments [26, 27].

Crustaceans: These are common in the Eocene North Sea basin [28, 29], but in the Dalum material they are only preserved in the reworked phosphorite concretions. The remains of Hoploparia sp. (Figures 7(6–8)) are significant as these lobsters could be largely responsible for the high degree of bioturbation in the glauconitic sands (Osteroden section, cf. Figures 3(d), 7(1)) and may have been responsible for the Ophiomorpha isp. burrows (Figure 7(1)), fragments of which are also found (as phosphorite concretion casts) within the gravels (Figures7(2) and 7(3)). Smaller burrows, especially the y-shaped forms (Figures 7(4) and 7(5)), may also be related to smaller crustaceans. Crab remains are more common and at least three different species can be identified from their pincers and carapaces: Xanthilithes sp. (Figures 7(11) and 7(12)), Zanthopsis (Figure 7(9)), andLitoricola/Portunites (Figure 7(10)).

Molluscs: Bivalves have been identified following the published works on the Belgian [30, 31] and English [7] sites. In the Dalum material almost all of these have been destroyed because of their aragonite shells, with only a few infaunistic forms such as Arca sp. and Striarca sp. (Figure 8(9)) preserved in phosphorite concretions, some epifaunistic forms such as Pteria sp. (Figure 8(10)), and the more abundant large oyster Pycnodonte sp. (only shell fragments preserved by metasomatic silica replacement), which seem to have settled on the gravels of the pebble beaches. The gastropods are more diverse with at least eight species represented (all reworked Eocene phosphorite steinkerns, Figures 8(1–8)), although most of the specimens are from the infaunistic Natica sp.; epifaunistic forms include Sconsia sp., Cyprea sp., Tectonatica sp., and Xenophora sp.

4.2.2. Vertebrates

The chimera, shark, and ray material was compared to the main important works on Paleocene-Eocene sharks of Europe [5, 7, 32–39]. The vertebrates remains are dominated by shark teeth (95%), followed by remains of rays (2%); the rest are chimera, bony fish, reptile, and (extremely rare) mammal remains (Table 2, Figure 5).

Chimera: Only one chimera species, Edaphodon bucklandi Agassiz, 1843, has been recorded to date; it is represented by jaw fragments (Figure 14(19)).

Rays: These are the second most common type of fish remains, represented mainly by broken teeth, but also by fin spines. At least four species can be recognised from their teeth including Aetobatus irregularis (Agassiz, 1843),Leidybatis jugosus (Leidy, 1876), Myliobatis dixoni (Agassiz, 1843), and one large unidentified species

(Figures14(14–17)), while Pristis lethami Galeotti, 1837, is represented by several rostral tooth fragments (Figure14(18)).

Sharks: Out of the total of 13,690 Paleogene shark teeth recovered from the Dalum site, 6,946 (more then 50%) have been identified at a species level (Table 3). This large number of identifications is a result of the large quantity of material and many complete and comparable teeth recovered, from different jaw positions in different species. The preliminary qualitative analysis herein includes only 19 of the larger Paleogene shark species and five ray species (>4 mm material), which can be found listed in Table 3 together with their time ranges and abundances. Preliminary sampling of the material below 4 mm indicates twice as much fish biodiversity, including of course the smaller species whose teeth are, however, quite rare. Large teeth from Otodus obliquus (Agassiz, 1843; Figure 9(1–3)) are all incomplete and mostly very rounded, but the sharp-edged teeth illustrated herein would appear to be more or less autochthonous (also rare sharp edged in abundant Striatolamia teeth), which is significant because this species now appears to extend from the Paleocene into the lower Middle Eocene in Europe, but not into the middle or upper parts of the Middle Eocene. All of the large teeth from Carcharocles subserratus (Agassiz, 1843; Figure 9(4-10)) are rounded, indicating most probably in all cases the reworking of horizons that appear to extend into the upper Paleocene or Lower Eocene. However, this material may partly belong to the most abundant type of white shark,Carcharodon auriculatus (Blainville, 1818), which is represented by both enrolled teeth and sharp serrated teeth, from various jaw positions (Figures 10(1–8)). The largest and most robust teeth recovered from the Dalum site are from the megatooth shark lineage—Carcharocles sokolovi (Jaeckel, 1895)—which is not well defined, and is rare in any material. All of the megatooth shark ancestor material will be discussed further in the future, when more teeth are available (only 13 have been recovered to date, and only two of these are in good condition: Figures 10(9) and 10(10)) and can be better distinguished from the other serrated teeth. Another, more abundant, species is Isurus praecursor (Lériche, 1905), with material from all jaw positions (Figure 11(1–10)). Jaeckelotodus is represented by two species, J. trigonalis (Jaeckel, 1895) from which all teeth are reworked (Figures 11(11–15)), and J. robustus (Lériche, 1921) which has both reworked and sharp-edged specimens (Figures 11(16–18)). The most abundant shark is the sand shark ancestor Striatolamia macrota(Agassiz, 1843), with teeth present from all the main jaw positions (Figure 12) and including teeth from juveniles right through to teeth from mature individuals. Only 0.1% of the teeth are sharp edged with little evidence of any transport. Smaller species are also abundant, such as Sylvestrilamia teretidens (White, 1931; Figures 13(1–12)), and Hypotodus verticalis (Agassiz, 1843; Figures 13(13–19)). Less abundant are teeth fromAnomotodon novus (Winkler, 1874; Figures 13(20–22)) which are all reworked, as is also largely the case with teeth from Brachycarcharias lerichei (Casier, 1946; Figures 14(1–6)). This latter species sometimes also has sharp-edged teeth, such as Carcharias sp. (Figures 14(7) and 14(8)). Rare teeth from some other species are also present (Table 3, Figures 14(9) and 14(13)).

Table 3: Shark species from the Dalum site based on the number of identifiable teeth >4 mm.

Sharks	Quantity	Time range
Striatolamia macrota	5,901	Paleocene-Eocene
Carcharodon auriculatus	455	Middle/Upper Eocene
Isurus praecursor	179	Lower/Middle Eocene
Sylvestrilamia teretidens	95	Eocene
Hypotodus verticalis	64	Eocene
Carcharocles subserratus	48	Lower Eocene
Otodus obliquus	47	Paleocene-Lower Eocene
Jaeckelotodus robustus	46	Lower/Middle Eocene
Brachycarcharias lerichei	36	Eocene
Jaeckelotodus trigonalis	33	Lower/Middle Eocene
Carcharocles sokolovi	13	Middle Eocene-Lower Oligocene
Carcharias sp.	12	Eocene
Anomotodon novus	5	Eocene
Isurolamia affinis	4	Eocene
Galeocerdo latidens	3	Eocene
Pachygaleus lefevrei	2	Eocene
Xiphodolamia ensis	1	Eocene
Usakias wardi	1	Eocene
Odontaspis winkleri	1	Eocene
Total	6,946	
Selachii Indet.	6,737	
Total number of teeth	13,683	

Bone Fish: Some vertebrae and teeth, mainly from Phylodus sp. (Figure 14(20)) and Pycnodus sp. (Figure 14(21)) and some rostrum fragments from Cylindracanthus sp. (Figure 14(22)) provide only a poor representation of the diverse fish fauna, which is normally estimated from the otoliths, but these are absent from Dalum.

Terrestrial Vertebrates: Only single vertebrate teeth have been described from Early to Middle Eocene marine deposits around the world [4, 7, 40], but large pachyostotic rib fragments from Dalum appear to be attributable to early marine Sirenian mammals when compared to Oligocene material [41], and these rare Middle Eocene marine mammal remains will be described in a future publication. The terrestrial vertebrate remains recovered from the Dalum material are of rare crocodile teeth, osteoderm fragments from Diplocynodon (Figures 15(1) and 15(2)), and freshwater turtle remains of Trionyx (Figure 15(3)), all of which are suggestive of freshwater lakes or swamps and therefore a possible deltaic environment. Some bone fragments are impossible to identify but appear to relate to mammals, judging by their spongious nature. A distal humerus fragment (Figure 15(4)) appears to belong to a medium-sized marine mammal, and a femur fragment (Figure 15(5)) may belong to a larger mammal.

5. DISCUSSION

5.1. Stratigraphy and Dating

The sediments from the 22-meter-thick sections at Dalum and Osteroden have a different lithology from the Lower to Middle Eocene Brussel and Lede Sand Formations of the Belgian Oosterzele and Balegem marine deposits [42]. They also contain less carbonate and no storm shell beds, but only massive clay and conglomerate layers, and hence a new formation name is proposed for these deposits: the Fürstenau Formation.

At the Dalum site only the older (Ypresian) Eocene layers are present, with initial marine sequences possibly overlain by a massive clay bed, although precise dating of this part of the section is not yet available. The conglomerate layer can be dated by the terrestrial mammal teeth (MP 11 zone), as at Balegem, in Belgium [4,43], and by shark teeth indicating the presence or absence of certain species (e.g., O. obliquus). The more carbonate-influenced Middle Eocene conglomerates (two horizons) of the Lede Sand Formation at Balegem and Osterzele [44] have been described as being of Middle Lutetian age, and may therefore possibly be slightly younger than those from northern Germany [1, 5]. The condensed phosphorite gravel beds at Dalum and Osteroden may be isochronous with the two bivalve-enriched condensed beds in the middle Middle Eocene Lede Sand Formation of Belgium [42, 45]. The conglomerates at Dalum and Osteroden also appear to be isochronous with the Lower Geiseltalium, and would therefore be of a similar age to the Messel and older Geiseltal mammal faunas [9].

The marine clay-sand parasequences of the Fürstenau Formation have been dated by nannoplankton into the middle Middle Eocene (Lutetian, NP 14 zone [46, 47]) and appear to be nonisochronous with the shark-tooth-rich layers at Balegem (NP 15 zone) [43, 48]. The lithology of the newly introduced Fürstenau Formation differs from that of the marine localities in the southern North Sea and Paris Basins in having a much lower carbonate content, an almost complete absence of sandstone beds, and an absence of the bivalve (dominated by the bivalve Megacardita) and gastropod shell beds.

At least two shark species appear to be from reworked Upper Cretaceous deposits, represented by some teeth from Cretolamna appendiculata and a single tooth from Squalicorax kaupi (Figures 4(24) and 4(26)). Upper Paleocene and Lower Eocene tooth material appears to be particularly well represented by Carcharocles obliquus and Otodus subserratus. While O. obliquus is typically present in the Late Paleocene and Lower Eocene of Europe (cf. London Clay in: [7]) it is represented here by sharp-edged teeth that show little signs of transport, and so its time range may extend into the Early Lutetian. In contrast, O. obliquus and C. appendiculata, are both absent from the diverse selachian fauna of Middle Lutetian age in Belgium [5]. TheCarcharodon auriculatus teeth from Dalum and Osteroden in Germany, and those from Balegem and Oosterzele in Belgium, differ in their serrations, especially on the lateral cusps which are more strongly developed in the younger material from Belgium [5]. The shark fauna from the Dalum and Osteroden gravels appears to be a little younger than the Middle Eocene Belgian faunas, which supports the dating of the Dalum and Osteroden gravels (but not all the teeth therein) into the Early Lutetian transgression (Early Middle Eocene [48], during which older Paleogene tooth material was reworked and concentrated.

5.2. Biodiversity and Facies

The mollusc population seems to be represented by forms from the sandy floor of a shallow sea, although only rudimentary taphonomic comparisons have been made to the Belgian and English sites [7, 30, 31]. There is anyway too little material available at this stage for a model to be made of the autochthonous benthic fauna on the sandy sea floor. The invertebrate fauna of the Middle Eocene beach gravel paleoenvironment is expected to be restricted, with a dominance of oysters [26], as has been demonstrated for the Oligocene pebble beaches of north-west Germany. Oysters such as Pycnodonte appear to have been attached to the gravels, but proof is yet to be found in the fragmented material from Dalum. The solitary corals are also in same state of preservation as the oysters and would not have survived any significant amount of transport, which means that they are also likely to have been attached to the gravels. A restriction of the marine invertebrate benthic fauna must also be expected due to the freshwater influence, but further analysis of the Osteroden site, with its well-preserved mollusc assemblages (Felker pers. com.) is still required.

The vertebrates are dominated by the ancestral sand shark Striatolamia macrota (Figure 12, Table 3), which appears to have also had nursery schools in those shallow waters, as indicated by the presence of teeth from all sizes of animals including many very small teeth from young individuals. Most interesting in the shark research are two different “white and megatooth shark” ancestors, one with massive and the other with slim tooth morphologies: the megatooth shark ancestor Carcharocles sokolovi (the massive-toothed form) [49], and the great white shark ancestor Carcharodon auriculatus (the smaller form) [49].

Two different “large serrated shark" forms can therefore be shown (from the well-dated and large quantities of teeth) to have already developed by the basal Middle Eocene, both of which will need to be further analysed in greater detail on the basis of the material from Dalum in order to further define the ancestry of the white/megatooth sharks and their evolutionary record, which remains as yet unclear, controversial, and a matter of some discussion [49, 50]. The Dalum material includes intermediate forms between Otodus obliquusand Carcharocles sokolovi, such as Carcharocles subserratus whose teeth seem to have all been reworked from Lower Eocene horizons, as have most of the O. obliquus teeth. The origin of the white/megatooth shark lineage will be investigated using about 1,000 serrated shark teeth from the two sites at Dalum and Osteroden, which is possibly the largest collection in general [51], and for the Eocene period in particular.

5.3. Paleogeographic Reconstruction and Paleoenvironment

Analysis of conglomerate pebbles and the reworked Mesozoic fossils has allowed a detailed paleogeographic map of the Osnabrück region, north of the Rhenish Massif, to be prepared, revising the former less refined models [13, 16, 20]. Previous pebble analysis work by Bartholomäus [2], in which the iron-carbonate clasts, ammonite fragments, and several pebble types were not recognised, has been considerably extended herein.

Large quantities of Paleozoic quartzite material (from Carboniferous sandstones and conglomerates) must have been transported, mainly from Schafberg area but also from the Piesberg area (see Carboniferous horst structures and Westphal C-D sandstones [21]), which appears to have been the result of major erosion under a tropical climate by an ancient (hypothetically reconstructed) river that flowed during the Middle Eocene between the mountain chains that today form the Wiehengebirge and Teutoburger Wald ranges (Figure 5). This river also may also have transported Paleozoic gravels (including lydite) from the Rhenish Massif. Only the hard silica-rich rock types could have survived long-distance transport. With such a river and a tropical climate an estuarine or deltaic situation can be expected at its mouth [52], as has been reconstructed herein (on a preliminary basis) using only the two Dalum and Osteroden sections and their differences in gravel thickness. Dalum would have been more proximal (80 cm thickness) than Osteroden (40 cm thickness) in the deltaic fan (see [52]). More sections and drill cores from the Fürstenau area will need to be studied in the future in order to obtain an improved understanding of this paleoenvironment. The conglomerates can anyway not be simply treated as tempestites or as being of purely transgressive origin [1, 13], as they are demonstrated herein to be a combination of marine transgressive and fluvial deltaic deposits. The Lower Cretaceous (Valanginian/Hauterivian) iron-carbonate concretions, the smaller burrows, and the many ammonite fragments, with all softer rock types showing burrows and boreholes from Eocene times (Figure 5), were transported during the Early Eocene (Ypresian [19]) and subsequently reworked during the Middle Eocene from the Early Eocene marine deposits, or possibly even reworked twice, and finally accumulated north of the Wiehengebirge mountain range (to the north-west of Osnabrück) in the southern pre-North Sea basin coastal pebble facies that were mixed with deltaic deposits (Figure 5).

The region north-west of Fürstenau (i.e., the Dalum/Osteroden region) represents a deltaic, fresh-water-influenced, low-carbonate, siliciclastic glauconite sand to pebble facies in the shallow marine coastal environment of the southern pre-North Sea basin, to the north of the Rhenish Massif. In contrast, the Belgian (Balegem/Osterzele) region was more carbonate influenced and was located in shallow marine areas between the southwestern pre-North Sea and the Paris Basins [44]. The Osterzele and Balegem facies had similarities with the Middle Eocene facies from north-west Germany, but had less terrestrial influences whereas in the Lower Eocene London Clay Formation the terrestrial input was even higher into the coastal paleoenvironments [14].

6. CONCLUSION

The Middle Eocene (Lutetian) transgressive conglomerates from Dalum and Osteroden in north-west Germany are a mixture that includes reworked Lower Cretaceous (Valanginian-Hauterivian) marine claystones, with the derived ammonite fragments and pebbles making up 20% of the total number of clasts. About 80% of these reworked Lower Cretaceous pebbles contain burrows or boreholes (e.g., Gastrochaenolithes boreholes made by bivalves). Boreholes have also been left in driftwood by the bivalve Teredo.

In addition, Paleogene phosphorite nodules sometimes contain benthic marine invertebrate fauna (and also burrow casts). These were transported southward to the pebble beach coast, as were all the Lower to basal Middle Eocene vertebrate remains and, in particular, the shark teeth that are highly concentrated within these gravels, with a density of about 5,000 per cubic meter of gravel.

The other main components of the conglomerate are Paleozoic (Carboniferous) siliciclastic sandstones, and the (mostly well-rounded) quartzites that make up 59% of the pebbles, but these must have been transported by river and come from the opposite direction, deriving from the mainland to the south. An ancient river between the Wiehengebirge and Teutoburger Wald mountain ranges must have flowed to the west of Osnabrück and the Carboniferous Schafberg where a river passed to the west of the main quartz source. Even the resistant sandstone rocks were affected by weathering in the harsh tropical climate. A "Fürstenau-Delta" is postulated, with a conglomerate thickness of about 80 cm in Dalum (proximal) and 40 cm in Osteroden (distal), suggesting a deltaic fan close to the western end of the mountain chains on the Rhenish Massif mainland. The coastline during the Early Eocene (Ypresian) was a little further south, and later moved some tens of kilometres north of the Wiehengebirge mountain range, where it was during the Middle Eocene (Lutetian). The northern Rhenish Massif (the Westphalian Cretaceous Basin/Bay) was not flooded during a medium high stand of the Early to Middle Eocene pre-North Sea Basin, nor was the area between the Wiehengebirge and Teutoburger Wald mountain ranges subjected to marine flooding. The area between the Wiehengebirgs and Teutoburger Wald mountain ranges was, however, subjected to flooding later during the Oligocene (Latdorfian to Chattian), when the old river drainage system of Eocene times was flooded and destroyed.

The shark fauna is mainly dominated by the sand shark Striatolamia macrota, for which a full range of ages is well represented. Otodus obliquus is also another common large shark. A "great white" shark ancestor is represented by teeth from Carcharodon auriculatus and a megatooth shark ancestor by the larger and more massively rooted teeth of Carcharocles sokolovi. Other moderately common sharks include Isurus praecursor,Jaeckelotodus robustus, and J. trigonalis. The Otodus obliquus, and C. subserratus teeth, in particular, appear to indicate reworking of older Early Eocene marine sediments, which are known to exist in north-west Germany. All of the other large shark species (teeth >4 mm), out of a total of 19 that have been identified, are more rare. The larger sharks show a moderate degree of biodiversity and do not indicate tropical marine water temperatures, nor does the absence of coral reefs and the low carbonate production, which is in contrast to the terrestrial tropical forest mammal fauna biodiversity of the Rhenish Massif Mainland.

ACKNOWLEDGMENTS

The excavation field project was sponsored by the "Geo- and Naturpark TERRA.Vita" (H. Escher), and the Landkreis, Osnabrück. The private company "PaleoLogic" was responsible for the field work and sponsoring of the scientific research. Special thanks are due to A. Felde from the Reuter company for his excellent work with the backhoe, and to the company for its full support. The Kuhlhoff Education Center leaders Mr. A. Bruns and Mr. W. Hollermann, and the mayor of Bippen, Mr. H. Tolsdorf, are thanked for their managements to install the field sieving laboratory and for any support. Finally, the author would also like to thank H. Felker, a local hobby paleontologist, who provided support, helped with sieving the material, and allowed access to his large collection (obtained over 25 years from both localities (Dalum and Osteroden) for comparisons.

REFERENCES

1. F. Hocht, "Eine Lagerstätte kreidezeitlicher und paläogener Chondrichthys-Reste bei Fürstenau (Niedersachsen)," Osnabrücker Naturwissenschaftliche Mitteilungen, vol. 6, pp. 35–44, 1979.

2. W. A. Bartholomäus, "Ein kondensiertes Knochenlager im marinen Alttertiär von Fürstenau/SW-Niedersachsen," Fossilien, vol. 2, no. 93, pp. 79–82, 1993.

3. W. Schwarzhans, "The otoliths from the Middle Eocene of Osteroden near Bramsche, north-western Germany," Neues Jahrbuch fur Geologie und Palaontologie, vol. 244, no. 3, pp. 299–369, 2007.

4. J. L. Franzen and T. Mörs, "Das nördlichste Vorkommen paläogener Säugetiere in Europa,"Palaontologische Zeitschrift, vol. 81, no. 4, pp. 447–456, 2007.

5. G. Van den Eeckhaut and P. De Schutter, "The Elasmobranch fauna of the Lede Sand Formation at Osterzele (Lutetian, Middle Eocene of Belgium)," Palaeofocus, vol. 1, pp. 1–57, 2009.

6. M. Lériche, "L'Eocene des Bassins Parisiennes et Belge, Livret-guide de la réunion extraordinaire de la Société Géologique de France," Bulletin de la Société Géologique de France, vol. 4, no. 12, pp. 692–807, 1912.

7. D. Rayner, T. Mitchel, M. Rayner, and F. Clouter, London Clay Fossils of Kent and Essex. Medway Fossil and Mineral Society, Caxton and Holmesdale Press Ltd, Kent, Ohio, USA, 2009.

8. J.L. Franzen, "The biostratigraphic and palaeoecologic significance of the Middle Eocene locality Geiseltal near Halle (German, Democratic Republic)," Münchner Geowissenchaftliche Abhandlungen A, vol. 10, pp. 93–100, 1987.

9. S. Schaal and W. Ziegler, Eds., Messel—ein Schaufenster in die Geschichte der Erde und des Lebens, Waldemar Kramer, Frankfurt, Germany, 1988.

10. F. W. Negendank, G. Irion, and J. Linden, "Ein eozänes Maar bei Eckfeld nordöstlich Manderscheid (SW-Eifel)," Mainzer Geowissenschaftliche Mitteilungen, vol. 11, pp. 157–172, 1982.

11. H. H. Schleich, "Neue Reptilfunde aus dem Tertiär Deutschlands. 13. Schildkröten- und Krokodilreste aus der eozänen Braunkohle des Untertargebaues Stolzenbach bei Borken (Hessen)," Courier des Forschungsinstitutes Senckenberg, vol. 173, pp. 79–101, 1994.

12. J.-L. Hartenberger, "Les Mammifères d'Egerkingen et l'histoire des faunes de l'Eocène d'Europe," Bulletin de la Société Geologique France, Serie VII, vol. 12, no. 5, pp. 886–893, 1970.

13. H. Hiltermann, "Tertiär," in Geologie des Osnabrücker Berglandes, H. Klassen, Ed., pp. 463–494, Osnabrück Germany, 1984.

14. C. King, "The stratigraphy of the London Clay Formation and Virginia Formation in the coastal sections of the Isle of Sheppy (Kent, England)," Tertiary Research, vol. 5, no. 3, pp. 121–160, 1984.

15. P. Mangelsdorf, Geologische Wanderkarte Landkreis Osnabrück, Landkreis Osnabrück, Hanover, Germany, 1984.

16. F. Bettenstaedt, "Paläogeographie des nordwestdeutschen Tertiärs," Erdöl und Tektonik in NW Deutschland, vol. 1941, pp. 143–172, 1949.

17. F. Bettenstaedt, "Tertiär Norddeutschlands," Leitfossilien der Mikropaläontologie, vol. 1962, pp. 339–379, 1962.

18. J. Rohling, "Beiträge zur Stratigraphie und Tektonik des Tertiärs in Südoldenburg," Decheniana A, vol. 100, pp. 1–103, 1941.

19. K. Staesche and H. Hilterman, "Mikrofaunen aus dem Tertiär NW Deutschlands," Abhandlungen der Reichsanstalt für Bodenforschung Neue Folge, vol. 201, pp. 1–26, 1940.

20. P. A. Ziegler, Geological Atlas of Western and Central Europe, Shell, Amsterdam, The Netherlands, 1990.

21. K.-H. Josten, K. Köwing, A. Rabitz, et al., "Oberkarbon," in Geologie des Osnabrücker Berglandes, H. Klassen, Ed., pp. 7–77, Osnabrück, Germany, 1984.

22. E. Kempe, Die Tiefe Unterkreide im Vechte-Dinkel-Gebiet, Dobler-Druck, Alfeld,Germany, 1992.

23. C. Diedrich, "Kieselgeoden als wichtiger Rohstoff für die Beilproduktion im Neolithikum von Nordwestdeutschland—erste Nachweise von Produktionsplätzen der Trichterbecherkultur," pp. 1–36, 2004, www.jungsteinsite.de/

24. W. Weitschat, "Lias und Dogger," in Geologie des Osnabrücker Berglandes, H. Klassen, Ed., pp. 335–385, Osnabrück, Germany, 1984.

25. M. Bertling and K. Hermanns, "Autochthone Muschelbohrungen im Neogen des Rheinischen Braunkohlenreviers und ihre sedimentologische Bedeutung," Zentralblatt für Geologie und Paläontologie I, vol. 1995, no. 1-2, pp. 33–44, 1996.

26. C. Diedrich, "Palaeoecology, facies and stratigraphy of the shallow marine macrofauna from the Upper Oligocene (Palaeogene) in the southern Pre- North-Sea Basin of Astrup near Osnabrück (NW Germany)," Cainozoic Research. In press.

27. M. Berndt and J. Welle, "Ahermatype Korallen aus dem Oberoligozän der Schächte Rheinberg und Sophia Jacoba 8 (Niederrheinische Bucht)," Neues Jahrbuch für Geologie und Paläontologie Abhandlungen, vol. 198, no. 1-2, pp. 47–68, 1995.

28. W. J. Quayle, "English Eocene Crustacea (Lobsters and Stomatopods)," Tertiary Research, vol. 30, no. 5, pp. 581–612, 1987.

29. J. S. H. Collins and J. Saward, "Three new genera and species of crabs from the Lower Eocene London Clay of Essex, England," Bulletin of the Mitzunami Fossil Museum, vol. 33, no. 2006, pp. 67–76, 2002.

30. M. Gilbert, "Les Bivalvia du Lédien (Eocène ,oyen supérieur) de la Belgique I, Paleotaxodonta, cryptodonta, Pteriomorpha," Bulletin d l`Institut royal des Sciences Naturelles de Belgique/Mededelingen van het Koninklijk Belgish Instituut voor Natuurwetenschappen, vol. 51, no. 3, p. 1, 1975.

31. M. Gilbert, "Les Bivalvia du Lédien (Eocène ,oyen supérieur) de la Belgique II. Heterodonta, et Anomalodesmata," Bulletin d l`Institut Royal des Sciences Naturelles de Belgique/Mededelingen van het Koninklijk Belgish Instituut voor Natuurwetenschappen, vol. 52, no. 5, pp. 1–80, 1980.

32. O. Jaeckel, Die Eozänen Selachier aus Monte Bolca, 1894.

33. O. Jaeckel, "Unter-Tertiäre Selachier aus Südrussland," Mémoires du Comité Géologique de St.-Pétersbourg, vol. 9, pp. 19–35, 1895.

34. M. Lériche, "Contribution à lètude des poisons fossils du Nord de la France et des regions voisines,"Mémoires de la Société Geologique du Nord, vol. 5, pp. 1–430, 1906.

35. E. Casier, "Contributions à l'étude des poisons fossiles de la Belgique VII—les Pristidés Eocènes," Bulletin de l'Institut Royal des Sciences Naturelles de Belgique/Mededelingen van het Koninklijk Belgisch Instituut voor Naturwetenshappen, vol. 25, no. 19, pp. 1–52, 1949.

36. E. Casier, "Fauna ichthyologique du London Clay," British Museum Natural History London, vol. 1966, pp. 1–496, 1966.

37. H. Cappetta, Chondrichthyes II. Mesozoic and Cenozoic Elasmobranchii. Handbook of Paleoichthyology, Gustav Fischer, Stuttgart, Germany, 1987.

38. H. Cappetta and D. Nolf, "Révision de quelques Odontaspididae (Neoselachii: Lamniformes) du Paléocene et de l'Eocène du Bassin de la mer du Nord," Bulletin de l'Institut Royal des Sciences Naturelles de Belgique/Mededelingen van het Koninklijk Belgisch Instituut voor Naturwetenshappen, Sciences de la Terre/ Aardwetenshappen, vol. 75, pp. 237–266, 2005.

39. S. Adnet, "Nouvelles faunes de Sélachiens (Elasmobranchii, neoselachii) de l'Eocène moyen des Landes (Sud-Ouest France)," Implications Dans la Connaissance des Communautés de Sélachians Dèaux Profonds. Palaeo Ichthyologica, vol. 10, pp. 1–128, 2006.

40. S. G. Lucas, "Fossil mammals and the Paleocene/Eocene series boundary in Europe, North Amrica, and Asia," in Late Paleocene-Early Eocene Climatic and Biotic Evolution, M.-P. Aubry, S. Lucas, and W. A. Berggren, Eds., pp. 451–500, Columbia University Press, New York, NY, USA, 1998.

41. C. Diedrich, "The food of the miosiren anomotherium langenwieschei (Siegfried)—indirect proof of seaweed or seagrass by xenomorphic oyster fixation structures in the Upper Oligocene (Neogene) of the Doberg, Bünde (NW Germany) and comparisons to modern Dugong dugon (Müller) feeding strategies, "Senckenbergiana Maritima, vol. 38, no. 1, pp. 59–73, 2008.

42. J. Hermann, E. Steurbaut, and N. Vandenberghe, "The boundary between the Middle Eocene Brussel Sand and the Lede Sand Formations on the Zaventem-Nederokkerzeel area (Northeast of Brussels, Belgium)," Geologica Belgica, vol. 3, no. 3-4, pp. 213–255, 2000.

43. T. Smith, B. De Wilde, and E. Steurbaut, "Primitive equoid and Tapiroid mammals: keys for interpreting the Ypresian/Lutatian transition in Belgium," Bulletin van het Koninklijk Belgisch Institut voor Natuurwetenshappen, Aardwetenschappen, Supplement, vol. 74, pp. 165–175, 2004.

44. B. Fobe and V. Spiers, "Sedimentology and facies distribution of the Lede Formation (Eocene) in Belgium and northern France," Contributions to Tertiary and Quaternary Geology, vol. 29, no. 1-2, pp. 9–20, 1992.

45. P. Jacobs and E. Sevens, "Middle Eocene sequence stratigraphy in the Balegem quarry (Western Belgium, Southern Bigth North Sea)," Bulletin de la Société belge de Géologie/Bulletin van de Belgische Vereniging voor Geologie, vol. 102, no. 1-2, pp. 203–213, 1994.

46. M.-P. Aubry, "Palaeocene/Eocene boundary events in space and time," Israel Bulletin of Earth Sciences, vol. 44, pp. 1–253, 1998.

47. H.-P. Lauterbacher, J. R. Ali, H. Brinkhuis, et al., "The paleogene period," in A Geological Time Scale, G. M. Gradstein, J. G. Ogg, A. Smith, et al., Eds., pp. 384–408, Cambridge University Press, Cambridge, UK, 2004.

48. W. A. Berggren and M.-P Aubry, "Late Paleocene-Early Eocene NW European and North sea magnetostratigraphic correlation network. A sequence stratigraphic approach," in Correlation of the Early Palaeogene in Northwestern Europe, O. R. B. Knox, R. C. Corfiled, and R. E. Dunay, Eds., vol. 101, pp. 309–352, 1996.

49. A. Martin, "Systematics of the Lamnidae and the Origination Time of Carcharodon carcharias inferred from the comparative analysis of mitochondrial DNA sequences," in Great White Sharks: The Biology of Carcharodon Carcharias, A. P. Kimley and D. G. Ainley, Eds., pp. 67–78, Academic Press, San Diego, Calif, USA, 1996.

50. S. P. Applegate and L. Espinosa-Arrubarrena, "The fossil history of Carcharodon and its possible ancestor,Cretolamna: a study in tooth identification," in Great White Sharks—The Biology of Carcharodon Carcharias. Ainley, A. P. Klimley and G. David, Eds., pp. 19–36, Academic Press, San Diego, Calif, USA, 1996.

51. D. J. Long and B. M. Waggoner, "Evolutionary relationships of White Shark: a phylogeny of lamniform sharks based on dental morphology," in Great White Sharks—The Biology of Carcharodon Carcharias, A. P. Klimley and G. David, Eds., pp. 37–47, Academic Press, San Diego, Calif, USA, 1996.

52. H. Füchtbauer and G. Müller, Sedimente und Sedimentgesteine, Schweizbart' sche Verlagsbuchhandlung, Stuttgart, Germany, 1970.

4

SPATIAL DISTRIBUTION OF SOUTHERN BROWN SHRIMP (FARFANTEPENAEUS SUBTILIS) ON THE AMAZON CONTINENTAL SHELF: A FISHERY, MARINE GEOLOGY AND GIS INTEGRATED APPROACH

Déborah E. G. Martins[1] , Maurício Camargo-Zorro[2] , Pedro Walfir M. Souza Filho[3] * , Israel Hidenburgo Aniceto Cintra[4] , Kátia C. A. Silva[4]

[1]Universidade Federal do Pará, Programa de Pós-Graduação em Ecologia Aquática e Pesca. Brasil Ministério da Pesca e da Aquicultura, Superintendência do Pará.

[2]Instituto Federal de Educação, Ciência e Tecnologia da Paraíba.

[3]Universidade Federal do Pará, Instituto de Geociências. Instituto Tecnológico Vale.

[4]Universidade Federal Rural da Amazônia, Instituto Socioambiental e dos Recursos Hídricos.

ABSTRACT

The spatial distribution of the southern brown shrimp *Farfantepenaeus subtilis* (Pérez-Farfante, 1967) was studied based on industrial fishing fleet activities and is associated with geological and oceanographic characteristics of the benthonic environments on the Amazon continental shelf. Using a geographical information system (GIS) this paper sought to calculate the relative abundance of brown shrimp based on catch per unit effort (CPUE) and compare it with bathymetry, type of sedimentary structure, sedimentation rate and bottom salinity. As a result, we have concluded that the relative abundance (in terms of CPUE) is not uniformly distributed in space. Spatial analysis indicates that commercial trawling efforts were made in the (foreset) region of the subaqueous Amazon delta at depths of 40 to 60 m. In this region, prawn are responsible for the bioturbation of the sediments and the creation of a sedimentary structure called mottled mud. In the foreset region, sedimentation rates progressively increased up to 10 cm.yr-1; re-suspension was reduced and bottom salinity was high (~ 36). It appears that all of these factors define a stable muddy area with intense bioturbation. This notable biological activity is to be explained by the occurrence of a high *F. subtilis* abundance that appears to originate in a microbial loop. We concluded that by combining fishery information with environmental data from a GIS, it was possible to identify abundance distribution patterns for southern brown shrimp and other economically important fishery resources and to understand how they change on a large spatial-scale.

KEYWORDS

Shrimp fishery; Spatial analysis; GIS; Oceanography

1. INTRODUCTION

During the last decade there has been an increasing interest in the development of geographical information systems (GIS) that are applied to fishery management in order to understand a variety of three-dimensional fields of knowledge (MORRIS; BALL, 2006; RIGHTON; MILLS, 2006). However, relatively few studies have been published to describe the use of GIS to map the activity of industrial fishing and its relationship with geological characteristics and oceanographic processes in continental shelf regions (WALUDA et al., 2001). Hence, this paper used the spatial ecology address to integrate geosciences and fishery engineering data in a GIS approach to jointly examine geographical science and ecological information (SAITOH et al., 2011;

SKIDMORE et al., 2011; LAFFAN et al., 2012). Knowledge of the spatial distribution of fishing resources is very important for fishery planning. For fishery managers, it is essential that all the information related to a given resource is presented succinctly and precisely in an easily understood format (CADDY; GARCIA, 1986). The spatial dimension is important for both the development and the planning of fishery resource exploitation (MEADEN, CHI, 1996; GRAAF et al., 2003; RIOLO, 2006).

Approximately 3.4 million tons of shrimp are caught worldwide every year. This resource is, therefore, the most important fishery commodity sold worldwide in terms of value, accounting for 17% of fishing product trading; the market is concentrated in the United States, Japan and Europe (GILLETT, 2008; FAO, 2009). In many countries, shrimp is the most valuable fishery export product (GILLETT, 2008). In Brazil, frozen shrimp were the second most exported fishery product in 2007, totaling 17,217 t and worth US$ 74.9 million (IBAMA, 2008). The northern coast of Brazil, which is 2500 km long and located between the mouths of the Parnaiba (Brazil) and the Orinoco Rivers (Venezuela), belongs to one of the most important shrimp fishery areas in the world. The southern brown shrimp *Farfantepenaeus subtilis* (PÉREZ-FARFANTE, 1967), which is the target of both artisanal and industrial fishing fleets (ISAAC et al., 1992), is among the harvested species.

In the northern portion of the western South Atlantic Ocean, which includes the Amazon continental shelf, no analysis has been performed that relates this information to the physical environmental characteristics or even to the spatialization of the harvested shrimp volume or the southern brown shrimp targeting efforts (DRAGOVICH, 1981). *Farfantepenaeus subtilis* occurs at depths of up to 190 meters, often on muddy and sandy bottoms and sometimes associated with mollusks (HOLTHUIS, 1980). Identifying areas of concentrated fishing efforts, quantifying the catch by area and assessing the relationships between captures and efforts are of interest to fishery resource managers (MEADEN, 2000). Understanding the relationships between the distribution of fishery resources and environmental factors provides information for establishing a fishery management plan that will make rational and sustainable harvesting of shrimp stocks possible.

Studies of the Amazon continental shelf have indicated that several overlapping factors, such as the location near the Equator, the great physical energy that is released by tides, ocean currents and winds and the enormous discharge of water, solutes and particulate matter by the Amazon River, play a major role in regional hydrological dynamics

(NITTROUER; DEMARTER, 1996). The Amazon River discharge forms a plume of low-salinity water that extends offshore and northwestward over the north Brazilian shelf (LENTZ; LIMEBURNER, 1995). Meteorological and physical processes (e.g., discharge, precipitation, winds, currents, tides, and waves) are among the principal factors responsible for spatial and temporal variations in salinity, turbidity, nutrient concentration, phytoplankton blooms and nekton along the Amazon coast (BARTHEM, 1985; NITTROUER; DEMASTER, 1986; DEMASTER et al., 1996; SANTOS et al., 2008; SILVA et al., 2010). However, only the spatial distribution of the southern brown shrimp is considered in this study. This study applies data from a geographical information system (GIS) to industrial southern brown shrimp fisheries on the Amazon continental shelf to identify possible spatial distribution patterns of stocks in areas where industrial fishing is active and to relate them to geological and oceanographic processes.

2. MATERIAL AND METHODS

Three main factors were considered when designing the sampling procedures: i) fishery areas, ii) catches and fishing effort, and iii) the accessibility of geospatial information related to continental shelf morphology, and bottom-salinity.

2.1 Study area

The Amazon River exhibits the largest discharge volume (5.5x1012 m3.year-1) and drainage basin size (6x106 km2) in the world (GIBBS, 1967; MEADE et al., 1979). The suspended matter transported by the Amazon is approximately 800x106 tons (MARTINEZ et al., 2009). The Amazon discharge is sufficiently large for ocean waters not to penetrate the river mouth even during the lowest flow period, which means that the estuarine circulation is displaced toward the shelf (GIBBS, 1970).

The Amazon continental shelf was defined by NITTROUER and DEMASTER (1986) as the region extending from the coastline out as far as the shelf break, which corresponds to the 100-m isobath. The region extends from the Pará River estuary to the 5ºN parallel and is limited by the frontier between Brazil and French Guyana (Figure 1).

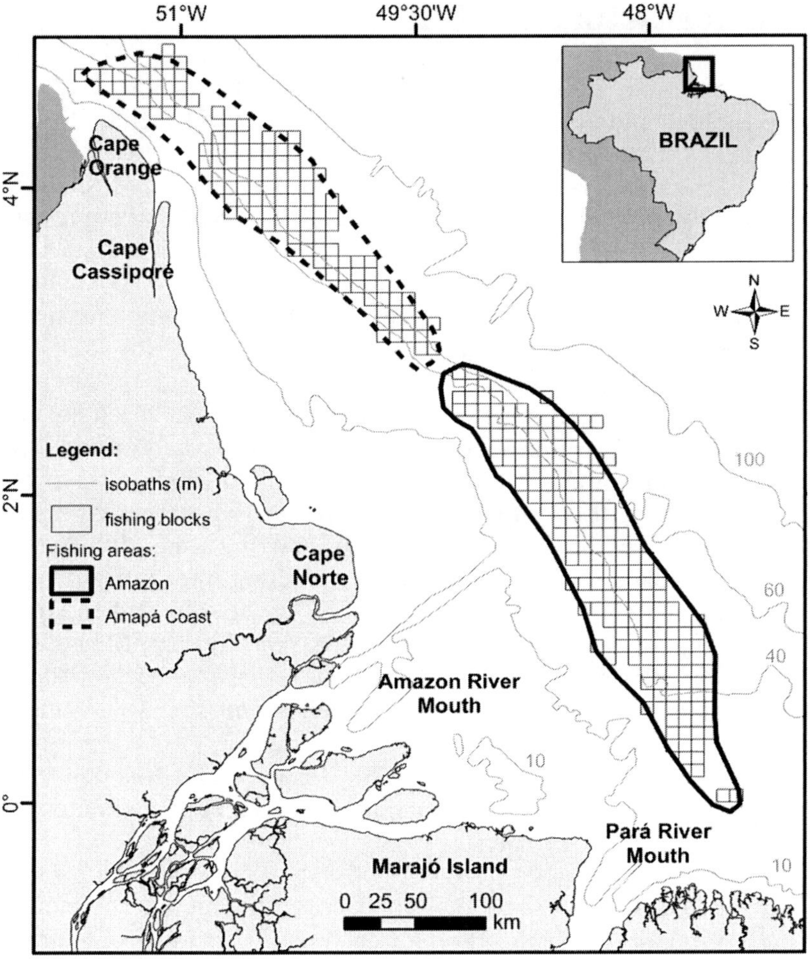

Figure 1 The Amazon continental shelf study area. The fishery sites and the representation of fishing blocks are shown.

Based on the typology of the mudbanks formed along the shelf, the region may be divided into two distinct areas: the Amazon and the Amapá coast. The Amazon coast is located southward of 2ºN latitude and is characterized by surface sand waves with heights of 3 to 6 m and wavelengths of 100 to 200 m. To the north, on the Amapá coast, the predominant features are large mudbanks with heights of 20 to 30 m and wavelengths of 6 to 8 km (NITTROUER; DEMASTER, 1996; NITTROUER et al., 1986).

2.2 Fishery data and CPUE

Data on catch and fishing efforts were extracted from the Project on Shrimp Biology and Fishing on the Northern Coast of Brazil undertaken by the Center for Research and Management of Fishery Resources of the Northern Brazilian Coast (*Projeto Biologia e Pesca de Camarões na Costa Norte do Brasil do Centro de Pesquisa e Gestão de Recursos Pesqueiros do Litoral Norte* - CEPNOR/ IBAMA) which collected monthly samples from at least two boats of the industrial fishing fleet. The data were collected from 2000 to 2004 at depths between 10 and 100 m along the Amazon continental shelf in daily fisheries.

The boats involved had steel hulls and an average total length of 22 meters. On average, these vessels performed 4 daily trawls with approximate durations of 5 to 6 hours using a double-rig system (ARAGÃO; SILVA, 1999). For each haul, onboard observers recorded a variety of information, such as the date, starting time, duration (h), initial latitude and longitude, depth (m), fishing site and amount of southern brown shrimp that were caught (kg). In addition to shrimp, catches on the Amazon continental shelf are typically composed of many other species that are generally discarded. Generally, part of this by-catch is retained depending on the species' economic value and the available on-board space. In the study area, the by-catch ratio was estimated as 7:1 (ARAGÃO; SILVA, 1999). However, for this paper, only the retained brown shrimp catch was registered and no estimates were made of the discarded by-catch.

The CPUE (capture per unit of effort) was adopted as the index of relative abundance and calculated for each haul. The CPUE is defined as the ratio between production (in kilograms) of southern brown shrimp and the haul duration (in hours), expressed in kg.h-1 (KING, 1995).

2.3 Geospatial information acquisition and analysis

Bottom salinity data for high water discharge at the Amazon River mouth (May) were spatialized based on LENTZ and LIMEBURNER (1995). The water discharge on May was chosen because the strongest seaward near-bottom salinity front occurs near the 20-m isobath at this time. Moreover, this tide-induced mixing influences the position and structure of the bottom salinity front that separates the well-mixed near-shore region from the stratified plume (GEYER et al., 1996).

Isobaths were obtained by digitizing nautical charts that were drawn by the Hydrography and Navigation Directorate (DHN) and the bathymetric chart that was prepared for the GEBCO Project (General Bathymetric Chart of the Oceans) (IOC et al., 2003).

The morphological classification of the Amazon continental shelf was determined in accordance with the characteristics proposed by NITTROUER et al. (1986), who divided the shelf into three distinct portions based on seismic stratigraphic data: (a) 10 to 40 m: wide and gently dipping inner shelf (< 1:3000), called the *topset*; (b) 40 to 60 m: middle shelf, with a relatively more abrupt inclination, known as the *foreset*, in which the inclination is 1:1000 on the Amazon, that is gentler than on the Amapá coast where the downward slope reaches 1:100; and (c) 60 to 100 m: outer slope, with a gentle downward slope (1:2000), called the *bottomset*.

The characterization of the Amazon continental shelf substrate was based on the descriptions of KUEHL et al. (1982); KUEHL et al. (1986b), who described the following regions according to the sedimentary structure: (a) interbedded mud and sand; (b) faintly laminated mud, and (c) mottled mud, characterized by sediments that had been considerably reworked by benthic macrofauna - and also according to the description of KUEHL et al. (1986a), who determined the annual sedimentation rate.

2.4 Shrimp fishery data analysis

By means of a GIS, the data that were organized in layers were integrated using the ArcGIS 9.3 program (ESRI, 2008). The bases for environmental factors (bathymetry, substrate and salinity) were digitized using the WGS-84 projection. The layers were integrated in pairs by applying the *Identity* command. Fishery data for the study period were analyzed simultaneously.

To identify areas with the highest CPUE values for southern brown shrimp, the fishing area was divided into 9-km x 9-km blocks with at least three hauls per block (Figure 1). The number of hauls and the median CPUE were calculated for each block and these results were then spatialized.

The CPUE spatial distribution patterns were based on 3,543 records of fishing operations (hauls) that had an average duration of 4.55 ± 1.3 hours, which represented 16,117.840 hours of total trawling. Descriptive CPUE statistical parameters were calculated for each depth interval and substrate type. Due to the bathymetric differences of each fishing area, the effect of depth on the CPUE was determined for each area. After negative normality tests had been performed on the CPUE data, the Kruskal-Wallis test was applied to the depth strata classification proposed by NITTROUER et al. (1986): 10 to 40 m and 40 to 60 m for the Amazon fishing area and 40 to 60 and 60 to 100 m for the Amapá Coast. To assess a possible correlation between depth and CPUE, the Spearman rank order correlation was applied (p = 0.05) for each fishing area.

3. RESULTS AND DISCUSSION

The boats of the industrial fishing fleet surveyed operated in a zone along the Amazon continental shelf that extended from the Amazon River mouth to Cape Orange. Figure 2a shows the spatial distribution of the hauls that were surveyed during the study period.

The median CPUE for the fishing blocks varied from 2 to 20 kg.h-1. Four areas presented the highest CPUEs for southern brown shrimp and these represent important fishing sites known by the local fishermen as Cabo Orange and Cascalho on the Amapá coast and Praia Grande and Lixeira in the Amazon region (Figure 2b).

3.1 CPUE distribution in relation to bathymetry

At the Amazon River mouth, the hauls were concentrated at depths of 40 to 60 m (67.83%), while on the Amapá coast, the largest haul frequency was at depths of 60 to 100 m (64.85%). The CPUE remained greater than 5.00 kg.h-1 except for the depth stratum of 60 to 100 m, where the median CPUE was 3.54 kg.h-1 (Table 1).

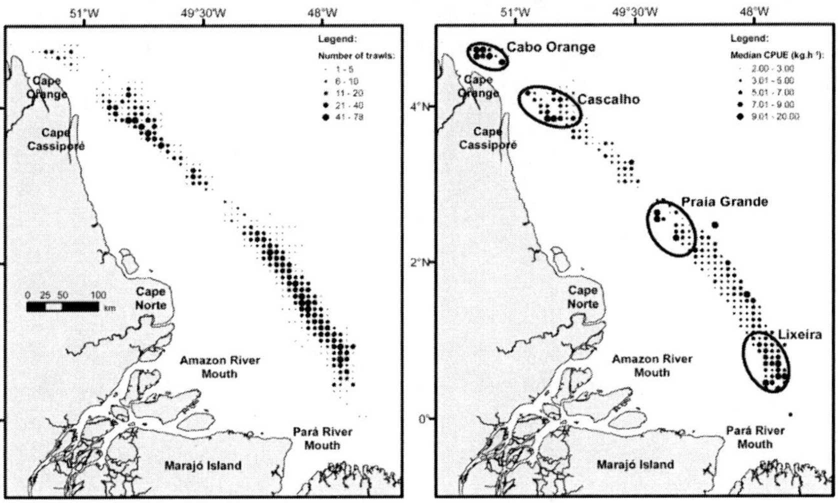

Figure 2 Number of hauls per fishing block (a) and spatial distribution of CPUE, which highlights the fishing sites for southern brown shrimp with the highest CPUE values (b).

On the Amapá Coast, the CPUE was significantly higher at depths of 40 to 60 m (H(1, N = 1098) = 87.64; $p < 0.05$). Moreover, for the Amazon, no significant differences were found between the different depth strata (H(1,N = 2313) = 0.36; $p = 0.55$) (Figure 3).

A low negative correlation was observed between the depth and CPUE on the Amapá Coast (rs = - 0.32; $p < 0.05$), while no correlation was found for the Amazon (rs = 0.03; $p = 0.13$) (Figure 4).

Even though catches of *Farfantepenaeus subtilis* occurred at depths from 12 to 100 m, there was clearly a higher incidence of hauls in the 40- to 60-m bathymetric stratum. Another common pattern on the Amapá coast was an evident reduction in the CPUE with increasing depth.

Most fishery efforts occur on the foreset of the Amazon delta, which overlaps the mottled mud sedimentary structure where the prawns form a dense population. We are convinced that the prawns are responsible for the bioturbation of the sediments and the creation of a sedimentary structure called mottled mud. Furthermore, the CPUE distribution on the Amazon shelf, with the largest CPUE areas adjacent to the mouths of the Amazon and the Pará Rivers, seems to be associated with the river discharges.

Table 1 Number of hauls (%) and CPUE (kg.h-1) of southern brown shrimp on the Amazon continental shelf according to the bathymetric and strata form data

Area	Bathymetric and strata form data*	Number of hauls (%)	Median CPUE	Minimum CPUE	Maximum CPUE
Amazon River Mouth	10-40 m, *topset*	32.2	5.00	0.03	53.33
	40-60 m, *foreset*	68.0	5.09	0.07	44.67
Amapá Coast	40-60 m, *foreset*	35.0	5.25	0.40	25.00
	60-100 m, *bottomset*	65.0	3.54	0.86	29.85

* The depth strata data were obtained from Nittrouer et al. (1986).

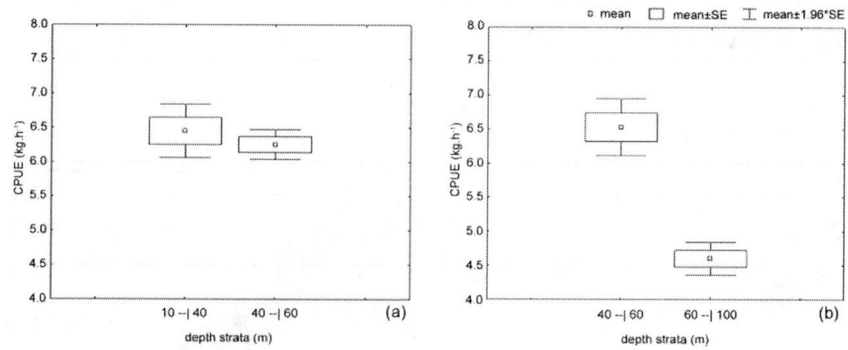

Figure 3 CPUE (kg.h-1) for southern brown shrimp for depth strata in the Amazon (a) and the Amapá coast (b) sites (SE: standard error).

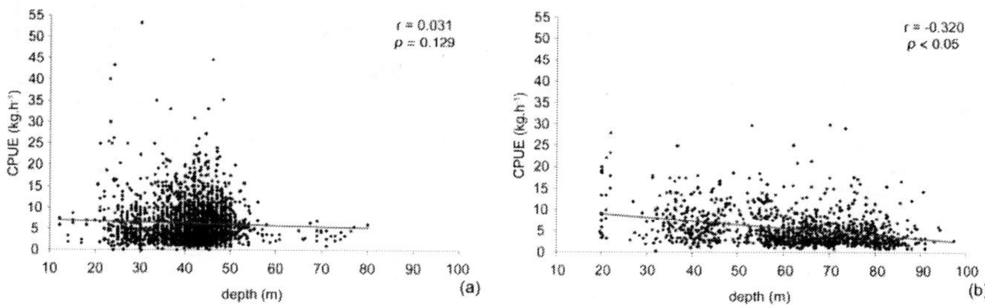

Figure 4 Relationship between the CPUE and depth for the mouth of the Amazon River (a) and the Amapá coast (b).

3.2 CPUE distribution in relation to substrate and bottom salinity

The hauls were concentrated in the mottled mud region (66.38%) (Figure 5b), which was followed by regions with faintly laminated mud (4.83%) and interbedded mud and sand (1.33%) in the region adjacent to the mouth of the Amazon River.

The largest median CPUEs were recorded in the bands of faintly laminated mud (6.86 kg.h-1) and interbedded mud and sand (6.00 kg.h-1); while in the mottled mud region the lowest median CPUE (4.42 kg.h-1) was observed. However, the number of hauls recorded for the latter region (2,352) was approximately ten times higher than the number of hauls that were recorded in the other two sediment bands (Table 2).

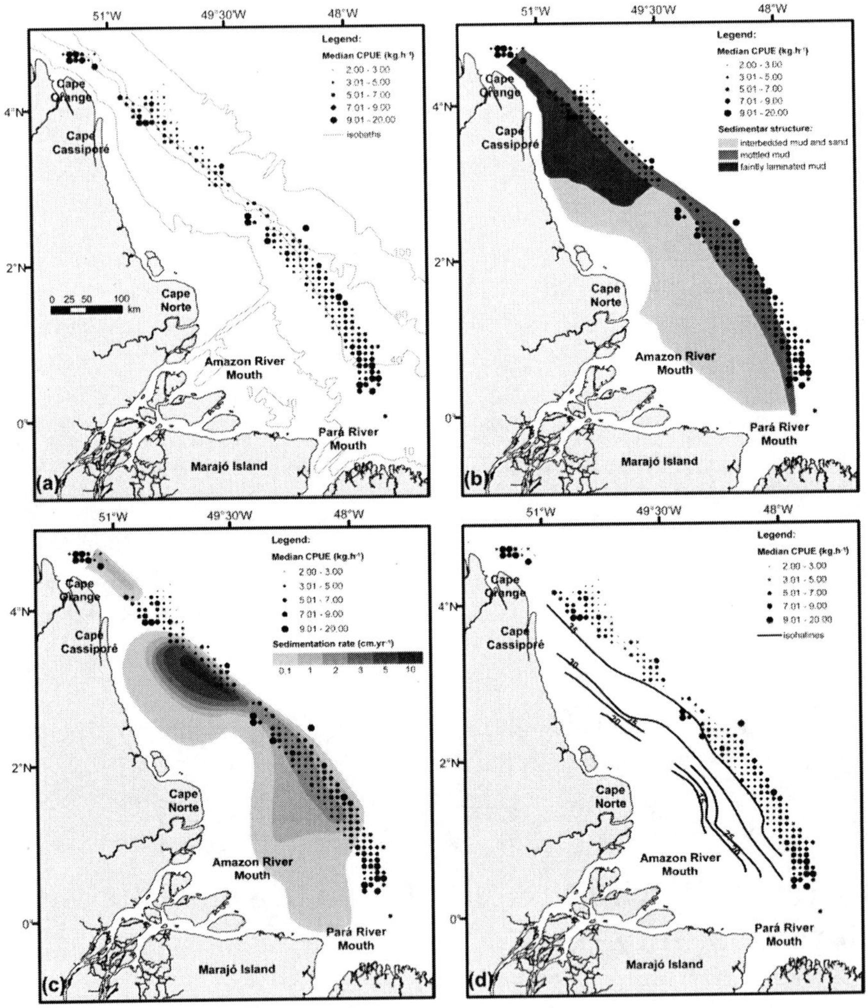

Figure 5 CPUE spatial distribution (kg.h-1) in relation to environmental characteristics: a) bathymetric chart (IOC et al., 2003); b) sedimentary structure (Nittrouer et al., 1986); c) sedimentation rate (Kuehl et al., 1986a); and d) bottom salinity (Lentz and Limeburner, 1995).

Table 2 Number of hauls (%) and CPUE (kg.h-1) for southern brown shrimp on the Amazon shelf for individual sediment bands

Sedimentary structure*	Number of hauls (%)	Median CPUE	Minimum CPUE	Maximum CPUE
Mottled mud	66.38	4.42	0.03	44.67
Interbedded mud and sand	1.33	6.00	1.00	43.33
Faintly laminated mud	4.83	6.86	1.25	28.00
Indeterminate	27.46	5.60	0.40	53.33

* The sedimentary structure data were obtained from Kuehl et al. (1982, 1986b).

The regions that exhibited the highest sedimentation rates (> 5 cm.year-1) related to the fishing blocks in which the CPUE was less than 5 kg.h-1 (Figure 5c). Some fishing blocks at the southern tip of the study area and as far north as 4oN latitude, close to Cape Cassiporé, occurred outside the boundaries of the study area of KUEHL et al. (1986a). A comparison between the sedimentation rate and the CPUE was not therefore viable.

According to the isohalines established by LENTZ and LIMEBURNER (1995), the majority of hauls were conducted in locations where the bottom salinity was always greater than 35 (Figure 5d).

What processes determine the patterns observed for *Farfantepenaeus subtilis* on the Amazon continental shelf? The Amazon River discharge forms a surface plume that advances 200 km into the ocean and 1000 km toward the northwest (GIBBS, 1970; LENTZ; LIMEBURNER, 1995). The river plume is a route for sediment deposits and determines an intense cycling of biologically relevant materials and the production of carbon that is exported to the sea as particulate organic carbon (POC), dissolved organic carbon (DOC) and nutrients (MCKEE et al., 2004). The Amazon continental shelf and the adjacent oceanic region are also directly affected by the North Brazil Current, which flows along and over the slope of the shelf break (RICHARDSON et al., 1994). Moreover, this current also drives the transport of sediment toward the northwest, outside the boundaries of the Brazilian shelf (KUEHL et al., 1986a).

The low salinity waters of the Amazon plume that extend along the coast have a surface effect, exhibiting average depths of 7.3 m and a transition boundary between river and oceanic waters at the 20-m isobath. Beyond that boundary, no significant seasonal variations occur in the 35 isohaline adjacent to the substrate (LENTZ; LIMEBURNER, 1995). This phenomenon means that the joint effect of the Amazon plume and the North Brazil Current causes a deposition and salinity gradient to form in deep waters, which may subsequently determine an environmental pattern for *Farfantepenaeus subtilis*. Given that the isohaline is displaced furthest seaward in May and the captures of *Farfantepenaeus subtilis* were concentrated at the 35 isohaline boundary, one may expect that the seasonal water discharge from the Amazon should not exhibit a straightforward relationship with the bottom spatial distribution of this species (MCKEE et al., 2004).

3.3 CPUE distribution in relation to sediment accumulation rates and bioturbation

It is likely that an intrinsic association between sediment accumulation rates and the spatial distribution of*Farfantepenaeus subtilis* has occurred. KUEHL et al. (1986a) found that on the inner shelf (< 15 m depth), a dense layer of sediment is present, which is continuously reworked by surface waves and by the strong tidal currents that limit the sedimentation rates. In the outer regions, known as the topset and foreset, and at approximately 80-100 kilometers from the coastline, the sedimentation rates increase progressively due to reduced sediment re-suspension. Finally, in the most oceanic areas, which correspond to the bottomset region, the sedimentation rates are generally < 1 cm.yr-1 in response to the reduced sediment contribution (Figure 6). These sediment stability categories, which were proposed by KUEHL et al. (1986a), may be controlling the *Farfantepenaeus subtilis*population gradients.

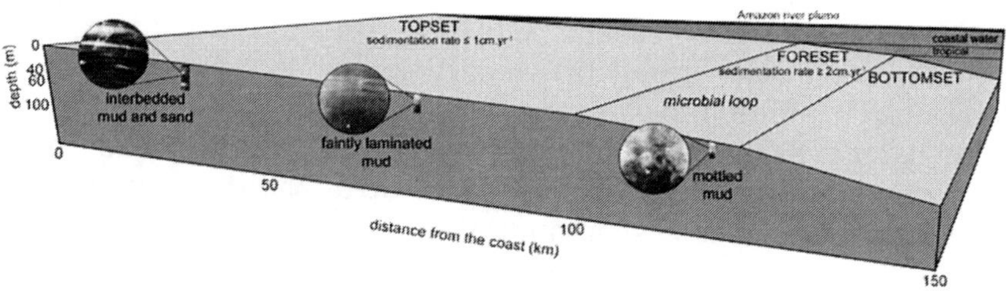

Figure 6 Cross-shelf transect showing the relationships between deltaic stratigraphy, sediment accumulation rates and the microbial loop activity hypothesis in relation to the *F. subtilis* spatial distribution and abundance.

What factor might also explain the density of *F. subtilis* in the foreset region? Analyses of the diet of several Penaeidae have shown that complex communities of bacteria, protozoa, fungi and algae encrusted in a matrix of extracellular polysaccharides secreted by bacteria (COSTERTON; IRVIN, 1981) or microbial material and detritus, plant debris and small animals are important sources of energy for these predators (DALL et al., 1990). In turn, more than 95% of the organic material in marine ecosystems consists of polymers and high molecular weight (HMW) compounds, such as proteins, polysaccharides and lipids, which are not assimilated by many of the marine organisms at the higher trophic levels (FENCHEL, 2008). It is possible that the bioturbation process that is conducted by various organisms in the foreset region is closely related to these rich and stable sources of energy that are stored in the substrate where the microbial loop process occurs (Figure 6). This microbial process proposed by POMEROY et al. (2007) can explain why industrial fishing of *Farfantepenaeus subtilis* is concentrated in the foreset region, corresponding to the location where the benthonic organisms revolve the muddy substrate and create the mottled pattern.

Furthermore, to understand exactly where the fishery resources are utilized is essential to fishery management (MEADEN; CHI, 1996). Therefore, this information will make it possible to determine areas that should be prohibited for southern brown shrimp capture and that require habitat recovery.

These results indicate that fishery data can also be obtained from commercial fishery sources. A GIS, such as the one developed in this study, can use information available in logbooks and vessel monitoring systems. However, this is only true if participation in these programs is an obligatory condition for an annual fishery license for the industrial shrimp fishing fleets that operate along the North Brazilian coast. Hence, new information would be generated based on integrated fishery, environmental and geographical data.

4. CONCLUSION

The concentration of industrial fishing fleet hauls for southern brown shrimp in the mottled mud band of the Amazon continental shelf suggests that the overlapping of environmental factors determines the greater occurrence of *F. subtilis* in the foreset region at depths of 40 to 60 m. In this region, bottom morphology, substrate type (mud), sedimentation rate (≥ 2 cm.y-1), and salinity (> 35) constitute the ideal habitat for adults of the species.

The abundance distribution patterns for southern brown shrimp and their spatial variability were determined using geo-referenced fishery information that was provided by commercial fishing fleets and combined with environmental data in a GIS framework. Therefore, in having depth, bottom-morphology and salinity data and the locations of higher and lower capture per unit effort values, it is possible to use this information to propose a more efficient fishery strategy along the Amazon continental shelf.

ACKNOWLEDGMENTS

The present research was supported by CAPES through Marine Science Announcement 2009. We would like to thank the Aquatic Ecology and Fishery postgraduate program of the Universidade Federal do Pará. PWM Souza Filho is grateful to CNPq for the research grant during this study. The authors also thank Stephen Ferrari for comments and the English review, and the anonymous reviewers for their inestimable contributions.

REFERENCES

1. ARAGÃO, J. A. N.; SILVA, K. C. A. National report of Brazil. In: National reports presented and stock assessment reports prepared at the CFRAMP/FAO/DANIDA. Stock Assessment Workshop on the Shrimp and Groundfish Fisheries on the Guiana-Brazil Shelf. Rome: FAO, 1999. p. 1-16.

2. BARTHEM, R. B. Ocorrência, distribuição e biologia dos peixes da baía de Marajó, estuário amazônico. Bol. Mus. Para. Emílio Goeldi, Ser. Zool., v. 2, n. 1, p. 49-69, 1985.

3. CADDY, J. F.; GARCIA, S. Fisheries thematic mapping: A prerequisite for intelligent management and development of fisheries. Oceanogr. Trop., v. 21, n. 1, p. 31-52, 1986.

4. COSTERTON, J. W.; IRVIN, R.T.; CHENG, K. J. The bacterial glycocalyx in nature and disease. Annu. Rev. Microbiol., v. 35, p. 299-324, 1981.

5. DALL, W.; HILL, B. J.; ROTHLISBERG, P. C.; STAPLES, D. J. The biology of the Penaeidae. In: BLAXTER, J. H. S.; SOUTHWARD, A. J. (Eds.).

Advanced Marine Biology, San Diego, Academic press. London: Academic Press, 1990. 489 p.

6. DEMASTER, D. J.; SMITH JR. W. O.; NELSON, D. M.; ALLER, J. Y. Biogeochemical processes in Amazon shelf waters: chemical distributions and uptake rates of silicon, carbon and nitrogen. Cont. Shelf. Res., v. 16, n. 5/6, p. 617-643, 1996.

7. DRAGOVICH, A. Guianas-Brazil shrimp fishery and related U. S. research activity. Mar. Fish. Rev., v. 43, n. 2, p. 9-18, 1981.

8. ESRI. What is ArcGIS? Redlands: Environmental Systems Research Institute, ESRI Software 9.3 Library, 2008. 130 p.

9. FAO. The state of world fisheries and aquaculture 2008. Rome: FAO Fisheries and Aquaculture Department, 2009. 176 p.

10. FENCHEL, T. The microbial loop - 25 years later. J. Exp. Mar. Biol. Ecol., v. 366, n. 1/2, p. 99-103, 2008.

11. GEYER, W.R., BEARDSLEY, R.C., LENTZ, S.J., CANDELA, J., LIMEBURNER, R., JOHNS, W.E., CASTRO, B.M., SOARES, I.D., 1996. Physical oceanography of the Amazon shelf. Cont. Shelf Res. 16, 575-616.

12. GIBBS, R. J. The Geochemistry of the Amazon River System: Part I. The factors that control the salinity and the composition and concentration of the suspended solids. Geol. Soc. Am. Bull., v. 78, n. 10, p. 1203-1232, 1967.

13. GIBBS, R. J. Circulation in the Amazon River estuary and adjacent Atlantic ocean. Mar. Fish. Rev., v. 28, n. 2, p. 113-123, 1970.

14. GILLET, R. Global study of shrimp fisheries. Rome: FAO, 2008. 331 p.

15. GRAAF, G.; MARTTIN, F.; AGUILAR-MANJARREZ, J.; JENNESS, J. Geographic Information Systems in Fisheries Management and Planning: Technical manual. FAO Fisheries Technical Paper No. 449. Rome: FAO, 2003. 162 p.

16. HOLTHUIS, L. B. Shrimps and prawns of the world: An annotated catalogue of species of interest to fisheries. FAO species catalogue. Vol.1. FAO Fisheries Synopsis. 125. Rome: FAO, 1980. 271 p.

17. IBAMA. Estatística da pesca 2007 Brasil: grandes regiões e unidades da federação. Brasília: IBAMA, 2008. 151 p.

18. IOC; IHO; BODC. Centenary Edition of the GEBCO Digital Atlas, General Bathymetric Chart of the Oceans. Liverpool: British Oceanographic Data Centre. 2003. <http://www.gebco.net/>

19. ISAAC, V. J.; DIAS-NETO, J.; DAMASCENO, F. G. Camarão-rosa da Costa Norte: biologia, dinâmica e administração pesqueira. Série de Estudos de Pesca, Coleção Meio Ambiente. Brasília: IBAMA, 1992. 187 p.

20. KING, M. Fisheries biology, assessment and management. Oxford: Fishing New Books, 1995. 342 p.

21. KUEHL, S. A.; NITTROUER, C. A; DEMASTER, D. J. Modern sediment accumulation and strata formation on the Amazon continental shelf. Mar. Geol., v. 49, p. 279-300, 1982.

22. KUEHL, S. A.; DEMASTER, D. J.;. NITTROUER, C. A Nature of sediment accumulation on the Amazon continental shelf. Cont. Shelf Res., v. 6, p. 209-225, 1986a.

23. KUEHL, S. A.; NITTROUER, C. A; DEMASTER, D. J. Distribution of sedimentary structures in the Amazon subaqueous delta. Cont. Shelf Res., v. 6, p. 311-336, 1986b.

24. LAFFAN, S. W.; SKIDMORE, A. K.; FRANKLIN, J. Geospatial analysis of species, biodiversity and landscapes: introduction to the second special issue on spatial ecology. Int. J. Geogr. Inf. Sci., v. 26, p. 2003-2007, 2012.

25. LENTZ, S. J.; LIMEBURNER, R. The Amazon river plume during AMASSEDS: spatial characteristics and salinity variability. J. Geophys. Res., v. 100, n. 2, p. 2355-2375, 1995.

26. MARTINEZ, J. M.; GUYOT, J. L.; FILIZOLA, N.; SONDAG, F. Increase in suspended sediment discharge of the Amazon River assessed by monitoring network and satellite data. Catena, v. 79, p. 257-264, 2009.

27. MEADE, R. H.; NORDIN, C. F.; CURTIS, W. F.; RODRIGUES, F. M. C.; VALE, C. M.; EDMOND, J. M. Sediment loads in the Amazon River. Nature, v. 278, p. 161-163, 1979.

28. MEADEN, G. GIS in Fisheries Management. GeoCoast, v. 1, n. 1, p. 82-101. 2000.

29. MEADEN, G. J.; CHI, T. Geographical Information Systems. Applications to Marine Fisheries. FAO Fisheries Technical Paper. Rome: FAO, 1996. 335 p.

30. MCKEE, B. A.; ALLER, R. C.; ALLISON, M. A.; BIANCHI, T. S.; KINEKE, G. C. Transport and transformation of dissolved and particulate materials on continental margins influenced by major rivers: benthic boundary layer and seabed processes. Cont. Shelf Res., v. 24, n. 7, p. 899-926, 2004.

31. MORRIS, L., BALL, D., 2006. Habitat suitability modelling of economically important fish species with commercial fisheries data. ICES J. Mar. Sci. 63, 1590-1603.

32. NITTROUER, C. A.; DEMASTER, D. J. Sedimentary Process on the Amazon Continental Shelf: Past, Present and Future Research. Cont. Shelf Res., v. 6, p. 5-30, 1986.

33. NITTROUER, C. A.; DEMASTER, D. J. , The Amazon shelf setting: tropical, energetic, and influenced by a large river. Cont. Shelf Res., v. 16, p. 553-573, 1996.

34. NITTROUER, C. A.; KUEHL, S. A.; DEMASTER, D. J.; KOWSMANN, R. O. The deltaic nature of Amazon shelf sedimentation. Geol. Soc. Am. Bull., v. 97, n. 4, p. 444-458, 1986.

35. PEREZ-FARFANTE, I. 1967. A new species and two new subspecies of shrimp of the genus Penaeus form western Atlantic. Proc. Biol. Soc. Wash. 80, 83-100.

36. POMEROY, L. R.; WILLIAMS, P. J. L.; AZAM, F.; HOBBIE, J. E. The microbial loop. Oceanography, v. 20, n. 2, p. 28-33, 2007.

37. RICHARDSON, P. L.; HUFFORD, G. E.; LIMEBURNER, R.; BROWN, W. S. North Brazil Current retroflection eddies. J. Geophys. Res., v. 99, n. 3, p. 5081-5093, 1994.

38. RIGHTON, D.; MILLS, C. Application of GIS to investigate the use of space in coral reef fish: a comparison of territorial behaviour in two Red Sea butterflyfishes. Int. J. Geogr. Inf. Sci., v. 20, n. 2, p. 215-232, 2006.

39. RIOLO, F. A geographic information system for fisheries management in American Samoa. Environ. Model. Softw., v. 21, n. 7, p. 1025-1041, 2006.

40. SAITOH, S-I.; MUGO, R.; RADIARTA, I. N.; ASAGA, S.; TAKAHASHI, F.; HIRAWAKE, T.; ISHIKAWA, Y.; AWAJI, T. In, T.; SHIMA, S. Some operational uses of satellite remote sensing and marine GIS for sustainable fisheries and aquaculture. ICES J. Mar. Sci., v. 68, n. 4, p. 687-695, 2011.

41. SANTOS, M. L. S.; MEDEIROS, C.; MUNIZ, K.; FEITOSA, F. A. N.; SCHWAMBORN, R.; MACEDO, S. J. Influence of the Amazon and Pará rivers on water composition and phytoplankton biomass on the adjacent Shelf. J. Coast. Res., v. 24, p. 585-593, 2008.

42. SKIDMORE, A. K.; FRANKLIN, J.; DAWSON, T. P.; PILESJÖ, P. Geospatial tools address emerging issues in spatial ecology: a review and commentary on the Special Issue. Int. J. Geogr. Inf. Sci., v. 25, n. 3, p. 337-365, 2011.

43. SILVA, A. C.; ARAUJO, M.; BOURLÈS, B. Seasonal variability of the Amazon River plume during REVIZEE program. Trop. Oceanogr., v. 38, n. 1, p. 76-87, 2010.

44. WALUDA, C.M., RODHOUSE, P.G., TRATHAN, P.N., PIERCE, G.J., 2001. Remotely sensed mesoscale oceanography and the distribution of Illex argentinus in the South Atlantic. Fish. Oceanogr. 10, 207-216.

5

ASSESSMENT OF LONGITUDINAL GRADIENTS IN NEMATODE COMMUNITIES IN THE DEEP NORTHERN GULF OF MEXICO AND CONCORDANCE WITH BENTHIC TAXA

Jyotsna Sharma1, Jeffrey G. Baguley2, Paul A. Montagna3, and Gilbert T. Rowe4

1Department of Biology, University of Texas at San Antonio, One University Circle, San Antonio, TX 78249, USA

2Department of Biology, The University of Nevada Reno, Mailstop 314, Reno, NV 89557, USA

3Harte Research Institute, Texas A&M University—Corpus Christi, 6300 Ocean Dr., Corpus Christi, TX 78412, USA

4Department of Marine Biology, Texas A&M University at Galveston, Galveston, TX 77553, USA

ABSTRACT

Meiobenthic nematode assemblages were examined at 16 stations along two transects on the eastern and western boundaries of the deep northern Gulf of Mexico (dNGOM) at depths of 212–3000 m. The highest abundance (297 individuals 10 cm−2) and number of genera (71) occurred at stations near the Mississippi River delta. Number of genera decreased with increasing depth, and showed differences in community composition between the east and west regions. The dominant family,

Comesomatidae, was represented by Sabatieria that was present at most shallow stations but absent at greater water depths. A significant difference in nematode feeding morphology was observed between depth groups but not between the two transects at different longitudes. Patterns of nematode community structure are congruent with harpacticoid copepods. Overall, the higher abundance and diversity of nematodes in the north-central Gulf of Mexico is consistent with findings of other benthic taxa and reflects organic material loading from the Mississippi River driving deep sea communities in the Gulf. The east-west gradient in composition of nematode communities suggests that nematode assemblages have well-defined distribution patterns similar to other meiobenthic taxa in the GOM but they are not aligned in the bathymetric zones observed in macrofauna, megafauna and demersal fishes.

1. INTRODUCTION

Deep-sea biodiversity remains woefully understudied, and it is estimated that only 1% of marine species is currently described [3]. Most studies of deep-sea benthic diversity are based on mega- or macrofaunal organisms, but studies of the meiobenthos and nannobenthos have revealed a diverse fauna [4–6]. Meiobenthic abundance and diversity exceed that of macrofauna and megafauna, and the biomass of the meiofauna community dominates that of macro- and megafauna with increasing water depth [7]. Comparison of biomass and respiration rates also suggests that meiofauna have greater rates of carbon flow through food webs in deeper waters in the Gulf of Mexico than the larger macrofauna and megafauna [8, 9].

As the numerically dominant meiofaunal taxon in deep-sea sediments, nematodes play significant roles in marine biogeochemical cycles through bioturbation and related processes that enhance sediment oxygenation and solute transport, thereby increasing rates of demineralization [10–12]. However, despite their numerical dominance, high diversity, and functional significance [13, 14], studies of this ubiquitous phylum are rare. Studies of nematode biodiversity in the deep sea conducted in the Pacific and Atlantic Oceans [15–19] reveal both latitudinal [20] and depth-related gradients [21–23]. Additionally, Danovaro et al. [24] observed a longitudinal gradient in nematode biodiversity in the Mediterranean Sea that was related to surface primary production and sediment organic matter. It appears that the spatial biodiversity and distribution of nematodes are affected by deep-sea habitat heterogeneity [6], further supporting the broader contention that habitat heterogeneity plays an important role in maintaining deep-sea species diversity [25] across taxa and size classes.

To date there is little information available on nematode distribution in the northern Gulf of Mexico, with most studies conducted in the intertidal zone. A recent checklist of the free-living marine nematodes of the Western North Atlantic and Gulf of Mexico [2] lists 190 species, all of which occur in shallow waters of the continental shelf. Of these 66% was endemic, and the identifications are also associated with poor taxonomic descriptions to confirm identifications [26]. Cobb [27], Chitwood [28], and Chitwood and Timm [29] described the free-living marine nematodes of various habitats from the Texas coast. Intertidal nematodes in the coastal northeastern Gulf of Mexico were studied by King [30], Hopper [31], and Keppner [32]. The first studies of subtidal nematodes in this region were conducted in the East Flower Garden Banks [33–36] at a brine seep at 70 m water depth. The nematode fauna of the Caribbean has been described by Boucher and Gourbault [37], de Jesus-Navarrete and Herrera-Gomez [38], and de Jesus-Navarrete [39]. Castillo-Fernandez and Lambshead [40] have also described nematodes from the southern GOM. However, nematode collections from the deep GOM have been limited to ecological studies of the meiofaunal community as a whole [41]. Due to the significant economic importance of the GOM and the inevitability of future oil and gas exploration in deep water, it is critical to understand the regional differences of its biodiversity.

Here we present the first ecological study of nematode assemblages in the deep northern GOM, specifically investigating patterns of nematode community structure over depth and longitude in the northern deep GOM. The north-central and northeastern GOM are characterized by strong river outflow of nutrients and organic matter that serve to enhance surface primary production and fuel-enhanced benthic secondary production. Surface productivity patterns have been well characterized [42], and physical oceanographic conditions create a favorable environment for primary productivity in an otherwise oligotrophic subtropical basin [43]. Specifically, in the northeastern region the Loop Current, Loop Current Eddies, and other smaller cyclonic and anticyclonic eddies transport nutrient-rich shelf waters onto the continental slope [43]. Comparatively, the northwestern GOM receives relatively less river outflow and is relatively more oligotrophic [42].

The analysis of Baguley et al. [41] revealed a strong influence of organic matter outflow from the Mississippi and Atchafalaya Rivers in northeast GOM on meiofauna communities. Additionally, Baguley et al. [41] reported a significant longitude by depth interaction of the meiofauna community abundance in the northern GOM and associated interactions with seafloor topographic features such as the Mississippi Trough, DeSoto Canyon, and Florida Escarpment. The distribution patterns of several macrofaunal taxa in the GOM are associated with surface phytoplankton production and show a horizontal distribution pattern that is related to the organic input of the Mississippi River [44]. Therefore, nematode assemblages may differ by longitude in the northern GOM in response to longitudinal variations in physical, chemical, and biological water column processes [42–44]. To test this hypothesis, we examined (1) the composition of the nematode community along depth and longitudinal gradients, (2) the structure of the nematode community in relation to water depth and longitude gradients, (3) the functional diversity of the nematode community as determined from the morphology of the buccal cavity, and (4) compared the community structure patterns of nematode spatial variability with that of Harpacticoida and Polychaeta. This study also provides a valuable baseline dataset for comparison with potential future environmental perturbations such as the Deepwater Horizon oil spill from April 20–July 15, 2010.

2. MATERIALS AND METHODS

2.1. Field and Laboratory Methods

Nematode biodiversity and community structure were analyzed at 16 stations in the northern GOM deep sea (Figure 1). Thirteen of 16 stations were along two transects in the northeastern (S39–S44) and northwestern (RW1–RW6, AC1, Alaminos Canyon) GOM that spanned a depth range of 212–3000 m. All stations were sampled as part of the Deep Gulf of Mexico Benthos (DGoMB) program during summer, 2000 [45]. Detailed field and laboratory sampling and sorting procedures are outlined by Baguley et al. [5, 41]. Briefly, samples were collected with a 0.2 m2 GOMEX boxcorer equipped with an array of smaller subsampling corers attached by hose clamps along a central stainless steel bar within the box. Meiofauna subcores had a 5.5 cm inner diameter and penetrated approximately 15 to 30 cm into the sediments. The top 3 cm of sediment was retained for examination, as it has been shown to contain over 80% of the meiofauna in deep-sea GOM samples [5]. Five replicate core samples were obtained from each of the sixteen stations. The taxonomic study of nematodes was initiated about 4 years after samples were collected and thus replicates from some stations were lost due to desiccation prior to taxonomic analysis. Only three replicates were analyzed from C1 and AC1. Only four replicates were analyzed from stations RW5, RW6, and S44.

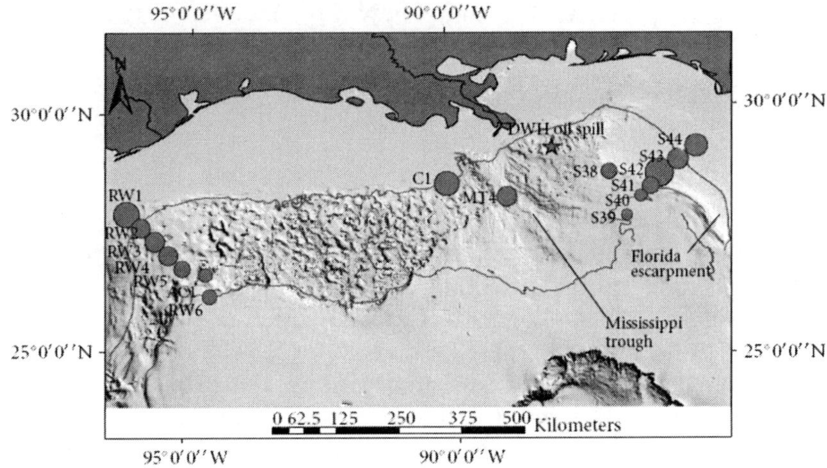

Figure 1: Spatial analysis of nematode abundance (N 10 cm−3) at dNGOM stations. Bubble size equals relative nematode abundance.

In the field, meiofauna were narcotized with 7% MgCl2 and preserved in 5% buffered formalin. In the laboratory, fine sediments were removed by rinsing samples through a 45 µm mesh sieve. Meiofauna retained on the 45 µm mesh sieve were extracted from coarser sediments and other debris using the Ludox centrifugation method of de Jonge and Bouwman [46]. Nematodes were hand picked under a Wild M3 stereoscope and transferred to ethanol and then cleared in glycerol by the method of Seinhorst [47] for identification on a Zeiss Axioskop.

2.2. Nematode Taxonomy

Standard identification keys and manuals [48] were used for taxonomic identification to genus level based on morphological characteristics. Data on nematode distributions at the species level are scarce and most deep-sea nematodes are undescribed, and thus it has been practical to identify nematode morphotypes to genus [49]. Many characters to distinguish species are based only on males and furthermore juveniles are indistinguishable beyond the genus level. Vanreusel et al. [6] note that several studies indicate that genus composition is an indicator of macroecological patterns. The samples will be archived at the National Museum of Natural History, Washington, DC, USA.

2.3. Feeding Groups

Genera were classified into four feeding groups based on buccal morphology, namely, 1A (selective deposit feeders), 1B (nonselective deposit feeders), 2A (epigrowth feeders), and 2B (omnivores and predators) [50]. Although Moens and Vincx [51] have proposed a broader classification into six feeding groups, the original categories proposed by Wieser [50] are used here because several genera from this region have not yet been described, and thus little information is available for them.

2.4. Statistical Analysis

Genus richness and evenness of nematodes were regressed against depth to determine the effect of this variable. Multivariate analyses were performed using PRIMER 6.1 [52]. Ordination of samples was performed using nonparametric multidimensional scaling (nMDS) to visualize the similarity patterns [53]. Jaccard similarity index was used to calculate similarity among samples after transformation for presence/absence of genera due to unequal sample size as a result of loss of some replicates and the ability of this transformation to downweigh the effect of common species [53]. Multivariate statistical differences in community structure over depth and longitude were determined using two-way crossed Analysis of Similarities (ANOSIM) where global $R = 1$ indicates complete separation of groups and global $R = 0$ indicates no separation [53]. Diversity analysis was also calculated using PRIMER 6.1 to determine the evenness of genera in each sample (Primer-E Ltd.). SIMPER analysis was used to investigate longitudinal differences in genus and family abundance and percent contribution to community structure. Similarity profile (SIMPROF) tests were performed on group average cluster analyses to test the null hypothesis that longitudinal differences in nematode community structure do not exist in the northern Gulf of Mexico deep sea.

2.5. Cross-Taxon Spatial Variability

The community structure patterns of nematode spatial variability were compared with that of Harpacticoida and Polychaeta using the RELATE procedure in Primer, which tests for matching among the station patterns computed using the Bray-Curtis similarity resemblance matrices. This is testing to determine if MDS patterns for the three groups are the same. This comparison can only be made where genus or species level identifications were performed and the organisms were found. This occurred only at the following 13 stations: AC1, C1, RW1, RW2, RW3, RW4, RW5, RW6, S35, S38, S39, S40, and S43.

3. RESULTS

3.1. Abundance

At the 16 stations investigated here for nematode community structure, abundance ranged from roughly 53–300 individuals 10 cm−2 (Table 1). Nematode abundance decreased with depth along both transects (Figure 2). The highest abundance, 267 individuals 10 cm−2, occurred at station C1 (depth 482 m) near the mouth of the Mississippi delta. The lowest abundance, 53 individuals 10 cm−2, occurred at station S39 in the eastern transect.

Table 1: DGoMB station locations, depth, and average meiofaunal abundance (average of five replicate cores) for pooled taxonomic groups.

Station	Latitude	Longitude	Depth	Abundance ($N/10cm^2$)
RW1	27.50014	−96.0029	212	277 ± 50
RW2	27.25403	−95.7468	950	145 ± 80
RW3	27.00836	−95.4924	1340	173 ± 16
RW4	26.75142	−95.2502	1575	164 ± 30
RW5	26.50753	−94.9967	1620	116 ± 50
AC1	26.39357	−94.5731	2440	92 ± 33
RW6	25.99730	−94.4956	3000	100 ± 40
S44	28.74999	−85.7477	212	221 ± 43
C1	28.05984	−90.2499	336	267 ± 134
S43	28.50294	−86.0768	362	209 ± 51
S42	28.25100	−86.4193	763	134 ± 33
MT4	27.82761	−89.1662	1401	172 ± 29
S38	28.27995	−87.3276	2627	97 ± 27
S40	27.83948	−86.7514	2972	60 ± 11
S41	28.01364	−86.5734	2974	65 ± 17
S39	27.48368	−86.9998	3000	53 ± 15

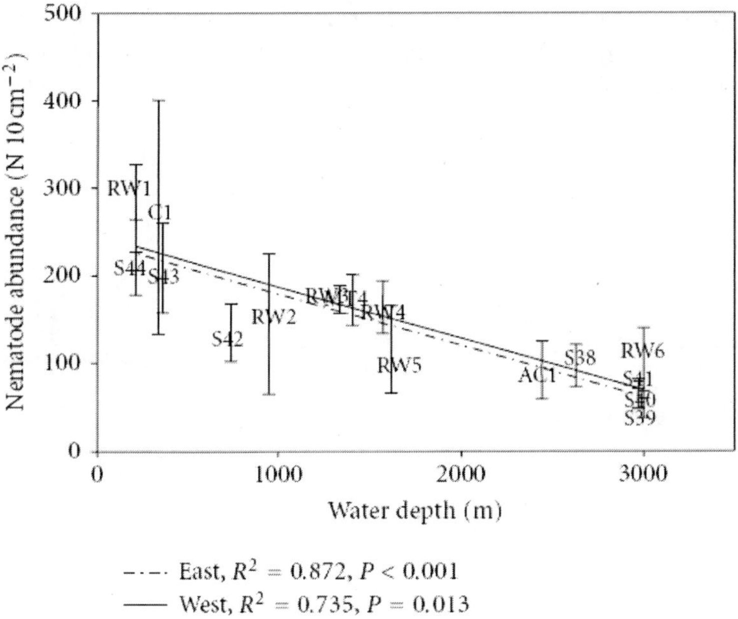

Figure 2: Abundance of nematodes as a function of depth at eastern and western transects in dNGOM.

3.2. Taxonomic Composition and Diversity

A total of 4764 nematode individuals were identified (Table 2) from the 16 stations, representing 128 nematode genera and 35 families. The highest number of genera (71) occurred at station S43 (depth 362 m) in the eastern transect and the lowest number of genera (23) occurred at station RW6 (depth 3000 m) in the western transect. The cumulative percent composition of 42 genera in the total nematode fauna examined is less than 0.1%, and 103 genera comprised less than 1% of the total fauna indicating the presence of many rare genera (Table 2). Our study is the first record of 56 genera that have not been recorded previously in the GOM [2].

Table 2: Nematode taxa in meiofaunal samples in the dGOM (listed by systematic scheme of de Ley and Blaxter [1]). Table includes order (in capital letters), family, genus, trophic guild assignment, and cumulative percent contribution of a genus to total nematode fauna. Genera not recorded previously in the GOM [2] are marked*.

Family/genus	Guild	Total abundance (cumulative % contribution)
ENOPLIA		
Benthimermithidae		
*Benthimermis**	1A	3 (0.06)
Rhaptothyreidae		
*Rhaptothyreus typicus**	1A	2 (0.04)
Thoracostomopsidae		
Enoplolaimus	2B	27 (0.57)
Enoploides	2B	8 (0.17)
Mesacanthion	2B	4 (0.08)
*Paramesacanthion**	2B	1 (0.02)
*Thoracostomopsis**	2B	1 (0.02)
Anoplostomatidae		
Anoplostoma	1B	2 (0.04)
*Chaetonema**	1B	1 (0.02)
Phanodermatidae		
Crenopharynx	1A	1 (0.02)
*Phanoderma**	1A	67 (1.41)
Phanodermopsis	1A	6 (0.13)
Anticomidae		
Anticoma	1A	8 (0.02)
Ironidae		
Dolicholaimus	2B	2 (0.04)
Syringolaimus	2B	49 (1.03)
Oxystominidae		
Halalaimus	1A	276 (5.79)
*Litinium**	1A	4 (0.08)

*Nemanema**	1A	4 (0.08)
*Oxystomina**	1A	17 (3.59)
*Thalassoalaimus**	1A	49 (1.03)
*Wieseria**	1A	3 (0.06)
Oncholaimidae		
Oncholaimus	2B	8 (0.17)
Viscosia	2B	17 (0.36)
Enchelidiidae		
Eurystomina	2B	21 (0.44)
*Bathyeurystomina**	2B	14 (0.29)
Pareurystomina	2B	9 (0.19)
Calyptronema	2B	19 (0.40)
Symplocostoma	2B	7 (0.15)
Rhabdodemaniidae		
*Rhabdodemania**	2B	8 (0.17)
Pandolaimidae		
*Pandolaimus**	2B	1 (0.02)
Trefusiidae		
*Rhabdocoma**	2B	7 (0.15)

Family/genus	Guild	Total abundance (cumulative % contribution)
Xenellidae		
Xenella	1A	3 (0.06)
CHROMADORIDA		
Desmoscolecidae		
Calligyrus	1A	28 (0.59)
Greefiella	1A	47 (0.99)
Desmoscolex	1A	121 (2.54)
Desmoscolex 2	1A	6 (0.13)
*Pareudesmoscolex**	1A	5 (0.10)
Quadricoma	1A	6 (0.13)
Tricoma	1A	(10.02)
Epsilonematidae		
*Epsilonema**	1A	11 (0.23)
Draconematodae		
*Apenodraconema**	2A	5 (0.10)
Chromadoridae		
Acantholaimus	2A	1 (0.02)
Actinonema	2A	12 (0.25)
Chromadora	2A	45 (0.94)
Chromadorina	2A	15 (0.31)
*Dichromadora**	2A	3 (0.06)
Euchromdora	2A	1 (0.02)
Hypodontolaimus	2A	5 (0.10)

Neochromadora	2A	8 (0.17)
*Neotonchus**	2A	8 (0.17)
Prochromadorella	2A	76 (0.02)
Ptycholaimellus	2A	1 (1.59)
Rhips	2A	1 (0.02)
Ethmolaimidae		
*Filitonchus**	1A	1 (0.02)
Selachinematidae		
Halichoanolaimus	2B	68 (1.43)
*Richtersia**	2A	4 (0.08)
*Selachinema**	2A	38 (0.80)
*Synonchiella**	2B	3 (0.06)
Cyatholaimidae		
Pomponema	2A	34 (0.71)
Paracanthonchus	2A	4 (0.08)
*Cyatholaimus**	2A	79 (1.66)
Longicyatholaimus	2A	12 (0.25)
Marylynnia	2A	3 (0.06)
Desmodoridae		
Acanthopharyngoides	2B	1 (0.02)
Desmodora	2A	73 (1.53)
Paradesmodora	2A	1 (0.02)
Spirinia	2A	66 (1.39)
Perspiria	2A	1 (0.02)

Family/genus	Guild	Total abundance (cumulative % contribution)
Microlaimidae		
Bolbolaimus	2A	6 (0.13)
Microlaimus	2A	187 (3.92)
*Molgolaimus**	2A	6 (0.13)
Paramicrolaimus	2A	1 (0.02)
Monoposthiidae		
Monoposthia	2A	1 (0.02)
Ceramonematidae		
*Ceramonema**	1A	90 (1.89)
Pselionema	1A	1 (0.02)
*Pterygonema**	1A	1 (0.02)
MONHYSTERIDA		
Xyalidae		
*Ammotheristus**	1B	97 (2.06)
*Amphimonhystera**	1B	161 (3.38)
*Cobbia**	2A	7 (0.15)
Daptonema	1B	126 (2.64)
Diplolaimella	1B	14 (0.29)
Elzalia	1B	83 (1.74)
Elzalia 2	1B	6 (0.13)
*Gnomoxyala**	1B	8 (0.17)
*Linhystera**	1B	21 (0.44)
Monhystera	1B	227 (4.76)
Paramonhystera	1B	21 (0.44)
*Prorhynchonema**	1B	11 (0.23)

Retrotheristus∗	1B	7 (0.15)
Rhynchonema	1B	17 (0.36)
*Steineria**	1B	2 (0.04)
*Stylotheristus**	1B	1 (0.02)
Theristus	1B	73 (1.53)
Xyala	1B	3 (0.06)
Scaptrellidae		
Scaptrella	2B	77 (1.62)
Sphaerolaimidae		
*Sphaerolaimus**	2B	144 (3.02)
*Metasphaerolaimus**	2B	63 (1.32)
Siphonolaimidae		
*Siphonolaimus**	2B	13 (0.27)
Linhomoeidae		
Didelta	1B	25 (0.52)
*Disconema**	1A	3 (0.06)
Terschellingia	1A	16 (0.34)
Desmolaimus	1B	1 (0.02)
*Eumorpholaimus**	1B	6 (0.13)
Linhomoeus	1B	80 (1.68)
*Paralinhomoeus**	1B	5 (0.10)

Family/genus	Guild	Total abundance (cumulative % contribution)
AREOLAIMIDA		
Axonolaimidae		
Axonolaimus	1B	13 (0.27)
Odontophora	1B	3 (0.06)
*Pseudolella**	1B	4 (0.08)
Comesomatidae		
*Cervonema**	1B	153 (3.21)
*Comesoma**	1B	7 (0.15)
Sabatieria	1B	1072 (22.5)
S. hilarula	1B	16 (0.34)
*Setosabatieria**	1B	1 (0.02)
*Paracomesoma**	1B	11 (0.23)
Mesonchium	1B	1 (0.02)
Metacomesoma	1B	1 (0.02)
*Dorylaimopsis**	2A	14 (0.29)
Diplopeltidae		
Araeolaimus	1A	49 (1.03)
*Campylaimus**	1B	15 (0.31)
*Diplopeltis**	1A	28 (0.59)
*Southerniella**	1A	2 (0.04)
Leptolaimidae		
Alaimella	1A	8 (0.17)
Camacolaimus	1A	3 (0.06)
Leptolaimus	1A	115 (2.41)

Plectide		
*Setoplectus**	1A	8 (0.17)
Aegialoalaimidae		
*Aegialoalaimus**	1A	11 (0.23)
*Cyartonema**	1A	11 (0.23)
*Diplopeltoides**	1A	17 (0.36)
Total Nematodes		4764

The average number of genera decreased with increasing water depth in both transects, but the correlation to water depth was stronger in the western transect (Figure 3). The average number of genera in the eastern transect was higher than in the western transect. The evenness of genera increased with water depth in both transects from 0.69 at 362 m to 0.94 at 2440 m depth (Figure 4).

A two-way crossed analysis of similarities of genera abundance indicated significant depth differences between all longitude groups (P = 0.001; global R = 0.925) but no significant difference between all depth groups (Table 3). A hierarchical cluster analysis based on Jaccard similarity coefficient of nematode genera shows a high level of similarity among the stations grouped by transects over depth that are located at different longitudes (Figure 5). The nMDS ordination of genera shows a clear separation of the east and west transects with stations RW6 and S38 most dissimilar (Figure 6). A separation of the nematode fauna in east and west transects is also noted by the high R values in the ANOSIM analysis of both the genera and family composition (Table 3).

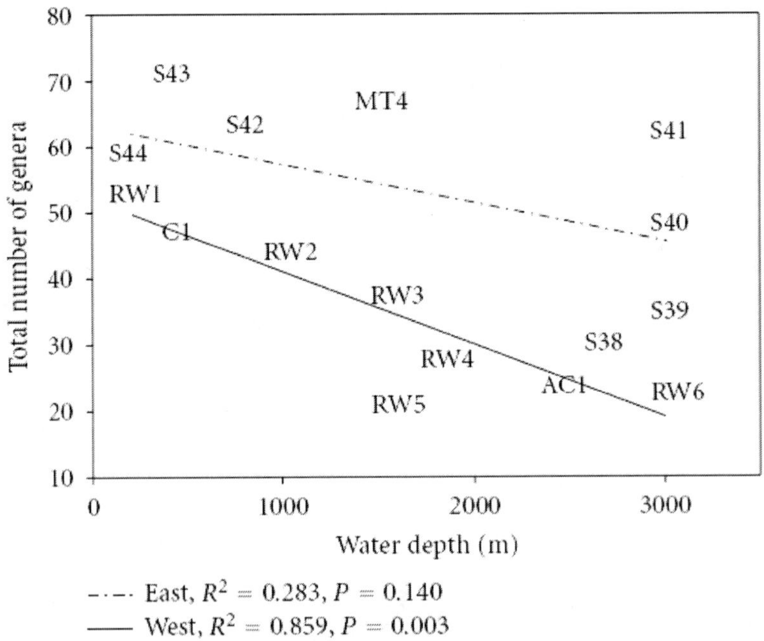

Figure 3: Number of nematode genera as a function of depth in eastern and western transect in dNGOM.

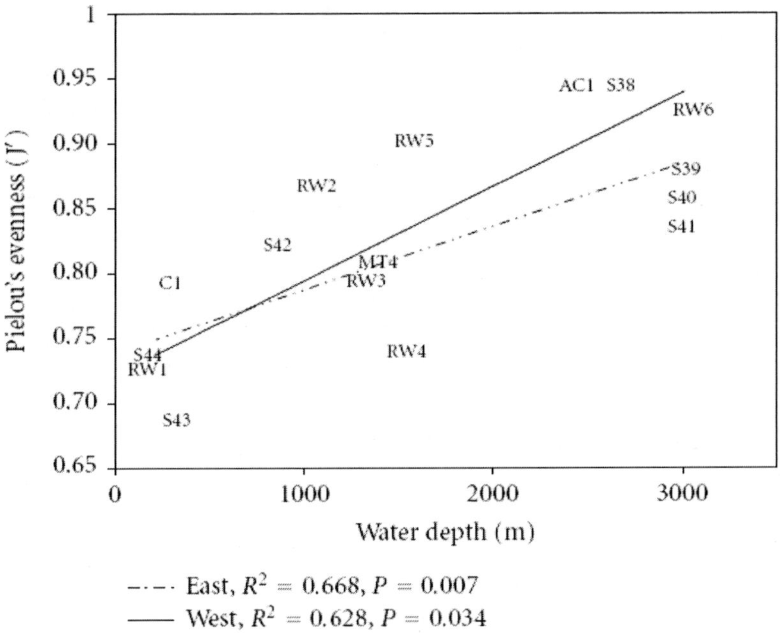

Figure 4: Evenness of nematode genera as a function of depth in eastern and western transects in dNGOM.

Table 3: R statistic values for ANOSIM test of the Bray-Curtis similarity matrices for pairwise comparison of longitudinal and depth differences in distribution of genera and family levels.

Pair-wise comparison	Global *R*	*P* value
Genus		
Longitude	0.925	0.001
Depth	0.326	ns
Family		
Longitude	0.650	0.003
Depth	0.338	ns
Feeding groups		
Longitude	0.269	ns
Depth	0.414	0.009

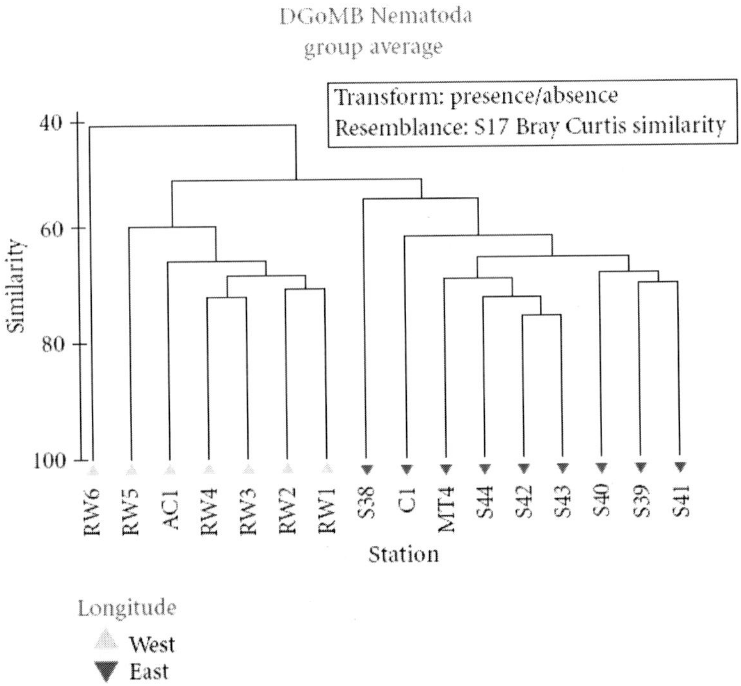

Figure 5: Hierarchical Cluster analysis with SIMPROF test on similarity of nematode genera at dNGOM stations and flower gardens brine seep.

A SIMPER analysis showed 69% dissimilarity between the genera of the eastern and western transects. Sabatieria was the most abundant genus in both transects though it was almost absent at the deeper stations (Table 4). The average similarity among genera of the eastern transect was 45%, while at the western transect it was 38%. The average abundance of most genera was higher in the eastern transect. Oxystomina was the only genus that had a higher abundance in the western transect.

Two families, Comesomatidae and Xyalidae, accounted for 43% of the total nematode fauna (Figures 7(a) and 7(b)). The Comesomatidae and Xyalidae were dominant in both transects. Among the most abundant families, only the Oxystominidae had a higher percent composition in the western transect. A SIMPER analysis showed a 45% average similarity among the families of the western transect and 44% average similarity among families of the eastern transect. Average dissimilarity between the families of the two transects was 59%.

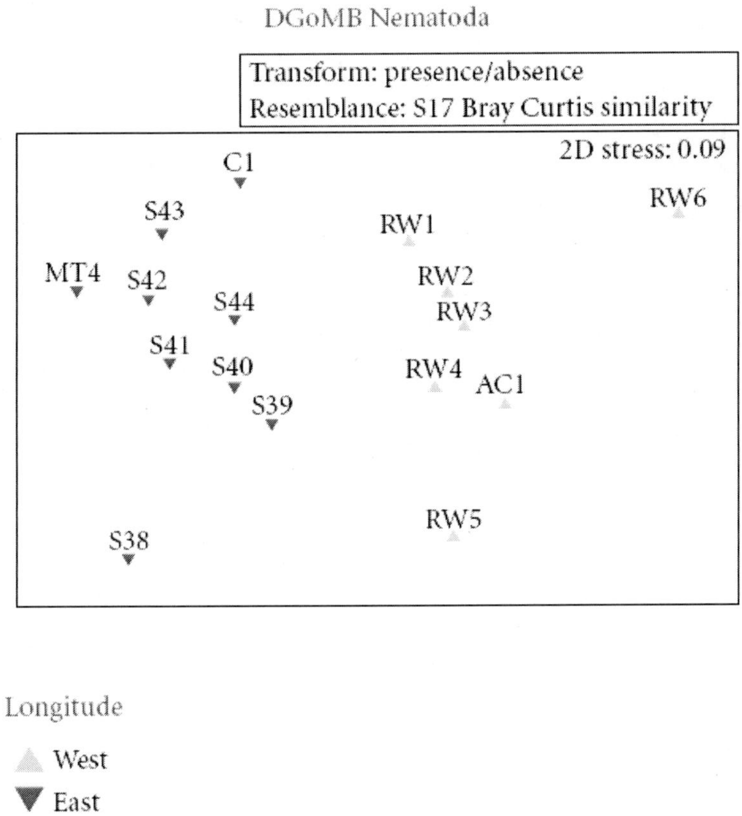

Figure 6: nMDS ordination of nematode genera abundance in dNGOM based on Bray-Curtis similarity of presence/absence data.

3.3. Functional Diversity

A regression of nematode feeding group abundance versus water depths illustrates that they were distributed fairly evenly across all depths (Figure 8). The shelf and inner slope stations were characterized by high numbers of nonselective deposit feeders such as Sabatieria. Deeper stations had higher numbers of selective deposit feeders such as Oxystomina and Halalaimus. The feeding groups showed a high similarity within the eastern and western transects. The greatest dissimilarity between the two transects was in the higher occurrence of the epigrowth feeders in the eastern transect (Table 5). The epigrowth feeders were less abundant at the slope stations.

Table 4: SIMPER (Similarity Percentages) analysis of 20 most common nematode genera at GOM eastern and western transects.

	Groups West and East		Average dissimilarity = 68.83
Genera	Group West Av.Abund	Group East Av.Abund	Contribution%
Sabatieria	48	81.78	19.56
Monhystera	0.29	25	7.23
Halalaimus	12.57	20.89	4.2
Oxystomina	14.86	7.44	3.28
Microlaimus	7.57	14.89	3.2
Cervonema	7.43	11.22	2.65
Sphaerolaimus	3.57	13.22	2.63
Spirinia	0	7.33	2.41
Amphimonhystera	9.43	10.56	2.38
Leptolaimus	7	7.33	2.29
Prochromadorella	0	8.44	2.16
Cyatholaimus	0	8.78	2.14
Daptonema	5.14	10	2.12
Scaptrella	1.14	7.67	2.04
Linhomoeus	3	6.56	2.01
Desmoscolex	5	9.56	1.79
Ceramonema	3.57	7.22	1.67
Theristus	1.57	6.89	1.64
Metasphaerolaimus	6.71	1.78	1.63
Elzalia	4.86	5.44	1.59

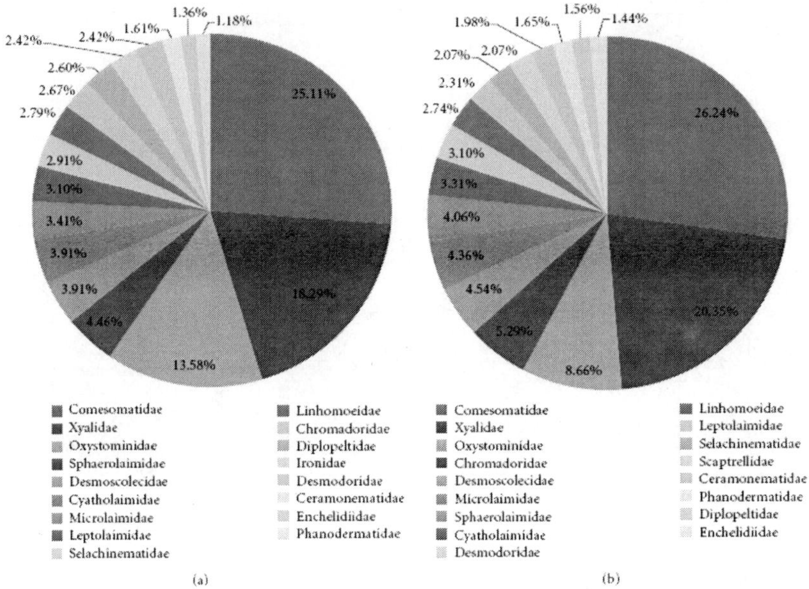

Figure 7: (a) Percent composition of nematode families in the western transect at dNGOM. (b) Percent composition of nematode families in the eastern transect at dNGOM.

Table 5: SIMPER (Similarity Percentages) analysis of nematode feeding groups at GOM eastern and western transects (1A nonselective deposit feeders; 1B selective deposit feeder; 2A epigrowth feeders; 2B omniovores and predators).

	Groups West and East		Average dissimilarity = 16.08
Feeding group	Group West Av.Abund	Group East Av.Abund	Contrib%
2A	1.92	2.21	36.68
1B	2.85	3.5	29.83
1A	2.67	2.92	19.29
2B	2.24	2.45	14.2

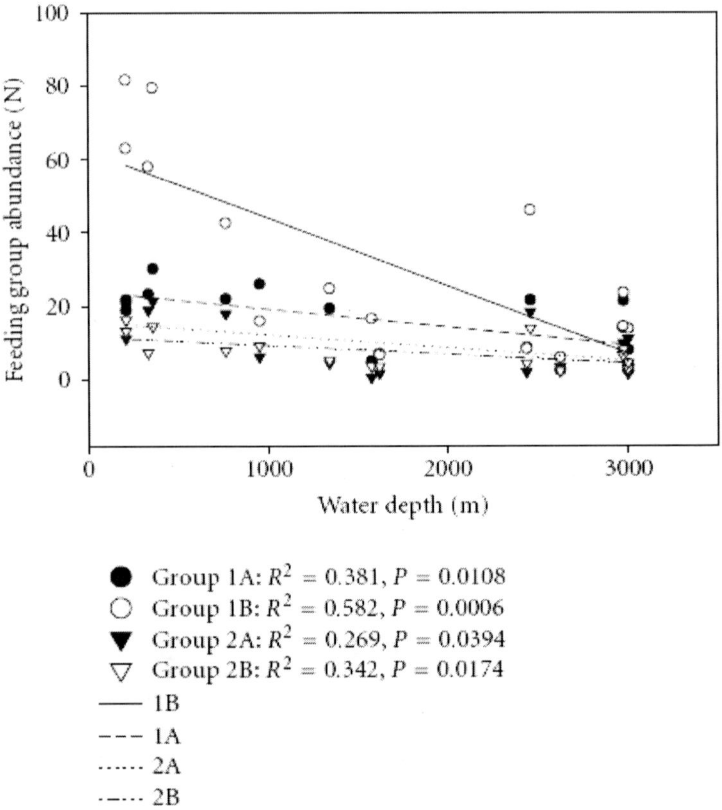

Figure 8: Regression of nematode feeding group abundances versus water depth in dNGOM. (1A nonselective deposit feeders; 1B selective deposit feeder; 2A epigrowth feeders; 2B omniovores and predators).

3.4. Cross-Taxon Spatial Variability

The Harpacticoida and Nematoda patterns match best (Rho = 0.249), but it is barely nonsignificant at . The patterns of the Polychaetes did not match the patterns for either nematodes or harpacticoids () (Table 6).

Table 6: Relationship between Nematoda, Harpacticoida, and Polychaeta spatial patterns. Values for the sample statistic (Rho) to test for matches in spatial patterns among stations and the significance probability level () for the match.

	Nematoda	Harpacticoida
Harpacticoida	Rho = 0.249 P = 0.057	—
Polychaeta	Rho = 0.028 P = 0.408	Rho = 0.159 P = 0.170

4. DISCUSSION

The GOM has significant economic importance and it is critical to understand the regional differences of its biodiversity. Our study confirms that nematodes are an abundant and diverse component of meiobenthos in the dNGOM and provides baseline data for a previously unexamined taxonomic group in this region. The proximity of eastern transect stations S38 and S40 to the Deepwater Horizon oil spill in April-July, 2011 reinforces the importance of this data for future environmental assessments.

4.1. Abundance

A significantly higher abundance of nematodes and all meiofauna groups on the eastern transect was noted by Baguley et al. [41] to be associated with high chlorophyll-a biomass in the overlying water column. The lack of a significant overall difference in abundance in our study may be attributed to the small subset of stations investigated here. The decrease in nematode abundance with increasing depth seen in the northern Gulf of Mexico reflects trends similar to those observed in previous studies [54, 55]. A comparative study of meiofauna from shelf and deep-sea sites off the S.E. coast of Brazil by Netto et al. [56] described the relative abundance of nematodes in the meiofauna that increased with depth. Vincx et al. [57] also note that exceptions to decreased density of meiofauna with increasing depth may be attributed to local physiographic variations. Such variation was also observed in our study where station S42 above the Florida Escarpment had a higher nematode abundance than other shallower stations. Baguley et al. [41] attributed a midbathyal meiofauna abundance maximum in the deep northeast GOM to interactions between seafloor topography, Mississippi River outflow, and the Loop Current. The overall decrease in nematode abundance with increasing water depth is related to the decreasing supply of organic matter away from land [4, 8, 58]. Morse and Beazley [59] also found a strong correlation between organic matter content and macrofauna biomass in the northern GOM. Exceptions to decreasing abundance and biomass with increasing water depth in the Mediterranean are associated with trophic gradients [60]. Though such trophic gradients are not defined in the GOM, the distribution of benthic organisms is related to surface phytoplankton production and associated particulate organic carbon flux to the benthos [44].

4.2. Diversity of Nematode Assemblages

The nematode fauna identified here represents a deep sea assemblage that is distinct from the intertidal and shallow subtidal studies of the GOM [32, 36]. The lack of significant differences in the composition of the nematode community between different deep sea sampling depths was also noted by Danovaro et al. [19]. They attributed the lower biodiversity at canyons to unique hydrodynamic conditions that were also noted by Baguley et al. [5]. The two stations that are less than 40% similar to their respective transects were RW6, located in the West lower slope and S38, located in the East Lower slope faunal zone classification of Wei et al. [44]. The nematode community at Alaminos canyon (station AC1) was closer to that of other stations in the western transect than that of RW6. The low similarity of the community at RW6 to other stations of this transect may be due to its depth at 3000 m. The Chromadoridae, Desmodoridae, and the genus Sabatieria, that were dominant at the slope stations, were absent at this station. The low similarity of the nematode community at station S38 to the other stations of this transect may be due to its separation from the Florida escarpment where impingement of the Loop current results in upwelling of nutrients [61]. Large scale comparisons of nematode community structure in other ocean basins indicate that continental shelf nematode communities were significantly different from slope communities, while there was no significant difference between slope and abyssal plain communities [6]. The lack of a significant difference between the genera and family depth groups suggests a more even distribution of nematode taxa at greater water depths (Table 3). This pattern supports the observations of harpacticoid diversity in the deep GOM, where there are proportionally more higher order taxa and therefore more phylogenetic or functional diversity with increasing water depth [5]. The dominant taxa in the deep GOM were the family Comesomatidae and the genus Sabatieria. These taxa also dominated the shelf and slope communities of the Mediterranean but were less prominent in deeper sediments in that study [21]. The Comesomatid, Cervonema, was found in deeper sediments in our study and suggests that nematode assemblages are specific for certain habitats in the deep GOM. The other two dominant families in our study, Xyalidae and Oxystominidae, were also associated with deep-sea sediments in temperate slope regions [62]. The Chromadoridae and Monhysteridae are the dominant families and Theristus andAcantholaimus are the dominant genera in the deep North Atlantic [16].

4.3. Functional Diversity

At all stations, the trend of the regressions of feeding group with depth is similar to that of taxonomic and phylogenetic diversity measures and indicates that there were proportionally more higher order taxa in the deep sea and therefore proportionally more functional diversity in the community. Although the number of genera was lower at greater water depths, the number of families was represented equally. This pattern indicates that there was more evenness among feeding types in deeper waters. In other words, one particular type (1B or 1A) was not dramatically dominant over the other two or three. This pattern also supports the observations of harpacticoid diversity [5] that in the deep GOM there are proportionally more higher order taxa and more phylogenetic or functional diversity. A study by Jensen [63] in the Norwegian Sea found that the deep sea fauna was dominated by deposit feeders and had few predators and scavengers. Deposit feeders were also the dominant group in the Puerto Rico trench and the Hatteras Abyssal plain [17]. Thistle and Sherman [16] found no evidence of feeding group correlation with environmental variables at 4625 m in the North Atlantic. Although, the proportion of predators and omnivores was low in the nematode meiofauna, their proportion was greater in nematodes collected in macrofauna [64]. Gambi et al. [65] suggested that the low prevalence of predatory and omnivorous nematodes in their study in the South Pacific Ocean may be attributed to the absence of freshly dead organisms.

4.4. Cross-Taxon Spatial Variability

The longitudinal eastwest gradient observed in the similarity of genera found in the GOM is comparable to patterns found in other semienclosed basins such as the Mediterranean [24]. This gradient was not observed in the distribution of larger macrofauna taxa [44]. The latter occurred in four depth-related zones that stretched across the entire northern GoM but with the two intermediate mid and deep slope zones being partitioned partially at intermediate longitudes. The similarity of the spatial pattern of nematode and harpacticoid copepods may be associated to their small size and dependence on similar food sources associated with detritus. The input of organic matter from continental margins in the Gulf of Mexico contributes significantly to benthic productivity [66]. The sedimentation of particulate organic matter from phytoplankton production in the surface waters ranged from about 100 mg C m−2 day−1 down to ca. 5 mg C m−2 day−1 on the abyssal plain, based on sediment community oxygen demand [9, 67] and satellite observations of ocean color [42, 44]. Similar estimates, however, were not made on the western transect. The pattern of macro- and megabenthic species richness has also been related to food, habitat, pollution, and location in the Gulf of Mexico [68]. The lack of similarity between the distribution patterns of nematodes and polychaetes may be due to utilization of different food sources because some polychaete species are suspension/deposit feeders [69]. In addition, polychaetes are macrofauna and are less important in terms of total biomass and carbon cycling in deeper waters [9, 44]. Nematodes occur in high abundance and diversity and thus should be included in multitaxon studies of ecosystems for evaluation of broad biodiversity patterns. In summary, our results indicate significantly different nematode community composition between the western and eastern transects in the deep northern GOM that is consistent with broad patterns observed among other taxa.

5. CONCLUSIONS

Free-living nematodes are an abundant and diverse component of the meiofauna in the deep Gulf of Mexico. They follow the general trends of decreasing nematode abundance and diversity with increasing depth observed among other benthic taxa in the northern GOM along the continental shelf and in the deep sea. Divergence from this general bathymetric pattern may be attributed to bottom topography and interacting physical, biological, and chemical oceanographic processes. With the exception of selective deposit feeders that decreased with increasing water depth, the trophic diversity of nematodes was fairly even at all depths, indicating their significance as consumers at all depths. Finally, our analysis of nematode community structure and biodiversity in the dNGOM corroborates previous meiobenthic community investigations in this region. These data reinforce the notion that the dNGOM is a heterogeneous habitat where the Loop Current interacts with the Mississippi River outflow to strongly influence regional variations in meiobenthic community structure and function. The east-west gradient in nematode distribution shows congruence with other meiobenthic taxa but not with macrofauna, megafauna, and demersal fishes [44, 70].

ACKNOWLEDGMENTS

The authors are grateful to Dr. Duane Hope, NMNH, for help in confirming identification of nematodes. The collection of the samples in this work was supported by the Minerals Management Service of the USA. Department of the Interior Contract 30991 to Texas A&M University. The authors appreciate the constructive comments and suggestions of the anonymous reviewers.

REFERENCES

1. P. de Ley and M. Blaxter, "Systematic position and phylogeny," in The Biology of Nematodes, D. L. Lee, Ed., pp. 1–30, Taylor & Francis, New York, NY, USA, 2002.

2. W. D. Hope, "An annotated checklist of the marine nematodes of the Western North Atlantic and Gulf of Mexico," Journal of Nematology, vol. 37, no. 4, pp. 1–208, 2005.

3. P. V. R. Snelgrove and C. R. Smith, "A riot of species in an environmental calm: the paradox of the species-rich deep-sea floor," Oceanography and Marine Biology, vol. 40, pp. 311–342, 2002.

4. H. Thiel, "Meiobenthos and nanobenthos of the deep sea," in The Sea, G. T. Rowe, Ed., vol. 8, pp. 167–229, John Wiley & Sons, New York, NY, USA, 1983.

5. J. G. Baguley, P. A. Montagna, W. Lee, L. J. Hyde, and G. T. Rowe, "Spatial and bathymetric trends in harpacticoida (Copepoda) community structure in the Northern Gulf of Mexico deep-sea," Journal of Experimental Marine Biology and Ecology, vol. 330, no. 1, pp. 327–341, 2006.

6. A. Vanreusel, G. Fonseca, R. Danovaro, et al., "The contribution of deep-sea macrohabitat heterogeneity to global nematode diversity," Marine Ecology—An Evolutionary Perspective, vol. 31, no. 1, pp. 6–20, 2010.

7. M. A. Rex and R. J. Etter, Deep-Sea Biodiversity Pattern and Scale, Harvard University Press, Cambridge, Mass, USA, 2010.

8. J. G. Baguley, P. A. Montagna, L. J. Hyde, and G. T. Rowe, "Metazoan meiofauna biomass, grazing, and weight-dependent respiration in the Northern Gulf of Mexico deep sea,"Deep-Sea Research Part II, vol. 55, no. 24–26, pp. 2607–2616, 2008.

9. G. T. Rowe, C. Wei, C. Nunnally et al., "Comparative biomass structure and estimated carbon flow in food webs in the deep Gulf of Mexico," Deep-Sea Research Part II, vol. 55, no. 24–26, pp. 2699–2711, 2008.

10. R. C. Aller and J. Y. Aller, "Meiofauna and solute transport in marine muds," Limonology and Oceanography, vol. 37, no. 5, pp. 1018–1033, 1992.

11. D. J. Cullen, "Bioturbation of superficial marine sediments by interstitial meiobenthos,"Nature, vol. 242, no. 5396, pp. 323–324, 1973.

12. J. Pike, J. M. Bernhard, S. G. Moreton, and I. B. Butler, "Microbioirrigation of marine sediments in dysoxic environments: implications for early sediment fabric formation and diagenetic processes," Geology, vol. 29, no. 10, pp. 923–926, 2001.

13. V. O. Mokievsky, A. A. Udalov, and A. I. Azovsky, "On the quantitative distribution of meiobenthos on the shelf of the World Ocean," Oceanology, vol. 44, no. 1, pp. 99–109, 2004.

14. D. M. Miljutin, G. Gad, M. M. Miljutina, V. O. Mokievsky, V. Fonseca-Genevois, and A. M. Esteves, "The state of knowledge on deep-sea nematode taxonomy: how many valid species are known down there?" Marine Biodiversity, vol. 40, no. 3, pp. 143–159, 2010.

15. Y. Shirayama, "The abundance of deep-sea meiobenthos in the western Pacific in relation to environmental factors," Oceanologica Acta, vol. 7, no. 1, pp. 113–121, 1984.

16. D. Thistle and K. M. Sherman, "The nematode fauna of a deep-sea site exposed to strong near-bottom currents," Deep-Sea Research, vol. 32, no. 9, pp. 1077–1088, 1985.

17. J. H. Tietjen, J. W. Deming, G. T. Rowe, S. Macko, and R. J. Wilke, "Meiobenthos of the hatteras abyssal plain and Puerto Rico trench: abundance, biomass and associations with bacteria and particulate fluxes," Deep-Sea Research Part A, vol. 36, no. 10, pp. 1567–1577, 1989.

18. K. Vopel and H. Thiel, "Abyssal nematode assemblages in physically disturbed and adjacent sites of the Eastern equatorial Pacific," Deep-Sea Research Part II, vol. 48, no. 17-18, pp. 3795–3808, 2001.

19. R. Danovaro, S. Bianchelli, C. Gambi, M. Mea, and D. Zeppilli, "α-, β-, γ-, δ-and ε-diversity of deep-sea nematodes in canyons and open slopes of NorthEast Atlantic and Mediterranean margins," Marine Ecology Progress Series, vol. 396, pp. 197–209, 2009.

20. P. J. D. Lambshead, C. J. Brown, T. J. Ferrero et al., "Latitudinal diversity patterns of deep-sea marine nematodes and organic fluxes: a test from the central equatorial Pacific,"Marine Ecology Progress Series, vol. 236, pp. 129–135, 2002.

21. K. Soetaert, C. Heip, and M. Vincx, "Diversity of nematode assemblages along a Mediterranean deep-sea transect," Marine Ecology Progress Series, vol. 75, no. 2-3, pp. 275–282, 1991.

22. T. Soltwedel, "Metazoan meiobenthos along continental margins: a review," Progress in Oceanography, vol. 46, no. 1, pp. 59–84, 2000.

23. S. Vanhove, H. Vermeeren, and A. Vanreusel, "Meiofauna towards the South Sandwich Trench (750–6300 m), focus on nematodes," Deep-Sea Research Part II, vol. 51, no. 14–16, pp. 1665–1687, 2004.

24. R. Danovaro, C. Gambi, N. Lampadariou, and A. Tselepides, "Deep-sea nematode biodiversity in the Mediterranean basin: testing for longitudinal, bathymetric and energetic gradients," Ecography, vol. 31, no. 2, pp. 231–244, 2008.

25. L. A. Levin, R. J. Etter, M. A. Rex, et al., "Environmental influences on regional deep-sea species diversity," Annual Review of Ecology and Systematics, vol. 32, pp. 51–93, 2001.

26. W. D. Hope, "Free-living marine nematoda of the Gulf of Mexico," in Gulf of Mexico Origin, Waters, and Biota: Volume I, Biodiversity, D. L. Felder and D. K. Camp, Eds., p. 1393, Texas A & M University Press, College Station, Tex, USA, 2009.

27. N. A. Cobb, "One hundred new nematodes," in Contributions to Science of Nematology, pp. 217–343, 1920.

28. B. G. Chitwood, "North American marine nematodes," Texas Journal of Science, vol. 4, pp. 617–672, 1951.

29. B. G. Chitwood and R. W. Timm, "Free-living nematodes of the Gulf of Mexico," Fishery Bulletin, vol. 55, no. 89, pp. 313–323, 1954.

30. C. E. King, "Some aspects of the ecology of psammolittoral nematodes in the Northeastern Gulf of Mexico," Ecology, vol. 43, no. 3, pp. 515–523, 1962.

31. B. Hopper, "Marine nematodes from the coast line of the Gulf of Mexico," Canadian Journal of Zoology, vol. 39, pp. 183–199, 1961.

32. E. J. Keppner, "New species of free-living marine nematode (Nematoda: Enoplidae) from Bay County, Florida, USA," Transactions of the American Microscopical Society, vol. 105, no. 4, pp. 319–337, 1986.

33. P. Jensen, "The nematode fauna in the sulphide-rich brine seep and adjacent bottoms of the East Flower Garden, NW Gulf of Mexico. I. Chromadorida," Zoologica Scripta, vol. 14, no. 4, pp. 247–264, 1985.

34. P. Jensen, "The nematode fauna in the sulphide-rich brine seep and adjacent bottoms of the East Flower Garden, NW Gulf of Mexico. II. Monhysterida," Zoologica Scripta, vol. 15, no. 1, pp. 1–11, 1986.

35. P. Jensen, "The nematode fauna in the sulphide-rich brine seep and adjacent bottoms of the East Flower Garden, NW Gulf of Mexico. III. Enoplida," Zoologica Scripta, vol. 15, no. 2, pp. 93–99, 1986.

36. P. Jensen, "Nematode fauna in the sulphide-rich brine seep and adjacent bottoms of the East Flower Garden, NW Gulf of Mexico IV. Ecological Aspects," Marine Biology, vol. 92, no. 4, pp. 489–503, 1986.

37. G. Boucher and N. Gourbault, "Sublittoral meiofauna and diversity of nematode assemblages off Guadeloupe Islands (French West Indies)," Bulletin of Marine Science, vol. 47, no. 2, pp. 448–463, 1990.

38. A. de Jesus-Navarrete and J. Herrera-Gomez, "Vertical distribution and feeding types of nematodes from Chetumal Bay, Quintana Roo, Mexico," Estuaries, vol. 25, no. 6, pp. 1131–1137, 2002.

39. A. de Jesus-Navarrete, "Nematodes from Isla Mujeres and Banco Chinchorro reefs, Quintana Roo, Mexico," Revista de Biologia Marina y Oceanografia, vol. 42, no. 2, pp. 193–200, 2007.

40. D. Castillo-Fernandez and P. J. D. Lambshead, "Revision of the genus Elzalia Gerlach, 1957 (Nematoda: Xyalidae) including three new species from an oil producing zone in the Gulf of Mexico, with a discussion of the sibling species problem," Bulletin of the British Museum (Natural History), Zoology Series, vol. 56, no. 1, pp. 63–71, 1990.

41. J. G. Baguley, P. A. Montagna, L. J. Hyde, R. D. Kalke, and G. T. Rowe, "Metazoan meiofauna abundance in relation to environmental variables in the Northern Gulf of Mexico deep sea," Deep-Sea Research Part I, vol. 53, no. 8, pp. 1344–1362, 2006.

42. D. C. Biggs, C. Hu, and F. E. Müller-Karger, "Remotely sensed sea-surface chlorophyll and POC flux at Deep Gulf of Mexico Benthos sampling stations," Deep-Sea Research Part II, vol. 55, no. 24–26, pp. 2555–2562, 2008.

43. A. E. Jochens and S. F. DiMarco, "Physical oceanographic conditions in the deepwater Gulf of Mexico in summer 2000–2002," Deep-Sea Research Part II, vol. 55, no. 24–26, pp. 2541–2554, 2008.

44. C. L. Wei, G. T. Rowe, G. Fain Hubbard et al., "Bathymetric zonation of deep-sea macrofauna in relation to export of surface phytoplankton production," Marine Ecology Progress Series, vol. 399, pp. 1–14, 2010.

45. G. T. Rowe and M. C. Kennicutt, "Northern Gulf of Mexico continental slope habitats and benthic ecology study: final report," Study MMS 2009-039, U.S. Department of the Interior Minerals Management Service, Gulf of Mexico OCS Region, New Orleans, La, USA, 2009.

46. V. N. de Jonge and L. A. Bouwman, "A simple density separation technique for quantitative isolation of meiobenthos using the colloidal silica Ludox-TM," Marine Biology, vol. 42, no. 2, pp. 143–148, 1977.

47. J. W. Seinhorst, "A rapid method for the transfer of nematodes from fixative to anhydrous glycerin," Nematologica, vol. 4, no. 1, pp. 67–69, 1959.

48. H. M. Platt and R. M. Warwick, Free Living Marine Nematodes. Part II. British Chromadorids, Synopses of the British Fauna, no. 38, The Linnean Society of London, 1988.

49. M. Mundo-Ocampo, P. J. D. Lambshead, N. Debenham et al., "Biodiversity of littoral nematodes from two sites in the Gulf of California," Hydrobiologia, vol. 586, no. 1, pp. 179–189, 2007.

50. W. Wieser, "Die beziehung zwischen munhohlengestalt, ernahrungsweise und vorkommen beifreilebenden marinen nematoden," Archives de Zoologie, vol. 2, pp. 439–484, 1953.

51. T. Moens and M. Vincx, "Observations on the feeding ecology of estuarine nematodes,"Journal of the Marine Biological Association of the United Kingdom, vol. 77, no. 1, pp. 211–227, 1997.

52. K. R. Clarke and R. N. Gorley, "PRIMER v6: User Manual/Tutorial," PRIMER-E, Plymouth, UK, 2006.

53. K. R. Clarke and R. M. Warwick, Change in Marine Communities: An Approach to Statistical Analysis and Interpretation, PRIMER-E, Plymouth, UK, 2nd edition, 2001.

54. K. Soetaert and C. Heip, "Nematode assemblages of deep-sea and shelf break sites in the North Atlantic and Mediterranean Sea,"

Marine Ecology Progress Series, vol. 125, no. 1–3, pp. 171–183, 1995.

55. T. Soltwedel, V. Mokievsky, and I. Schewe, “Benthic activity and biomass on the Yermak Plateau and in adjacent deep-sea regions northwest of Svalbard,” Deep-Sea Research Part I, vol. 47, no. 9, pp. 1761–1785, 2000.

56. S. A. Netto, F. Gallucci, and G. F. C. Fonseca, “Meiofauna communities of continental slope and deep-sea sites off SE Brazil,” Deep-Sea Research Part I, vol. 52, no. 5, pp. 845–859, 2005.

57. M. Vincx, B. J. Bett, A. Dinet, et al., “Meiobenthos of the deep Northeast Atlantic,” inAdvances in Marine Biology, J. H. S. Blaxter and A. J. Southward, Eds., vol. 30, pp. 2–88, Academic Press, London, UK, 1994.

58. J. H. Tietjen, “Abundance and biomass of metazoan meiobenthos in the deep sea,” inDeep-Sea food Chains and the Global Carbon Cycle, G. T. Rowe and V. Pariente, Eds., pp. 45–620, Kluwer Academic, Dodrecht, The Netherlands, 1992.

59. J. W. Morse and M. J. Beazley, “Organic matter in deepwater sediments of the Northern Gulf of Mexico and its relationship to the distribution of benthic organisms,” Deep-Sea Research Part II, vol. 55, no. 24–26, pp. 2563–2571, 2008.

60. S. Bianchelli, C. Gambi, D. Zeppilli, and R. Danovaro, “Metazoan meiofauna in deep-sea canyons and adjacent open slopes: a large-scale comparison with focus on the rare taxa,”Deep-Sea Research Part I, vol. 57, no. 3, pp. 420–433, 2010.

61. K. M. Yeager, P. H. Santschi, and G. T. Rowe, “Sediment accumulation and radionuclide inventories (239,240Pu, 210Pb and 234Th) in the Northern Gulf of Mexico, as influenced by organic matter and macrofaunal density,” Marine Chemistry, vol. 91, no. 1–4, pp. 1–14, 2004.

62. A. W. Muthumbi, A. Vanreusel, G. Duineveld, K. Soetaert, and M. Vincx, “Nematode community structure along the continental slope off the Kenyan Coast, Western Indian Ocean,” International Review of Hydrobiology, vol. 89, no. 2, pp. 188–205, 2004.

63. P. Jensen, “Nematode assemblages in the deep-sea benthos of the Norwegian Sea,” Deep Sea Research Part A, vol. 35, no. 7, pp. 1173–1184, 1988.

64. J. Sharma, J. Baguley, B. A. Bluhm, and G. Rowe, “Do meio- and macrobenthic nematodes differ in community composition and body weight trends with depth?” Plos ONE, vol. 6, no. 1, Article ID e14491, 2011.

65. C. Gambi, A. Vanreusel, and R. Danovaro, “Biodiversity of nematode assemblages from deep-sea sediments of the Atacama Slope and Trench (South Pacific Ocean),” Deep-Sea Research Part I, vol. 50, no. 1, pp. 103–117, 2003.

66. G. T. Rowe and M. C. Kennicutt, “Introduction to the deep Gulf of Mexico Benthos program,” Deep-Sea Research Part II, vol. 55, no. 24–26, pp. 2536–2540, 2008.

67. G. T. Rowe, J. Morse, C. Nunnally, and G. S. Boland, “Sediment community oxygen consumption in the deep Gulf of Mexico,” Deep-Sea Research Part II, vol. 55, no. 24–26, pp. 2686–2691, 2008.

68. R. L. Haedrich, J. A. Devine, and V. J. Kendall, “Predictors of species richness in the deep-benthic fauna of the Northern Gulf of Mexico,” Deep-Sea Research Part II, vol. 55, no. 24–26, pp. 2650–2656, 2008.

69. K. Fauchald and P. A. Jumars, “Diet of worms: a study of polychaete feeding guilds,”Oceanography and Marine Biology, vol. 17, pp. 193–284, 1979.

70. C. L. Wei, G. T. Rowe, E. Escobar-Briones, C. Nunnally, and Y. S. Soliman, “Standing stocks and body size of deep-sea macrofauna: predicting the baseline of 2010 deepwater horizon oil spill in the Northern Gulf of Mexico,” Deep-Sea Research Part II. In press.

6

PHYTOPLANKTON BIOMASS DYNAMICS AND ENVIRONMENTAL VARIABLES AROUND THE ROCAS ATOLL BIOLOGICAL RESERVE, SOUTH ATLANTIC

Marina Cavalcanti Jales1, Fernando Antônio do Nascimento Feitosa1 , Maria Luise Koening1 , Manuel de Jesus Flores Montes1, Moacyr Cunha de Araújo Filho1 , Rodolfo Araújo da Silva1

1Universidade Federal de Pernambuco. (Av. Arquitetura S/N, Departamento de Oceanografia, Cidade Universitária, Recife - Pernambuco, Brasil)

ABSTRACT

The Rocas Atoll Biological Reserve is located in the Atlantic Ocean, at 3º 51' S and 33º 49' W. It lies 143 nautical miles from the City of Natal, Rio Grande do Norte (Brazil). The purpose of this study was to analyze the hydrology, water masses, currents and chlorophyll *a* content to determine the dynamics of phytoplankton biomass around the Rocas Atoll. Samples were collected in July 2010 in the area around the Atoll, using the Research Vessel Cruzeiro do Sul of the Brazilian Navy. Two transects were established according to the surface currents, one of which at the southeast of the Atoll (SE) and the other at norwest (NW). Three collection points were determined on each of these transects. Samples were collected at different depths (surface and DCM - Deep Chlorophyll Maximum) and different times (day and night). According to PCA (Principal Component Analysis), the nutrients analyzed, DIN (dissolved inorganic nitrogen), DIP (dissolved inorganic phosphorus) and silicate, were inversely correlated with temperature and dissolved oxygen. Most environmental variables showed a significant increase due to the turbulence on the Northwest transect. There was an increase in the concentration of chlorophyll *a* and nutrients when the temperature and oxygen in the mixed layer was reduced due to the influence of the SACW (South Atlantic Central Water). Despite the increase observed in some variables such as nutrient salts and chlorophyll *a*, the temperature in the mixed layer attained a mean value of 23.23 ºC due to the predominance of Tropical Water. The increase of the phytoplankton biomass on the NW transect was, therefore, caused by the "island effect" and not by upwelling.

KEYWORDS

Reef environment; Chlorophyll a; Hydrology; Surface current

1. INTRODUCTION

Tropical oceans, in general, can be considered "blue deserts", or areas of low productivity and biomass, but high planktonic diversity (LONGHURST; PAULY, 2007). However, the submarine relief with its mountains and islands, as well as local currents, can influence the hydrological processes and also the physical processes such as transport and mixing, which are reflected in the concentration of chlorophyll *a* in surface water (SOUZA et al., 2013). According to LEITE et al. (2008), this increase in planktonic biomass around these areas, makes these islands and seamounts the main targets of ocean fisheries off northeastern Brazil.

According to MACÊDO et al. (2009), at low latitudes the temperature variation is not very great and it does not, therefore, cause any marked changes in the biological rhythms of organisms. The thermocline, which forms an ecological barrier, reduces nutrient turnover between deep and superficial layers, limiting primary production. The waters found beneath this "barrier" exhibit more stable physical and chemical variables and are deeper and colder. This is due to the fact that they do not suffer direct interference from environmental factors such as wind, rain, sunlight and evaporation. In tropical regions this thermocline is located at a depth of between 50 and 170 m (SOUZA et al., 2013).

While nutrients are limited to the surface layer, the vertical penetration of light is the limiting factor for primary productivity below the thermocline. Thus, only the strata near the thermocline are not limited by light and nutrients, favoring photosynthesis. This explains the existence of a characteristic Deep Chlorophyll Maximum (DCM), close to the base of the mixed layer (METZLER et al., 1997). CORDEIRO et al. (2013) reported that these mixing processes between the water masses above and below the thermocline may occur on a scale of dozens of meters.

Some studies suggest that the significant increase of plankton biomass and nutrient mixture near ocean islands is associated with the "island mass effect." This term refers to the effects of the turbulence which changes the dynamics of ocean circulation around these areas, causing moderately deep water, rich in nutrients, to rise into the photic zone (DOTY; OGURI, 1956). Some studies, such as those of GILMARTIN and REVELANTE (1974) in Hawaii;PALACIOS (2002) in Galapagos; MARTINEZ and MAAMAATUAIAHUTAPU (2004) in French Polynesia; MELO et al. (2012) in the Archipelago of *São Pedro e São Paulo*, and LIRA et al. (2014) in Fernando de Noronha Archipelago, reported these occurrences and the ecological influence that the "island mass effect" can have.

However, despite the recognized ecological importance of the island ecosystem, studies on plankton in the Rocas Atoll are still scarce, and may be mentioned PINTO et al. (1997); FEITOSA and PASSAVANTE (2004); MAFALDA JÚNIOR et al. (2009); CHAVES et al. (2008); NOGUEIRA and SASSI (2011) and SOUZA et al. (2013). This demonstrates the need for further investigation into the dynamics of plankton and oceanographic variables around the Rocas Atoll in the South Atlantic Ocean.

The present study aims to analyze the hydrology, water masses, currents and chlorophyll *a* content to determine the dynamics of the phytoplankton biomass around the Rocas Atoll Biological Reserve. The null hypothesis to be tested was: the chlorophyll *a* content and the nutrients analyzed in the study area are very low, thus demonstrating an absence of upwelling, but a significant difference between the transects analyzed regarding the predominant surface current.

2. MATERIAL AND METHODS

The Rocas Atoll is located in the Atlantic Ocean, south of the equator, at 3º 51' S and 33º 49' W. It lies 143 nautical miles from the city of Natal, Rio Grande do Norte, Brazil (GHERARDI; BOSENCE, 2001). It was formed on the flattened top of a volcano whose base lies at 4000 m depth on the seafloor (KIKUCHI; LEÃO, 1997; GHERARDI; BOSENCE, 2005).

According to MEDEIROS et al. (2009), the fracture process that led to the creation of the Fernando de Noronha Archipelago also created the Rocas Atoll and a series of banks found along the coast of Ceará and Rio Grande do Norte States.

The temperature may reach 42 ºC in the tidal pools, with the surface salinity ranging between 35 and 39 (PINTO et al., 1997). The mesotide regime is semidiurnal, with a maximum variation of 2.7 m, exposing the reef flat at low spring tide (GHERARDI; BOSENCE, 1999).

The climate is considered hot and tropical with a mean annual temperature of 26 ºC, with a maximum absolute temperature of 32 ºC and a minimum of 18 ºC. The annual rainfall varies between 1250 and 1500 mm, and the rainy season occurs between March and July. The most intense rainfall occurs in April while October is considered the driest month. There is a high relative humidity of 80% or more all year round (ANDRADE, 1960; IBAMA, 1989) *apud*SCHULZ-NETO (2004).

East and southeast winds are prevalent throughout the year, with an average frequency of 45%. Between June and August, winter in the Southern Hemisphere, 35% of winds come from the SE and 15% from the E. During summer in the Southern Hemisphere, between December and April, the occurrence of SE winds is of about 20%. The wind speed varies from 6 to 10 m/s, dominant throughout the year. During the winter, however, wind speeds between 11 and 15 m/s are common. Speeds exceeding 20 m/s are recorded more frequently during the summer months (KIKUCHI; LEÃO, 1997).

The water masses reported for the study area are as follows: Tropical Water (TW), South Atlantic Central Water (SACW), Antarctic Intermediate Water (AIW), North Atlantic Deep Water (NADW) and Antarctic Bottom Water (ABW) (MEDEIROS et al., 2009).

The Rocas Atoll Biological Reserve (ReBio) was established on June 5, 1979 and is comprised of the atoll itself and the surrounding waters as far out as the 1,000 m isobath. Prior to the creation of ReBio, the atoll was a target of intense fishing activity and extraction of coral and sand (MORAES et al., 2003). Currently the reserve is used strictly for scientific and educational activities.

The field work was conducted around the Rocas Atoll Biological Reserve in July 2010. Data were collected both day and night at each sampling point. The Research Vessel Cruzeiro do Sul (NH 38), belonging to the Brazilian Navy, was used for this.

Water-column profiling was undertaken at all the stations with a conductivity, temperature and depth system (CTD); in addition to these sensors, a fluorescence sensor was also connected to the CTD. Water samples were collected with 5 L Niskin bottles coupled to a rosette system at previously determined levels.

In the insular oceanic area, two transects were made with three sampling points on each (Figure 1): one SE (Southeast) and one NW (Northwest) transect, in relation to the prevailing surface current. The predominant direction of the currents from 0 m to 350 m was, therefore, determined, by means of an ADCP hull, before the sampling was begun in the study area.

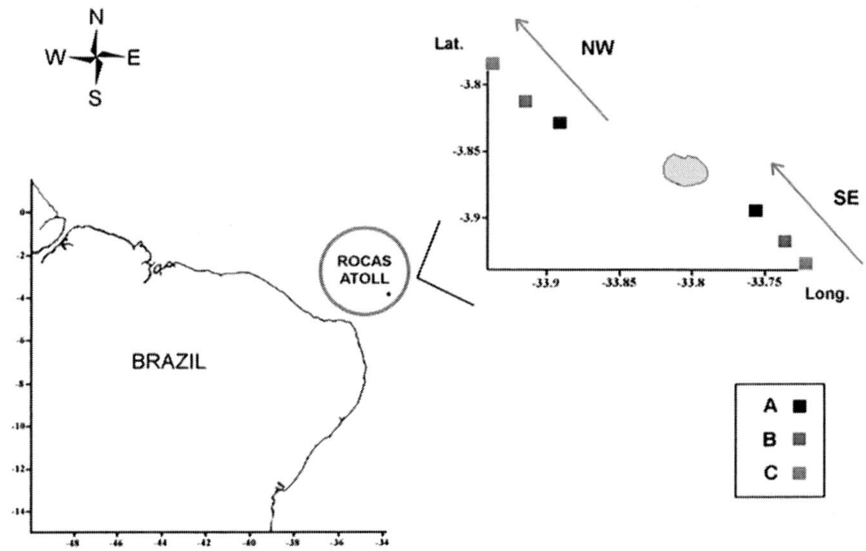

Figure 1 Map of the study area with sampling points and direction of the surface current, Southeast (SE) and Northwest (NW) in the Rocas Atoll (South Atlantic).

The thermocline depth and DCM were used as references for the sampling. For these *in situ* measurements, a CTD coupled with a fluorometer was used. However, in shallow areas, the deepest sample occurred at 75% of the location depth.

The rainfall data for the month of July during the last 10 years were acquired through the meteorological station of Fernando de Noronha and provided by Pernambuco State Agency for Water and Climate (APAC).

Niskin bottles and a rosette coupled to CTD were used for the water samples taken for the analysis of dissolved oxygen, salinity and dissolved inorganic nutrients such as ammonia, nitrite, nitrate, phosphate and silicate.

While the water samples were still on board, the dissolved oxygen content was determined by the modified Winkler method described by STRICKLAND and PARSONS (1972). The saturation rate was calculated by correlating the temperature and salinity data using the UNESCO Table (1973).

For the analysis of dissolved inorganic nutrients, the water samples were stored in plastic bottles (300 ml) and immediately frozen and transported to the Chemistry Laboratory of the Oceanography Department of UFPE. In the laboratory, analyses were performed as described by STRICKLAND and PARSONS (1972) and GRASSHOFF et al. (1983).

Samples for phytoplankton biomass analysis were taken on the surface and at DCM (80 to 100 m) and 3 L of water were filtered for each depth. Membranous filters of cellulose acetate 47mm in diameter with porosity of 0.45 µm (Schleicher and Schüll) were used. After drying, the filters were wrapped in aluminum foil, packed in paper envelopes properly identified and kept in the freezer at -18 ºC until their analysis. The method for determining the chlorophyll *a* was the spectrophotometric analysis of UNESCO (1966).

In order to extract chlorophyll *a*, test tubes of 10 ml were used, with 90% acetone, leaving them in a freezer at a temperature of -18 ºC for 24 hours in order to make it possible to extract these pigments. After this period, the material was centrifuged for 10 minutes at 3000 rpm and the supernatant was placed in optical cuvettes of 1 cm3, and the respective absorbance readings were made on a spectrophotometer. To calculate the concentration of chlorophyll *a*, PARSONS and STRICKLAND (1963) equation was applied.

2.1 Data Analysis

For the significance test, the data were assessed for normality using the Shapiro-Wilk test, and normal data were tested by factorial ANOVA, in which the values of $p \leq 0.05$ were considered significant. A comparison was made between the environmental variables using the following factors: transects SE and NW; time, comparing the diurnal and nocturnal data, and depth, comparing the concentration of chlorophyll *a* on the surface and at DCM. For these univariate analyses, the program STATISTICA 8.0 was used. For multivariate analysis, Principal Component Analysis (PCA) was performed, using the software Plymouth Routines in Multivariate Ecological Research (PRIMER 6).

3. RESULTS

The direction of the currents in this region during the campaign period were Southeasterly (SE) to Northwesterly (NW) (Figure 1). In these vertical profiles the deepening of the mixed layer was observed at the stations on the NW transect, with respect to the SE transect (Figure 2 to 4), where the region, with a depth between approximately 70 and 150 m, appears more mixed due to the turbulence generated by the currents as they cross the Atoll.

For the rainfall analysis, the data from a meteorological station on Fernando de Noronha Archipelago were used. The ten-year mean value (from 2003 to 2012), for the month of July, was 170.05 mm of rainfall. However, in July 2010, an above average value of 396 mm of rainfall was observed.

Regarding the temperature, a significant difference could only be observed at the depth analyzed (Figure 5). On the SE transect, the mean temperature ranged from 26.23 to 27.84 ºC. On the NW transect, the temperature decreased significantly, exhibiting lower values at the depth of the DCM, reaching a mean value of 23.23 ºC.

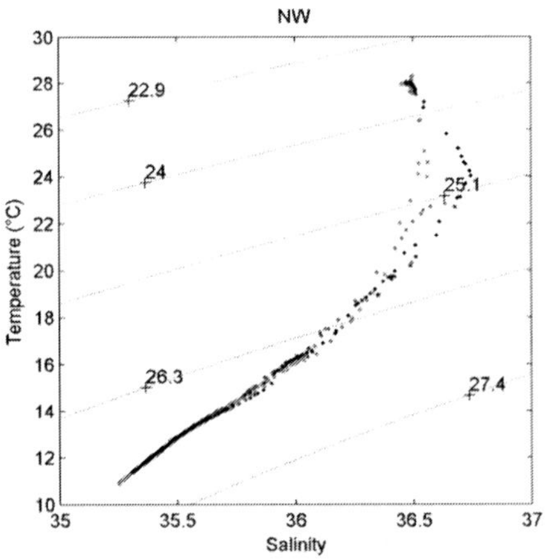

Figure 2 TS curve (Temperature-Salinity) around the Rocas Atoll (South Atlantic) in 2010.

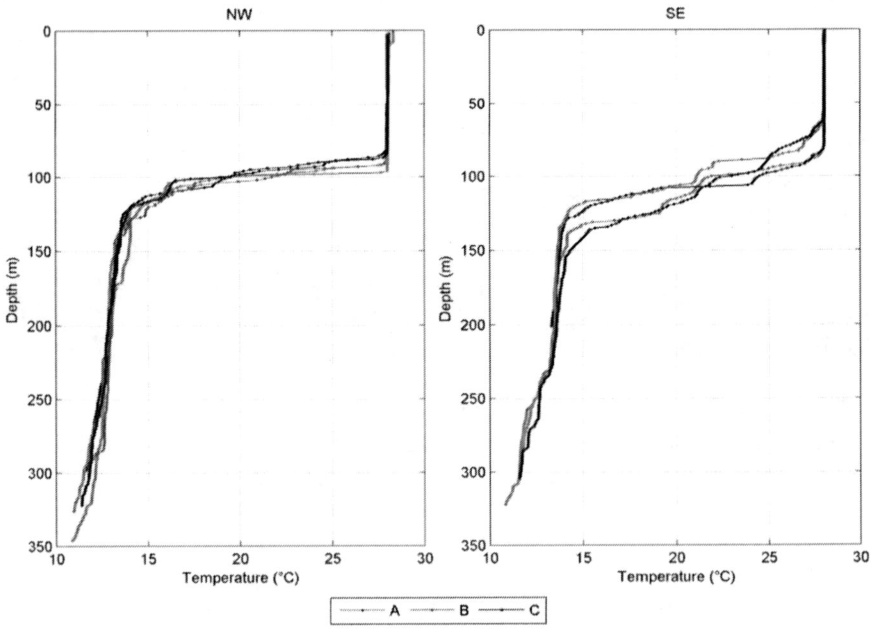

Figure 3 Vertical pattern of temperature (ºC) at points A, B and C around the Rocas Atoll (South Atlantic) in 2010.

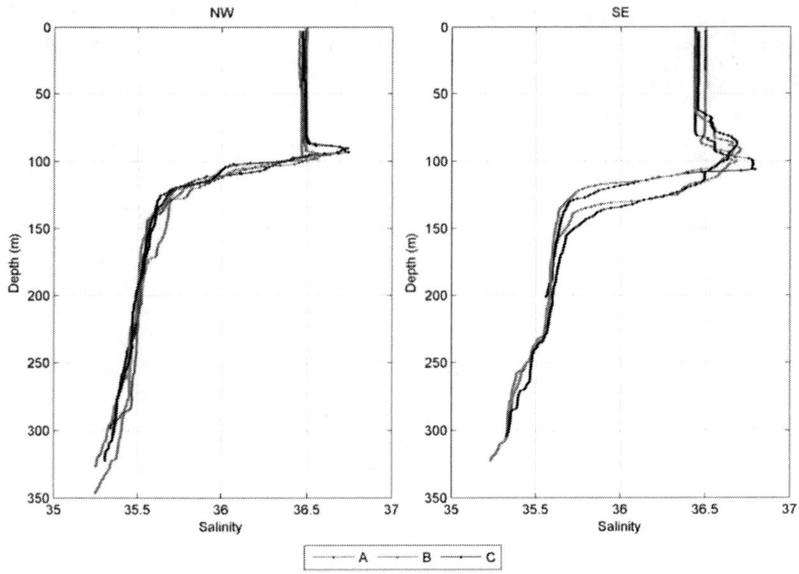

Figure 4 Vertical pattern of salinity at points A, B and C around the Rocas Atoll (South Atlantic) in 2010.

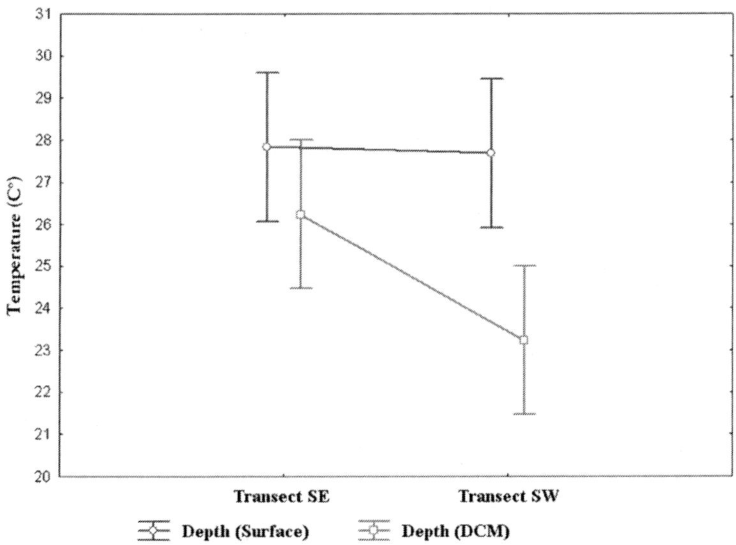

Figure 5 Spatial and vertical variation of temperature (ºC) around the Rocas Atoll (South Atlantic) in 2010.

The salinity showed no significant difference in any of the factors analyzed. The means of the SE and NW transects were 36.1 and 36.2, respectively (Figure 6).

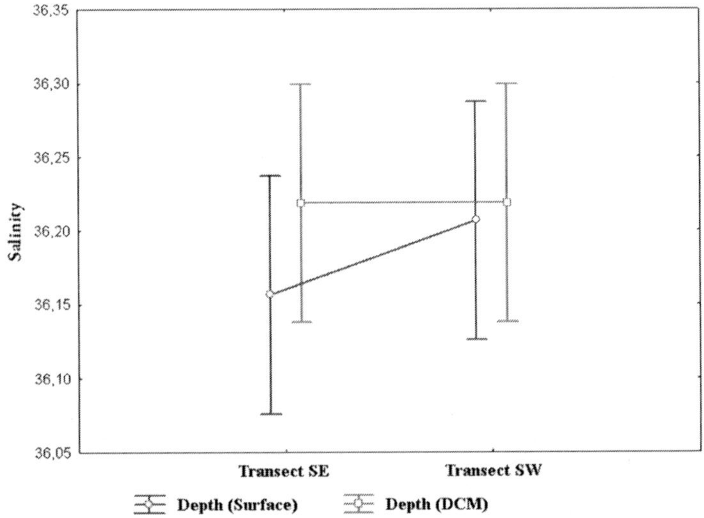

Figure 6 Spatial and vertical variation of salinity around the Rocas Atoll (South Atlantic) in 2010.

The dissolved oxygen showed significant differences in terms of depth. On both transects, the values of dissolved oxygen and its saturation rate were lower at the depth of DCM (Figure 7). On the NW transect lower values were more evident. The mean was 3.95 ml.L-1 with a saturation of 81.7% at the DCM, whereas on the SE transect, the mean was 4.47 ml.L-1 with a saturation of 92.8%.

The research showed a significant difference in the DIN (dissolved inorganic nitrogen) in terms of depth as well as transect (Figure 8). The highest mean concentration was found on the NW transect and at the depth of DCM, with a mean of 4.26 μM. Although no significant difference in surface waters was observed, the mean content of DIN was also higher on the NW transect with the value of 0.84 μM, while on the SE transect, the mean value was 0.56 μM.

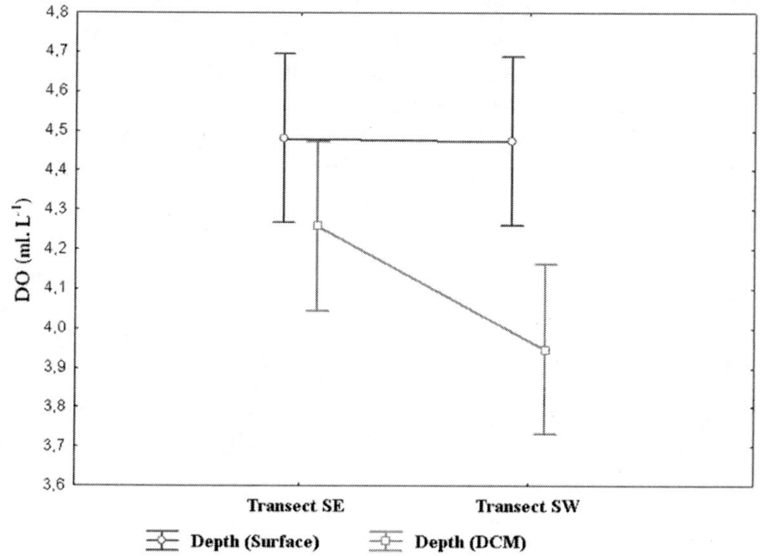

Figure 7 Spatial and vertical variation of dissolved oxygen (ml.L-1) around the Rocas Atoll (South Atlantic) in 2010.

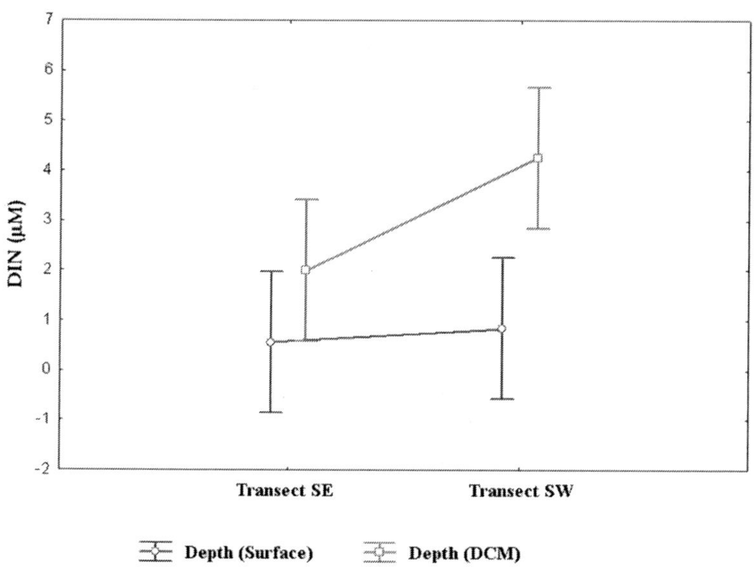

Figure 8 Spatial and vertical variation of DIN nitrite, nitrate and ammonia (µM) around the Rocas Atoll (South Atlantic) in 2010.

Only the depths showed significant differences in the concentration of DIP (dissolved inorganic phosphorus) (Figure 9). The contents of DIP were higher on both transects at the depth of DCM than at the surface. Mean values on the SE transect ranged from 0.16 to 0.18 μM while the difference in mean values was more evident on the NW transect with 0.06 and 0.36 μM at the surface and DCM, respectively.

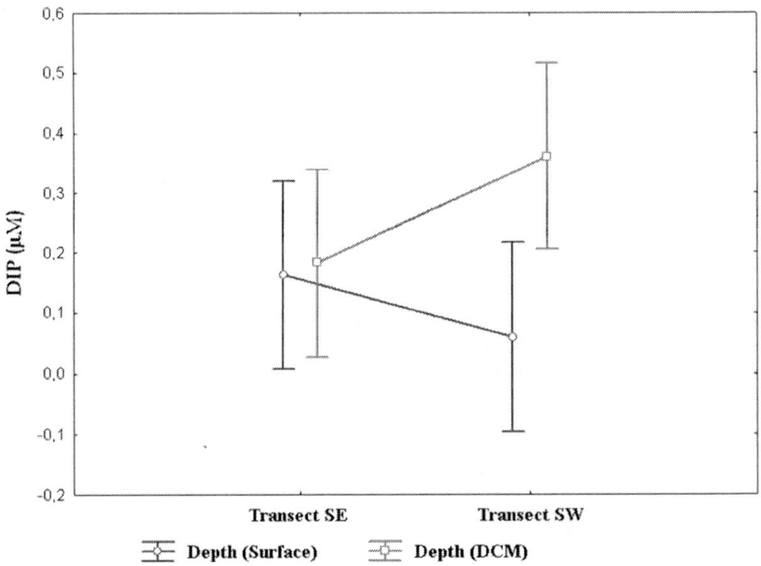

Figure 9 Spatial and vertical variation of DIP (μM) around the Rocas Atoll (South Atlantic) in 2010.

The Si02 (Silicate) content showed no significant difference between depths or transects (Figure 10). Besides the depth and transect factors, the difference in the concentration between the daytime and nighttime periods (time factor) was also analyzed. Regarding this factor, a significant difference was noted only in variable Si02, with mean values of 1.62 and 1.02 μM during the day and night, respectively (Figure 11).

The contents of chlorophyll *a* showed no significant difference in any of the factors analyzed (Figure 12). However, mean values were highest at the depth of DCM, with values ranging from 0.79 to 0.87 mg.m-3, on the NW and SE transects, respectively.

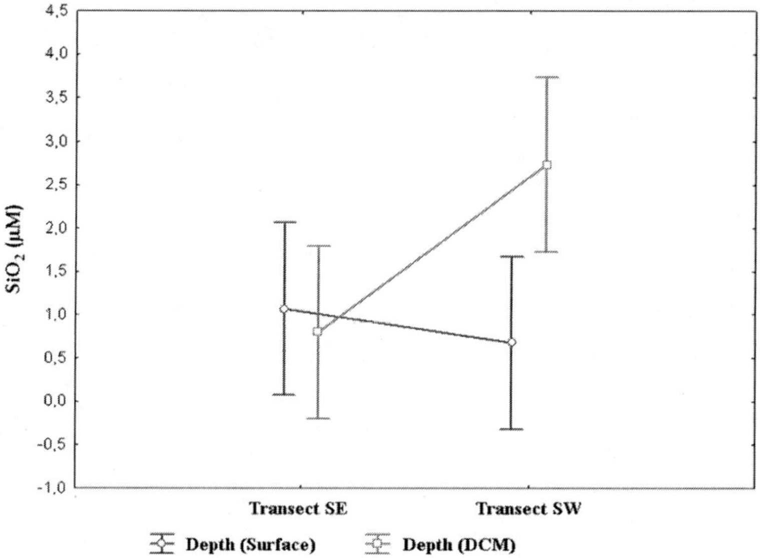

Figure 10 Spatial and vertical variation of silicate (µM) around the Rocas Atoll (South Atlantic) in 2010.

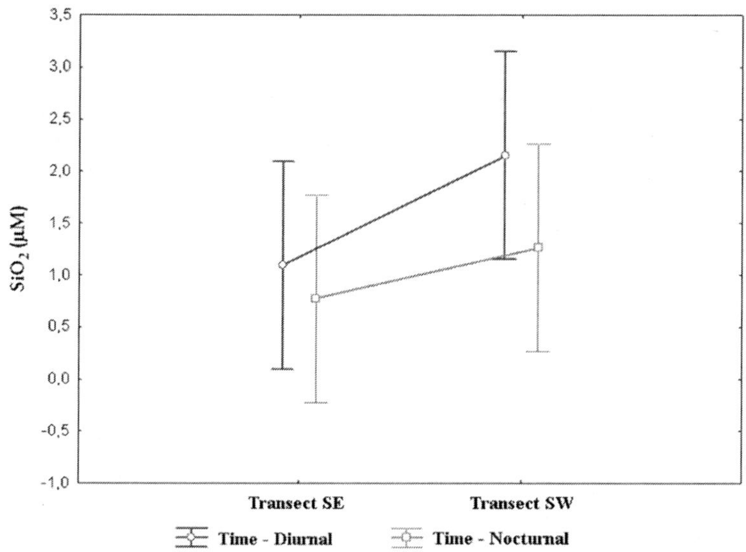

Figure 11 Temporal and vertical variation of silicate (µM) around the Rocas Atoll (South Atlantic) in 2010.

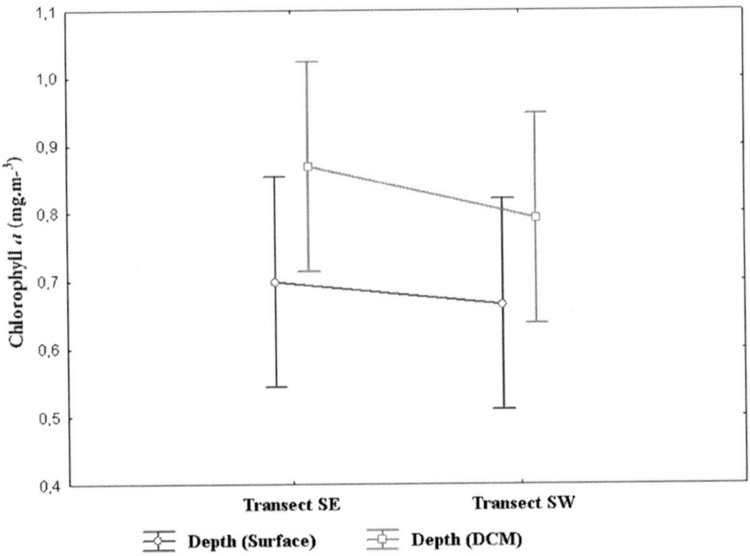

Figure 12 Spatial and vertical variation of chlorophyll *a* (mg.m-3) around the Rocas Atoll (South Atlantic) in 2010.

3.1 PCA - Principal Component Analysis

Based on the data a PCA was performed, by which it was found that three factors explained 86% of the variation in environmental variables, with the factor being represented by 58%, 15% and 13% respectively (Table 1).

Factor 1 showed a direct correlation between the variables dissolved oxygen and temperature, and an inverse correlation of these variables with the nutrients DIP, DIN and SiO2. Factor 2 evidenced the salinity and factor 3, the chlorophyll *a*. However, the variables found in factors 2 and 3 were not correlated with any others (Figure 13).

4. DISCUSSION

The surface layer of the tropical Atlantic is dominated by the Tropical Surface Water (TSW) (STRAMMA; SCHOTT, 1999), with a low concentration of dissolved inorganic nutrients and low biological productivity, as is characteristic of the tropical oceans (SAMPAIO, 1998).

Table 1 Principal Component Analysis around the Rocas Atoll (South Atlantic) in 2010

ENVIRONMENTAL VARIABLES	**Factor 1 (58%)**	**Factor 2 (15%)**	**Factor 3 (13%)**
Dissolved Oxygen (ml.L^{-1})	0.472	-0.035	0.160
Temperature (°C)	0.476	-0.130	0.060
Salinity	-0.166	0.836	0.285
DIN (μM)	-0.432	0.131	-0.145
DIP (μM)	-0.421	-0.220	0.110
SiO_2 (μM)	-0.351	-0.347	-0.272
Chlorophyll *a* (mg.m^{-3})	-0.190	-0.312	0.885

According to STRAMMA and SCHOTT (1999), the South Atlantic Central Water (SACW) is transported mainly within the South Equatorial Current (SEC) to the Brazilian Continental Shelf and is found below the TWS. Two types of SACW are found: the less dense type from the southwest subtropical South Atlantic that circulates in the subtropical gyre, and the denser type that probably originates in the southern South Atlantic.

Therefore, the oceanic current that most influences the study area is the SEC, and through the TS curve diagram (temperature and salinity) the direction of this current which was bathing the Atoll (Southeast - SE) and other after the island (Northwest - NW), was determined.

In the Atoll a greater disruption of the thermohaline structure and the vertical displacement of isotherms on the NW transect were registered, and an increase in the depth of the variables temperature and salinity was observed. This was due to the process of turbulence generated after the passage through the Atoll, i.e., the factor known as "island mass effect".

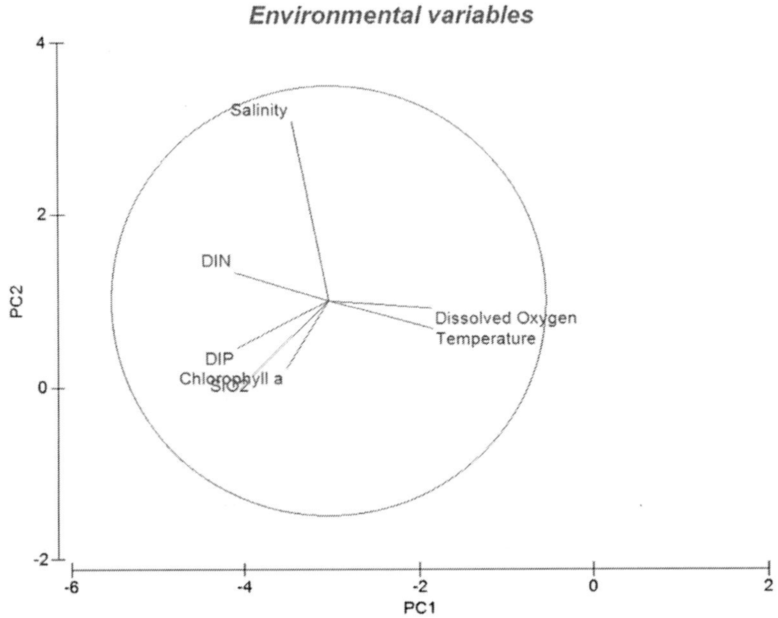

Figure 13 Principal Component Analysis of environmental variables around the Rocas Atoll (South Atlantic) in 2010.

In studies conducted by the Evaluation of the Sustainable Potential of Life's Resources in the Exclusive Economic Zone Program (REVIZEE), MEDEIROS et al. (2009) also observed differences in the thermohaline structure and the change in depth of isotherms, particularly close to the Seamounts of the North Chains and along the slope.

Despite the identification of mixture in the water mass, there was no evidence of a cold spot in the surface waters surrounding the atoll, the temperature was always over 26 ºC. This indicates the improbability of the occurrence of the upwelling of deeper waters, rich in nutrients. In a study conducted by SOUZA et al. (2013), a similar result was observed in the Fernando de Noronha Archipelago and in the Seamounts of the North Chains. However, in the same study, the *São Pedro e São Paulo* Archipelago presented colder waters on the surface, which probably caused an enrichment of surface waters. Only the depths of the DCM, that range from 83 to 100m, presented temperatures below 20 ºC, thus showing that they are very close to the barrier between the TSW and SACW water masses.

Surface temperatures are similar to those reported by MEDEIROS et al. (1999) and SOUZA et al. (2013) in the Tropical Atlantic and FEITOSA and PASSAVANTE (2004) in the Rocas Atoll, but slightly below the values found by PINTO et al. (1997), who found a thermal variation from 28.00 to 32.20 ºC. The same happened with salinity, for which PINTO et al. (1997) found values between 35 and 39 whereas in the present study the salinity remained above 36 but did not exceed 36.43.

Temperature is however, according to the results of PCA, directly correlated with the concentration of dissolved oxygen. In the surface layer, a greater dissolution of oxygen can be observed due to the influence of wind and thus it presents a similar pattern to most of the ocean. The same happens with the saturation rate, with lower values at the DCM.

Despite the fact that generally the decrease in temperature causes an increase in the dissolving capacity of gases, this study showed a decrease in the rate of dissolved oxygen below the surface layer. This fact can be explained by the consumption of dissolved oxygen by the biological activity of organisms and the oxidization of organic matter. Similar patterns occurred in the study performed in the JOPS (Joint Oceanographic Projects) byMEDEIROS et al. (1999).

All nutrients showed significant differences in some of the factors analyzed and were higher at the DCM, a fact also observed by SOUZA et al. (2013) in the Tropical Atlantic.

However, the contents of the DIN presented the greatest differentiation, as they showed not only differences as between depths but also between the transects, thus emphasizing the disturbance of the thermohaline structure and the vertical displacement of isotherms on the NW transect, as quoted above. On the other hand, the DIP showed no differentiation between transects, but like most other parameters, it presented differences between depths.

Both DIN and DIP attained a higher concentration in the Rocas Atoll, reaching a maximum of 5.72 and 0.61 μM, respectively, when compared with the Tikehau Atoll, Tuamotu Archipelago, French Polynesia (CHARPY; CHARPY-ROUBAUD, 1990) and the Tyrrhenian Sea in the

Mediterranean (MISIC et al., 2012). In the Tikehau Atoll, the authors observed maximum values of 0.09 and 0.016 µM and in the Tyrrhenian Sea, values of 4.40 and 0.20 µM for nitrogen and phosphorus, respectively.

In a study conducted in the waters of Northeastern Brazil by MEDEIROS et al. (1999), the distribution pattern of nutrients was similar, though the lower concentrations in the surface layer increased with depth. However, the values were higher, reaching concentrations greater than 10 and 1 µM for nitrogen and phosphorus, respectively.

In the Rocas Atoll, the maximum value of silicate was 5.49 µM, being slightly higher than that in the Tyrrhenian Sea in the Mediterranean, which showed a maximum value of 4.18 µM (MISIC et al., 2012). In other studies conducted in the Rocas Atoll, the maximum values of silicate found were 16.51 and 17.52 µM, by, respectively,FEITOSA and PASSAVANTE (2004) and SOUZA et al. (2013). The silicate was the only parameter that showed a significant difference between the time periods, with the lowest values occurring during the night. This fact is probably due to the consumption of silicate by the nanoplankton. As observed by KOCHHANN (2011), zooplanktonic radiolarians have broad spatial and temporal distribution in the ocean basins. The skeleton of radiolarians is impregnated with silicate and their vertical distribution in the water column of the Central Pacific is at a maximum in the layer between 70-110 m depth.

According to the results of the PCA, the DIN, DIP and silicate showed inverse correlations to temperature. This can be explained by the fact that the samplings were performed in the DCM, which coincided with the meeting of the water masses.

In oceanic areas, the processes that occur in the first hundred meters are essential for the heat exchanges with the atmosphere and for organic production. In this layer, the search for light and nutrients is undertaken simultaneously by the producers.

According to MARGALEF (1978), the characteristics of planktonic organisms that do not change easily, as far as the morphological, mechanical and general physiological properties of the cells are concerned, are less sensitive to temperature, salinity and shades of light, and much more dependent on turbulence and general nutrient availability. However, according to GOLDMAN (1988), turbulence and other small scale physical processes may be limiting a wide range of biological processes such as the absorption of nutrients by the phytoplankton.

One of the best known phytoplankton features is the maximum subsurface chlorophyll *a*, which develops during periods of stratification in different regions of the ocean (GIANESELLA, 2000).

According to LALLI and PARSON (2006), regions with low concentrations of essential nutrients, and therefore, low primary productivity, are called oligotrophic. These areas typically have chlorophyll *a* concentrations lower than 0.05 mg.m-3 at the surface and a maximum of 0.10 to 0.50 mg.m-3 at a depth of 100 to 150 m. Eutrophic waters contain high concentrations of nutrients and high phytoplankton densities that are expressed by chlorophyll concentrations from 1.0 to 10 mg.m-3 in the surface layer. Mesotrophic is a term sometimes applied to intermediate waters with concentrations ranging from 0.60 to 0.90 mg.m-3.

However, chlorophyll *a* concentrations and macronutrients cannot be the only factors to define oligotrophic or eutrophic areas. Over 20% of the world's open-ocean surface waters are characterized by the presence of adequate nitrate, phosphate and silicate in the euphotic zone, but have a relatively low corresponding phytoplankton biomass. Thus, biological productivity is often limited by a lack of micronutrients such as iron (MARTIN et al., 1994).

The DCM are common in situations of oligotrophy, such as those found in the large gyre of the Atlantic and Pacific oceans and in temperate areas during the summer. The upper layers of the photic zone have nutritional depletion and a subsurface maximum of chlorophyll *a* occurs within the pycnocline associated with the nutricline (GIANESELLA, 2000).

The chlorophyll *a* around the Rocas Atoll showed low concentrations, with values ranging from 0.30 to 1.07 mg.m-3, the same pattern occurring in studies conducted in oceanic regions by FEITOSA and PASSAVANTE (2004),MISIC et al. (2012) and SOUZA et al. (2013), with values from 0.64 to 1.10 mg.m-3, 0.01 to 0.93 mg.m-3 and 0.46 to 1.65 mg.m-3, respectively.

In Southeastern Brazil, GIANESELLA (2000) states that the greater variability between the depth of maximum chlorophyll *a* and nitracline was mainly due to the high nitrate concentrations observed at the surface without a corresponding increase in phytoplankton biomass.

Turbulent processes in the mixed layer or the unavailability of any micronutrient due to the presence of chelating agents in water were the hypotheses offered to explain this low biomass.

Regarding the water surrounding the Atoll, there was a relative increase in the chlorophyll *a* content in the mixed layer due to the influence of a nutrient rich water mass, the SACW. Most of the environmental variables analyzed showed a significant increase due to turbulence. Therefore, the content of chlorophyll *a*, in spite of presenting an increase, was not significant. This could possibly be due to the negative effect of turbulence, as explained by GOLDMAN (1988), or to the lack of any micronutrient corroborating with GIANESELLA (2000).

The present study indicates that low concentrations of chlorophyll *a* in the study region show that the productivity of phytoplankton is low and that the most plentiful food supply for the next link in the trophic web (zooplankton) is provided by the productivity of phytobenthos, mucus and aggregate organic matter produced by the benthic community of the Atoll, just as in the study undertaken by NOGUEIRA and SASSI (2011). According toMITRA et al. (2014), the microbial loop surely makes its contribution due the existence of protist mixotrophs. However, the realisation that these organisms are major players in the planktonic food web, contributing substantially to the flow of carbon and other nutrients in aquatic ecosystems, is new and requires more research for the adequate simulation of these events.

Despite the increase observed in some variables such as nutrient salts and chlorophyll *a*, the temperature in the mixed layer reached a mean value of 23.23 ºC due to the predominance of Tropical Water. The increase of phytoplankton biomass on the NW transect was, therefore, caused by the "island effect" and not by upwelling.

ACKNOWLEDGEMENTS

The authors wish to express their thanks to the Ministry of Science and Technology (MCT), the Chico Mendes Institute for Biodiversity Conservation (ICMBio), the Directorate of Hydrography and Navigation (of the Brazilian Navy), the Federal University of Pernambuco and the Fund for the Support of Science and Technology of Pernambuco (FACEPE), for their logistical support and the undertaking of the research collections.

REFERENCES

1. CHARPY, L.; CHARPY-ROUBAUD, C. J. A model of the relationship between light and primary production in an Atoll lagoon. J. Mar. Biol. Ass. U. K., v. 70, p. 357-369, 1990.

2. CHAVES, T. B. C.; MAFALDA JR, P. O.; SANTOS, C.; SOUZA, C. S.; MOURA, G.; SAMPAIO, J.; MELO, G. Biomassa planctônica e hidrografia na Zona Econômica Exclusiva do Nordeste do Brasil. Trop. Oceanogr., v. 36, p. 14-27, 2008.

3. CORDEIRO, T. A.; BRANDINI, F. P.; ROSA, R. S.; SASSI, R. Deep Chlorophyll Maximum in Western Equatorial Atlantic - How does it Interact with Islands Slopes and Seamounts? Mar. Sci., v. 3, n. 1, p. 30-37, 2013.

4. DOTY, M. S.; OGURI, M. The island mass effect. J. Cons. Int. Explor. Mer., v. 22, n. 1, p. 33-37, 1956.

5. FEITOSA, F. A. N.; PASSAVANTE, J. Z. O. Produtividade fitoplanctônica e hidrologia do Atol das Rocas (Brasil). In: CHELLAPPA, N. T.; CHELLAPA, S.; PASSAVANTE, J. Z. O. (Eds.). Ecologia aquática tropical. Natal: ServGraf, 2004. p. 143-156.

6. GHERARDI, D. F. M.; BOSENCE, D. W. J. Composition and community structure of the coralline algal reefs from Atol das Rocas, South Atlantic, Brazil. Coral Reefs, v. 19, n. 3, p. 205-219, 2001.

7. GHERARDI, D. F. M.; BOSENCE, D. W. J. Modeling of the ecological succession of encrusting organisms in recent coralline-algal frameworks from Atol das Rocas, Brazil. Palaios, v. 4, p. 145-158, 1999.

8. GHERARDI, D. F. M.; BOSENCE, D. W. J. Late Holocene Reef growth and relative sea-level changes in Atol das Rocas, equatorial South Atlantic. Coral Reefs, v. 24, n. 2, p. 264-272, 2005.

9. GIANESELLA, S. M. F. Variabilidade da camada de clorofila máxima na região de quebra da Plataforma Continental Sudeste Brasileira. 2000. 166f. Tese (Livre-Docência) - Instituto Oceanográfico, Universidade de São Paulo, São Paulo, 2000. http://www.teses.usp.br/teses/disponiveis/livredocencia/21/tde-04052010-103957/pt-br.php

10. GIANESELLA, S. M. F.; SALDANHA-CORRÊA, F. M. P. Sustentabilidade dos oceanos. São Paulo: Blucher press, 2010. 200 p.

11. GILMARTIN, M.; REVELANTE, N. The "island mass" effect on the phytoplankton and primary production of the Hawaiian islands. J. Exp. Mar. Biol. Ecol., v. 16, p.181-204, 1974.

12. GOLDMAN, J. C. Spatial and temporal discontinuities of biological process in pelagic surface waters. In: ROSTHCHILD, B. (Ed.). Toward a Theory of Biological and Physical Interaction in the world ocean. Amsterdam: Kluwer, 1988, p. 273-296.

13. GRASSHOFF, K.; EHRHARDT, M.; KREMLING, K. Methods of seawater analysis. 3rd ed. Weinheim: Verlag Chemie, 1983. 419 p.

14. HOPPENRATH, M; ELBRÄCHTER, M.; DREBES, G. Marine phytoplankton: Selected microphytoplankton species from the North Sea around Helgoland and Sylt. 1st ed. Stuttgart: Schweizerbart Science Publishers press, 2009. 264 p.

15. KIKUCHI, R. K. P.; LEÃO, Z. M. A. N. Rocas (Southwestern Equatorial Atlantic, Brazil): an atoll built primarily by coralline algae. In: 8th INTERNATIONAL CORAL REEF SYMPOSIUM, 1997, Panamá. Proceedings. International Society for Reef Studies. Panamá, 1997. p. 731-736.

16. KOCHHANN, K. G. D. Radiolários: estado do conhecimento e aplicações às geociências. Terra. Didat., v. 7, n. 1, p. 18-28, 2011.

17. LALLI, C. M.; PARSON, T. R. Biological Oceanography: an introduction. 2nd ed. Amsterdam: Elsevier, 2006. 314 p.

18. LEITE, T. S.; HAIMOVICI, M.; LINS, J. E. A pesca de polvos no Arquipélago de Fernando de Noronha, Brasil. Bol. Inst. Pesca, v. 34, n. 2, p. 271-280, 2008.

19. LIRA, S. M. A.; TEIXEIRA, I. A; LIMA, C. D. M; SANTOS, S.; LEITÃO, S. N.; SCHWAMBORN, R. Spatial and nycthemeral distribution of the zooneuston of Fernando de Noronha, Brazil. Braz. J. Oceanogr., v. 62, n. 1, p. 35-45, 2014.

20. LONGHURST, A. R.; PAULY, D. Ecologia de oceanos tropicais. 2nd ed. São Paulo: EDUSP, 2007. 419 p.

21. MACÊDO, S. J.; MONTES, M. J. F.; COSTA, K. M. P. Oceanografia Biológica: Condições hidrológicas da Zona Econômica Exclusiva (ZEE) do Nordeste. In: FÁBIO, V. H. (Org.). Programa REVIZEE - Score Nordeste. Fortaleza: Martins & Cordeiro, 2009. p. 104-136.

22. MAFALDA JÚNIOR, P. O.; MOURA, G.; MELO, G.; SAMPAIO, J.; FEITOSA, F. A. N.; PASSAVANTE, J. Z. O.; SOUZA, M.O.; SAMPAIO, C. Oceanografia Biológica: Biomassa Fitoplanctônica na ZEE da Região Nordeste do Brasil. In: FÁBIO, V. H. H. (Org.). Programa REVIZEE - Score Nordeste. Fortaleza: Martins & Cordeiro, 2009. p. 11-23.

23. MARGALEF, R. Life-forms of phytoplankton as survival alternatives in an unstable environment. Oceanol. Acta, v. 1, n. 4, p. 493-509, 1978.

24. MARTIN, J. H.; COALE, K. H.; JOHNSON, K. S.; FITZWATER, S. E.; GORDON, R. M.; TANNER, S. J.; HUNTER, C. N.; ELROD, V. A.; NOWICKI, J. L.; COLEY, T. L.; BARBER, R. T.; LINDLEY, S.; WATSON, A. J.; VAN SCOY, K.; LAW, C. S.; LIDDICOAT, M. I.; LING, R.; STANTON, T.; STOCKEL, J.; COLLINS, C.; ANDERSON, A.; BIDIGARE, R.; ONDRUSEK, M.; LATASA, M.; MILLERO, F. J.; LEE, K.; YAO, W.; ZHANG, J. Z.; FRIEDERICH, G.; SAKAMOTO, C.; CHAVEZ, F.; BUCK, K.; KOLBER, Z.; GREENE, R.; FALKOWSKI, P.; CHISHOLM, S. W.; HOGE, F.; SWIFT, R.; YUNGEL, J.; TURNER, S.; NIGHTINGALE, P.; HATTON, A.; LISS, P.; TINDALE, N. W. Testing the iron hypothesis in ecosystems of the equatorial Pacific Ocean. Nature, v. 371, p. 123-129, 1994.

25. MARTINEZ, E.; MAAMAATUAIAHUTAPU, K. Island mass effect in the Marquesas Islands: Time variation. Geophys. Res. Lett., v. 31, p. 1-4, 2004.

26. MEDEIROS, C.; ARAÚJO, M.; FREITAS, I.; ROLLNIC, M. Massas d'água da região oeste do Atlântico tropical. In: HAZIN, F. H. V. (Org.). Meteorologia e Sensoriamento Remoto, Oceanografia Física, Oceanografia Química e Oceanografia Geológica. Programa REVIZEE - Score Nordeste. Fortaleza: Martins & Cordeiro, 2009. p. 56-69.

27. MEDEIROS, C.; MACÊDO, S. J.; FEITOSA, F. A. N.; KOENING, M. L. Hydrography and phytoplankton biomass and abundance of North-East Brazilian waters. Arch. Fish. Mar. Res., v. 47, n. 2/3, p. 133-151, 1999.

28. MELO, P. A. M. C.; DIAZ, X. F. G.; MACEDO, S. J.; NEUMANN-LEITÃO, S. Diurnal and spatial variation of the mesozooplankton community in the Saint Peter and Saint Paul Archipelago, Equatorial Atlantic. Mar. Biodivers. Rec., v. 5, p. 1-14, 2012.

29. METZLER, P. M.; GLIBERT, P. M.; GAETA, S. A.; LUBLAN, J. New and regenerated production in the South Atlantic of Brazil. Deep-Sea Res., v. 44, n. 3, p. 363-384, 1997.

30. MISIC, C.; BAVESTRELLO, G.; BO, M.; BORGHINI, M.; CASTELLANO, M.; HARRIAGUE, A. C.; MASSA, F.; SPOTORNO, F.; POVERO, P. The "seamount effect" as revealed by organic matter dynamics around a shallow seamount in the Tyrrhenian Sea (Vercelli Seamount, western Mediterranean). Deep-Sea Res., v. 67, p. 1-11, 2012.

31. MITRA, A.; FLYNN, K. J.; BURKHOLDER, J. M.; BERGE, T.; CALBET, A.; RAVEN, J. A.; GRANÉLI, E.; GLIBERT, P. M.; HANSEN, P. J.; STOECKER, D. K.; THINGSTAD, F.; TILLMANN, U.; VAGE, S.; WILKEN, S.; ZUBKOV, M. V. The role of mixotrophic protists in the biological carbon pump. Biogeosciences, v. 11, p. 995-1005, 2014.

32. MORAES, F. C.; VILANONA, E. P.; MURICY, G. Distribuição das esponjas (Porifera) na reserva biológica do Atol das Rocas, Nordeste do Brasil. Arq. Mus. Nac. (Rio de J.), v. 61, n. 1, p. 13-22, 2003.

33. PALACIOS, D. M. Factors influencing the island-mass effect of the Galapagos archipelago. Geophys. Res. Lett., v. 29, n. 23, p. 21-34, 2002.

34. PARSONS, T. R.; STRICKLAND, J. D. H. Discussion of spectrophotometric determination of marine plankton pigments, with revised equations of ascertaining chlorophyll a and carotenoids. J. Mar. Res., v. 21, n. 3, p. 155-163, 1963.

35. PINTO, N. C. T.; MAFALDA, P.; SANTOS, A. T. Caracterização do zooplâncton da Reserva Biológica do Atol das Rocas na Campanha de março-1991(verão). Trab. Oceanog. Univ. Fed. PE, v. 25, p. 31-46, 1997.

36. SAMPAIO, J. A. A. Distribuição horizontal e vertical da clorofila a fitoplanctônica em águas da Zona Econômica Exclusiva do Nordeste Brasileiro. João Pessoa: Relatório Técnico Final, 1998. 21 p.

37. NOGUEIRA, E. M. S.; SASSI, R. Nychthemeral variations of tintinnina (Ciliata: Oligotrichida) near the Rocas Atoll (South Atlantic) and relationship with other microzooplanktonic components. Arq. Ciênc. Mar., v. 44, n. 1, p. 5-11, 2011.

38. SCHULZ-NETO, A. Aves marinhas do Atol das Rocas. In: Branco, J. O. (Ed.) Aves marinhas e insulares brasileiras: bioecologia e conservação. Itajaí: UNIVALI, 2004, p. 169-192.

39. SOUZA, C. S.; LUZ, J. A. G.; MACÊDO, S. J.; MONTES, M. J. F.; MAFALDA, P. Chlorophyll a and nutrient distribution around seamounts and island of the tropical south-western Atlantic. Mar. Fish. Rev., v. 64, p. 168-184, 2013.

40. STRAMMA, L.; SCHOTT, F. The mean flow field of the tropical Atlantic Ocean. Deep-Sea Res. II, v. 46, p. 279-303, 1999.

41. STRICKLAND, J. D. H.; PARSONS, T. R. A practical handbook of sea water analysis. Bull. Fish. Res., v.167, p. 1-311, 1972.

42. UNESCO. International Oceanographic Table. 2nd ed. Wormly: National Institute of Oceanography of Great Britain & UNESCO, 1973.

43. UNESCO. Determination of photosynthetic pigments in sea water. Paris: Monography on Oceanography Methodology, 1966. 69 p.

7

ENVIRONMENTAL INFLUENCES ON CARIBBEAN FISH CATCH

Mark R. Jury1,2

1Department of Physics, University of Puerto Rico Mayagüez, PR 00681, Puerto Rico

2University of Zululand, KwaDlangezwa 3886, South Africa

ABSTRACT

A Caribbean fishery index is used to identify contrasting environmental conditions in atmosphere and ocean reanalysis fields associated with high- and low-catch years over the period 1971–2004. A number of composite features are noted: cooler surface temperatures and warmer, drier weather across the southern half of the Caribbean favors higher catch rates. There is a ridge of elevated sea level on 16°N. South of the ridge the Caribbean Current is strengthened while north of the ridge anomalous eastward currents flow past the Antilles Islands. The atmospheric Hadley circulation weakens in years of high catch and tropical cyclones are rare. This paper uncovers basin-scale forcing of aggregate fish catch, reflecting a north-south gradient in land-atmosphere hydrology, and composite oceanographic differences based on reanalysis data. In addition, seasonal and interannual cycles of ocean productivity are investigated using satellite ocean color. Summertime upwelling along the coast of Venezuela joins fresher waters from the North Brazil Current to infuse the Caribbean ecosystem with higher nutrients.

1. INTRODUCTION

The Caribbean Sea is a semienclosed basin bounded on the south and west by South and Central America and on the east and north by the chain of Antilles Islands (8°N–25°N, 85°W–55°W). Persistent trade winds, solar insolation and tropical water exchanges limit seasonal variations in surface temperatures (25°–28°C). The thermocline is ~200 m deep and the upper 1200 m is stratified [1–3]. Several gaps between eastern islands (Grenada, St Vincent, St Lucia) are passes for ocean currents connecting with the Atlantic Ocean [4, 5]. From there, water flows across the Caribbean Sea as a meandering westward current reaching a surface velocity of 0.7 m/s along 14°N [6–9]. The current passes the Nicaragua Rise southwest of Jamaica, with a branch forming the counterclockwise Panama Gyre [5, 10, 11] and the bulk exiting via the Yucatan Channel. The Gulf Stream near Florida is joined by the Antilles Current that transports Atlantic waters northwestward around the Caribbean [12–14]. Schmitz and Richardson [15] and Johns et al. [4] find ~28 Sv of throughflow in the Caribbean from both North and South Atlantic sources [16]. South Atlantic waters are fresher and more oxygenated than North Atlantic waters of the same density, and enter the Caribbean north of Trinidad next to the Orinoco plume [17]. While the mean structure of water masses and currents in the Caribbean is well known and processes underlying year-to-year fluctuations are being uncovered [18], basin-scale impacts on fish catch are relatively unknown.

Tropical fishery abundance has been linked to coral reef environments with adequate food sources [19–21]. Trade winds that blow across the Caribbean establish basin-scale currents that are modulated by islands, offering chances for recirculation and biological retention [22–27]. Greater currents and turbulence in exposed open seas tends to scatter food sources [28–32]. In the Caribbean most spawning is during spring when sea surface temperatures (SST) are cooler [33–36], particularly along the Caribbean's southern edge, where coastal upwelling contributes to higher productivity ([34, 37–39] cf. Figures 1(d), 1(e), and 1(f)).

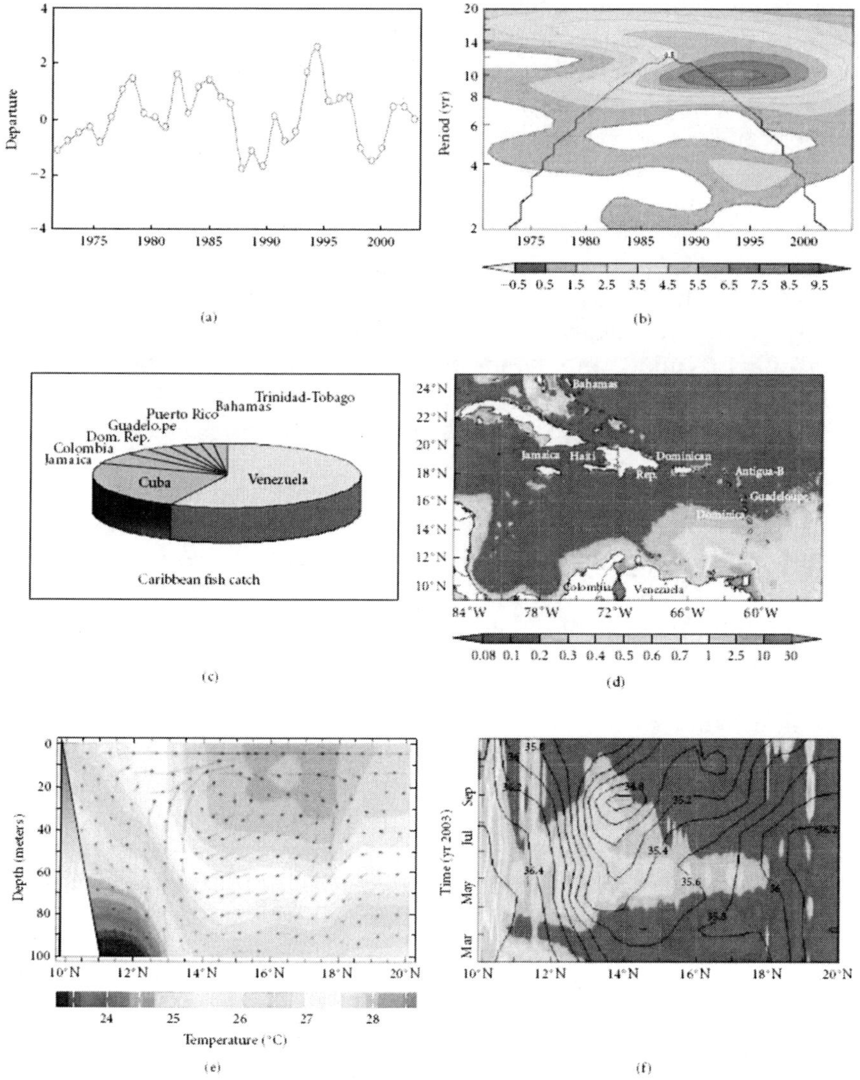

Figure 1: (a) Caribbean fish catch index time series, where 1 T/yr and (b) its wavelet spectrum with cone of validity and power shaded. (c) Caribbean country fish catch ranked by weight-excluding lesser, (d) satellite mean chlorophyll (mg/m3) with countries in the catch index labelled. (e) Depth section of sea temperature and vertical overturning circulation north of Venezuela averaged March-October 2003, (f) latitude-time hovmöller of chlorophyll (shaded, as in (d)) and salinity (g/kg) for the same period averaged 65–70 W.

Low salinity and high turbidity associated with river discharges can have localized impacts on fisheries. The Orinoco plume in eastern Venezuela spreads across the southern Antilles area (>106 km2, Muller-Karger et al. [17, 40, 41]). At varying distances downstream river plumes can be a source of nutrients or pollutants, and a mechanism for transport and stratification, producing convergence fronts that aggregate plankton and fish [42–44]. While long-term mean ocean currents act as pathways for population connectivity, they exhibit eddies and gyres [45–48]. Drifters and profiling floats reveal much flow variability in the Caribbean, often moving across or looping around the mean flow [49]. Leeward of oceanic islands nutrient enrichment is often seen due to runoff and wind-induced upwelling, leading to greater zooplankton biomass [50–52] that provide food sources in oligotrophic waters such as the northern Caribbean.

This literature survey highlights some of the factors influencing tropical fisheries. Winds, currents, water stratification and nutrient levels vary on different time and space scales, and can produce irregular cycles in marine resources, notwithstanding changes in fishing effort and consumption. To understand how the environment modulates fish catch it is necessary to analyze the subsurface ocean, and with the advent of global ocean monitoring and reanalysis products, new tools are available. While past work on the fishery biology is usually focused at local level over the lifespan of individual species, the approach here is to uncover basin-scale influences on the aggregate fishery during the catch year. Given the uneven distribution of historical ocean observations, evidence of biologically influential conditions may be elusive unless suitable methods are employed. Here a composite technique is used to quantitatively contrast reanalysis fields in high- and low-catch years. The key questions which are addressed include: What are the key environmental features that distinguish high- and low-catch years? Are these features significant in the context of natural variability? Is ocean-atmosphere coupling evident? Do conditions relate to catch effort or nutrification? Are the key features having north-south or east-west gradients? and, Does the surface hydrology play an important role?

2. DATA AND METHODS

A fishery index was developed by gathering catch records for all Caribbean countries (excluding Central America) from the website: http://www.fishbase.us/ [53] over the period 1971–2004. There are known reporting problems associated with commercial catch records [54, 55], so annual data as an aggregate for all species were extracted for each country by weight and also as standardized departures: subtracting the mean and dividing the standard deviation for each. The total-by-weight favors countries with longer coastlines and bigger catch (e.g., Venezuela, cf. Figure 1(c)), while standardized departures ensure equal representation (spreading the risk of uneven reporting). Individual country data were then cross-correlated with the two series and those with significant values (Table 1) were retained to form a Caribbean fish catch index by averaging the standardized departures. Next the linear upward trend related to increasing effort [56] was removed. The annual detrended standardized fish catch values were then ranked. Years with fish catch $>+1$ and <-1 were identified: seven of each. The mean difference between high- and low-catch is ~200 000 T/yr, about half the total. To aggregate the environmental data and reduce the number of maps/sections for interpretation, composite averages were made for the high catch years and the low-catch years, and the low field was subtracted from the high field to contrast environmental conditions. The composites draw 14 years of observations into a single field-limiting problems of sparse subsurface observations.

Atmospheric conditions are represented by the National Center for Environmental Prediction (NCEP) reanalysis product [57]. These monthly gridded fields at 2° horizontal and 500 m vertical resolution are based on surface ship and coastal observations, radiosonde profiles and aircraft reports, and satellite remote sensing of clouds, temperature, humidity and winds; all of which are optimally interpolated using a numerical weather prediction model. Because the ocean is affected by fresh water inputs from the land and atmosphere; rainfall, river gauges and run-off estimates [58] were analyzed.

To describe the oceanographic environment, the monthly SODAv2.4 ocean reanalysis products of the University of Maryland at 0.5° horizontal and ~20 m depth resolution are employed [59]. These make use of all hydrographic data (XBT, CTD, ADCP, etc.), optimally interpolated using a numerical ocean model that employs European Community Medium-range Weather Forecast (ECMWF) wind stress and surface fluxes, coastal stations, moored buoys, drifters, and satellite infrared SST (1979+), altimetry (1985+) and wind scatterometry (1998+).

Table 1: Correlation values of individual country fish catch with respect to the all-Caribbean total by weight and by standardized departure over the period 1971–2004. For , significant at 90% confidence (bold number and country).

By weight	By sigma	Country
0.36	**0.37**	**Antig-Barbuda**
−0.74	0.06	Aruba
0.86	**0.30**	**Bahamas**
−0.53	0.08	Barbados
0.61	**0.42**	**Colombia**
0.28	0.25	**Dominica**
0.65	0.06	**Dominican Rep.**
−0.36	−0.33	Grenada
−0.43	**0.52**	Cuba
0.61	**0.49**	**Guadeloupe**
0.44	0.08	**Haiti**
0.68	0.23	**Jamaica**
0.21	0.02	Martinique
0.05	0.14	Neth.Antilles
−0.35	0.25	Puerto Rico
−0.82	**0.39**	StKitts/Nevis
−0.17	**0.31**	St. Lucia
0.29	0.12	St. Vincent
−0.18	**0.36**	Trinidad-Tob.
−0.26	0.05	US Virg.Isl.
0.95	**0.31**	**Venezuela**

Contrasts in the environmental conditions in years of high- and low-catch were analyzed by field subtraction via the IRI Climate Library website http://iridl.ldeo.columbia.edu/SOURCES/.CARTON-GIESE/SODA/.v2p0p2-4/ and differences were evaluated for: SST, sea surface height, wind stress, and in depth sections or as depth-averages for temperature, salinity, currents and vertical motion. In addition, composite tracks of tropical cyclones (hurricanes) in the Caribbean were mapped with respect to high- and low-catch years, based on the

National Hurricane Center reanalysis product. The annual fish catch index in the period 1971–2004 was analyzed for evolving spectral cycles using wavelet transform via the website:http:// ion. researchsystems.com/IONScript/wavelet/. To determine whether differences in high- and low-catch years are modulated by a naturally occurring mode of ocean climate variability, the environmental fields were objectively analyzed by empirical orthogonal function (EOF) decomposition of the covariance matrix, with variability reduced to modes represented by spatial loadings and time scores. The domain of analysis covers the Caribbean from 8°–25°N, 85°–55°W with a focus on the leading mode of 0–200 m zonal currents and comparison with patterns that relate to the Caribbean fishery.

Given that the fish catch index does not account for changes in effort and may not coincide with abundance, an understanding of environmental modulation using an objective measure of productivity is needed. For this, monthly satellite chlorophyll estimates from SeaWifs and Modis were obtained from reason.gsfc.nasa.gov/Giovanni and analyzed in the period 1998–2007 for seasonal cycle and consecutive anomalies. Seasonal cycles were also constructed for rainfall, winds, currents, vertical motion, temperature and salinity, and anomaly time series were smoothed with a 7-month running mean. Further ocean data on chlorophyll, nitrate and mixed layer depth were obtained from the NOBM model [60] averaged over the Caribbean, with a focus on the seasonal cycle. The NOBM model uses a similar reanalysis system as SODAv2.4 and offers additional ocean chemistry and productivity outputs. The environmental variables were regressed onto chlorophyll using the forward stepwise method and an optimal multivariate solution was determined using an objective Statgraphics routine. Statistical significance was evaluated with 8 degrees of freedom (df) for the seasonal cycle and 18 df for the smoothed anomalies.

3. RESULTS

3.1. Fish Catch, Seasonal Cycle and Productivity

A statistical intercomparison of individual country fish catch with the Caribbean indices over the period 1971–2004 is listed in Table 1. Those significantly correlated include: Antigua-Barbuda, Bahamas, Colombia, Dominica, Dominican Republic, Guadeloupe, Haiti, Jamaica and Venezuela. Thus 9 of 21 contribute to the Caribbean fish catch index

(Figure 1(a)) subsequently considered as a basis for composite analysis. The catch index wavelet spectrum exhibits significant oscillations at 3-4 years and 8–12 years, particularly since 1985 (Figure 1(b)). High and low years tend to occur in succession (e.g., 1971–75, 1977–78, 1979–81, 1984–87, 1988–90, 1994–95, 1999–2001). Figure 1(d) is the SeaWifs chlorophyll content averaged 1998–2007. It reveals high productivity in the southern Caribbean and nutrient-poor waters in the subtropical west Atlantic. Around the larger islands and coastal headlands are zones of enhanced phytoplankton accumulated from upwelling and nutrients carried in river run-off. Although the ocean color record is too short for composite analysis with respect to fish catch, it is studied to understand how productivity is modulated.

The seasonal cycle of environmental variables averaged over the Caribbean in the period 1998–2007 is presented in Figure 2. Both rainfall and near-surface temperature gradually increase to a maximum in late summer, with rain showing a secondary peak in spring (Figure 2(a)). Mixed layer depth remains near 50 m with a minimum in late summer. Currents are consistently westward with a slight northward component that is maximum in June and minimum in late summer. Zonal currents are slightly stronger in summer and winter than in transition seasons. There is mean downwelling across the Caribbean that inhibits productivity, particularly in winter (Figure 2(b)). Vertical motions are near zero in late summer. Wind stress is persistently westward with a weak southward component (Figure 2(c)). Trade winds are weakest in the hurricane season (August–October). Near-surface salinity is greatest at the end of the dry season (March-April) and falls rapidly in August as river run-off and rainfall increase. Chlorophyll content in both satellite and model display a single peak in mid-summer (June-July) when trade winds and south coast upwelling are enhanced (Figure 2(d)). Model-derived nitrate on the other hand, gradually declines during the year to a minimum in the hurricane season, inversely to rainfall. An environmental model for the seasonal cycle of chlorophyll was objectively formulated using mixed layer depth, temperature and meridional wind stress (Table 2(a), Figure 2(e)). The model explains two-thirds of the variance (significant at 90% confidence) but the peak is delayed 1-2 months. Productivity is more variable during summer, suggesting increased sensitivity to environmental forcing then. Developing an objective multivariate model for smoothed consecutive monthly chlorophyll anomalies (Table2(b), Figure 2(f)), 45% of variance is explained (significant at 90% confidence). The model tracks interannual fluctuations using rainfall, temperature, meridional current and zonal wind stress. The association of cooler temperatures and enhanced trade winds with higher productivity is highlighted. There is a chlorophyll event in the summer of 2003 (cf. Figure 2(f)) related to the spread of upwelled water from Venezuela northward into a zone of lower salinity (cf. Figures 1(e) and 1(f)), where the Caribbean Current carries it westward.

3.2. Composite Environmental Differences

Figure 3 presents contrasts in the ocean environment with respect to fish catch. SST differences (Figure 3(a)) exhibit a cool plume extending across the southeastern Caribbean (8°–15°N, 55°–65°W). In contrast the Bahamas and northern Cuba experience warmer SST. Thus a warm/north—cool/south gradient is superimposed during years of high fish catch. The composite sea surface height field displays a key feature: an east-west 0.05 m ridge along 16°N stretching 2000 km from the Cayman Islands to Guadeloupe. It has a narrow north-south dimension of 200 km, a frontal feature [61]. Apart from this ridge there are negative sea surface height differences off eastern Venezuela and in the Panama gyre. Anomalous currents tend to flow anticyclonically around the zonal sea surface ridge (Figure 3(c)), with enhanced westward flow on 15°N and weakened westward flow (eastward anomalies) on 18°N. This weakening of currents near the Antilles improves chances for the retention of food sources, larvae and juvenile fish [62]. The North Brazil Current is enhanced and may entrain nutrient-rich waters from the east Atlantic. Wind stress is anomalous westward particularly north of Venezuela (Figure 3(d)); the trade winds are stronger and more southeasterly than usual and promote northward Ekman transport and cooler SST. The temperature and salinity sections (Figures 3(e) and 3(f)) indicate the Caribbean is cool and fresh in the upper 100 m with respect to years of high fish catch. There is a warm salty layer that slopes downward to the north below 100 m. Meridional currents and vertical motion differences reflect overturning near the South American coast (Figure 3(f)) and broad northward flow that spreads productive waters from Venezuela across the Caribbean.

Table 2: Multi-variate regression of environmental variables onto the seasonal cycle of chlorophyll averaged over the Caribbean 1998–2007.

(a) Using mixed layer depth, temperature and meridional wind stress. Multi-variate regression onto smoothed consecutive chlorophyll anomalies averaged over the Caribbean.

Adj R Sq	0.66	Regression statistics for **seasonal** Chl model			
Std Error	0.04				
ANOVA	deg.fr.	SS	MS	*F*	
Regression	3.0	0.04	0.01	7.99	
Residual	8.0	0.01	0.00		
	Coeff	Std Err	*t* Stat	*P*-value	
Intercept	−6.17	1.78	−3.47	.01	
mixL.D.	0.02	0.01	3.01	.02	
Temp	0.23	0.06	3.73	.01	
Tau Y	7.18	1.94	3.70	.01	

(b) Using rainfall, temperature, meridional currents and zonal wind stress. Refer to Figures 2(e) and 2(f).

Adj R Sq	0.45	Regression statistics for **anomaly** Chl model		
Std Error	0.01			
ANOVA	deg.fr.	SS	MS	*F*
Regression	4.0	0.02	0.01	26.14
Residual	18.0	0.02	0.00	
	Coeff	Std Err	*t* Stat	*P*-value
Intercept	0.002	0.00	1.45	.15
Rain	0.0005	0.00	4.46	.00
Temp	−0.04	0.00	−8.09	.00
V cur	−1.97	0.38	−5.22	.00
Tau X	−2.08	0.28	−7.51	.00

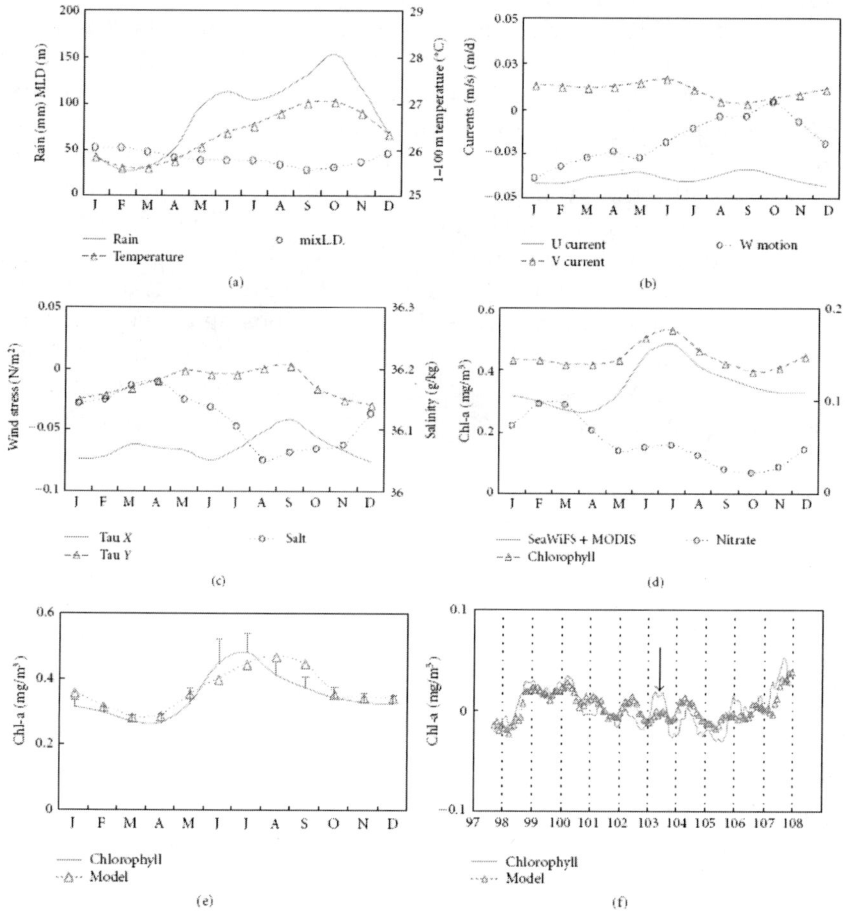

Figure 2: Caribbean area seasonal cycle for (a) rain, mixed layer depth and 0–100 m temperature, (b) 0–200 m zonal/meridional currents and vertical motion, (c) zonal/meridional wind stress and 0–50 m salinity, (d) SeaWifs chlorophyll, chlorophyll and nitrate from NOBM model, (e) chlorophyll with deviations and multivariate model. (f) Smoothed monthly SeaWifs chlorophyll anomaly and multivariate model as in Table 2, with pointer to event in Figures 1(e) and 1(f).

Atmospheric contrasts associated with fish catch are illustrated in Figure 4. An upper level high pressure prevails over the tropical Atlantic while an upper level low pressure prevails over the east Pacific (not shown). The result is a meridional overturning atmospheric circulation

(Figure 4(a)) that represents a weakening of the normal Hadley Cell with respect to high catch years; with upward motion north of the Antilles and sinking motion over South America. Atmospheric precipitable water is reduced over South America and enhanced in the northern Caribbean (Figure 4(b)). The climatic scenario corresponding with high fish catch is a wet/north—dry/south climatic gradient consistent with the warm/north—cool/south SST field. In the atmospheric height sections (Figures 4(c) and 4(d)) air temperatures and humidity over Venezuela are anomalous warm and dry, while in the Antilles zone (Hispanola) it is cool and moist with respect to high fish catch. The sea surface ridge on 16°N bisects opposing climate regimes. Warmer air temperatures in the 800 hPa layer ~11°N cap the trade winds, enhancing cyclonic curl and upwelling along the Venezuela coast. Thus ocean-atmosphere coupling is indicated in the composite difference fields.

4. DISCUSSION

A fishery index developed from nine country records has been used to contrast environmental differences between high- and low-catch rates in the Caribbean. The index has pointed to a number of key features, most notably a cool SST plume and changes in the atmospheric Hadley cell which could suppress tropical cyclones. In Figures 5(a) and 5(b) hurricane tracks are given for the seven high- and low-catch years. Few hurricanes pass through the area in high catch years, while in low-catch years the Caribbean is inundated by hurricanes. Hurricanes influence "catchability" and "effort" due to the rough seas that are generated. Hurricanes typically bring swells >6 m and can deepen the mixed layer >100 m, dispersing plankton and inhibiting feeding. Composite differences were found to be negligible over the Gulf Stream and Bermuda (not shown), so hurricanes do not "disappear" in years of high fish catch, they just shift out of the Caribbean due to cooler SST and corresponding changes in climate. It is noteworthy that Caribbean hurricanes have fluctuated with an 8-year cycle since 1970 [63, 64]. Years with more hurricanes coincide with weaker trade winds and diminished coastal upwelling that inhibit the spread of productive waters.

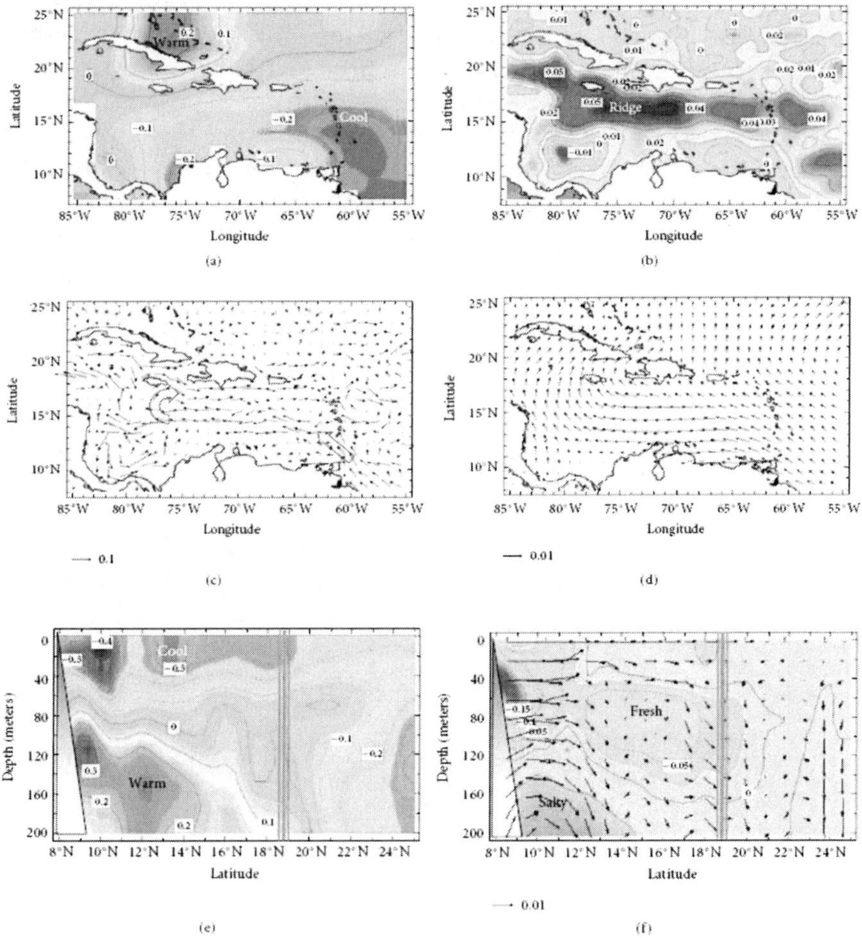

Figure 3: Composite difference maps for (a) SST (°C), (b) sea surface height (m), (c) 0–200 m vector currents (m/s) in respect of high minus low fish catch, and (d) vector wind stress (N/m2). Composite vertical N-S sections of (e) temperature (°C), (f) salinity (g/kg) and V W currents (m/s), with vertical motion exaggerated. Vertical bar represents Antilles, South America is left-side of sections. Key vectors are given.

The current patterns with respect to fish catch seen in Figure 3(c) reflect a southward shift of the Caribbean Current and anomalous eastward (decreased westward) flow in the Antilles islands. A question that arises is whether this current pattern is specific to fluctuations of the fishery, or is it a natural mode of ocean variability? This is addressed by decomposition of environmental fields into empirical orthogonal functions (EOF). The spatial loading pattern for the leading mode of zonal currents is illustrated in Figure 5(c) and corresponds with the vector field of Figure 3(c). The Caribbean Current alternates N-S of its mean axis (−U on 14°N, +U on 18°N), modulating the flow around Jamaica, Hispanola and the Anegada Passage, and thus chances for biological retention. Its scatterplot with fish catch (Figure 5(d)) indicates 25% of variance explained, significant at 90% confidence with 34 df.

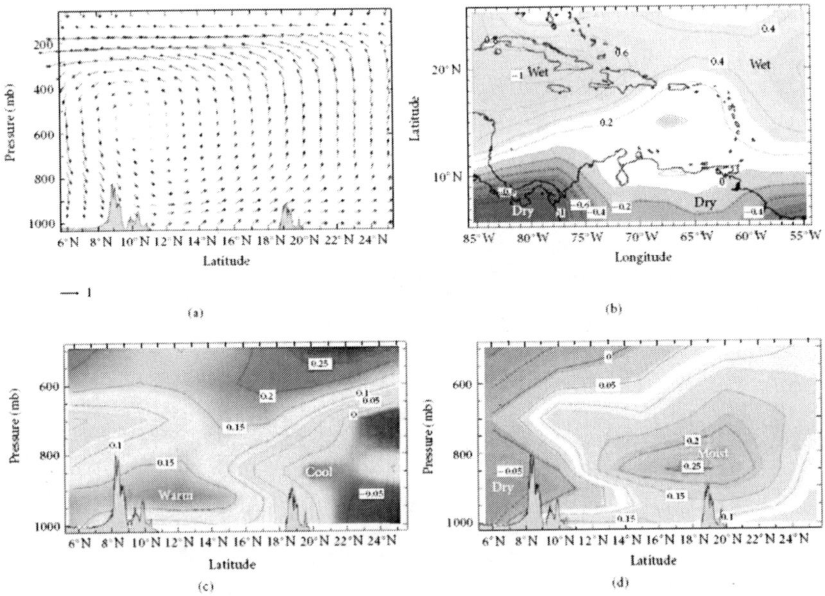

Figure 4: Composite difference height sections of (a) meridional winds and vertical motion (m/s), (c) air temperature (°C), and (d) specific humidity (g/kg) with land represented. Composite difference map for (b) precipitable water (mm).

In summary, a key feature uncovered in this work based on a Caribbean fishery index is an anomalous sea surface ridge on 16°N that reflects anticyclonic recirculation during years of higher catch. South of the ridge, the westward Caribbean Current and North Brazil Current are strengthened. North of the ridge, anomalous eastward currents are found in the axis of the Antilles Islands (Jamaica, Hispanola, Puerto Rico, Guadeloupe). The ridge divides warm dry weather over Venezuela from cool wet weather over the northern Antilles, set up by an anomalous reversal of the atmospheric Hadley circulation. Trade winds and upwelling are stronger along the Venezuela coast in high catch years. The meridional ocean circulation (cf. Figure 3(f)) advects waters northward, infusing the Caribbean with more nutrients. Recirculation gyres along the north edge of the Caribbean Current (cf. Figure 3(c)) inhibit the loss of productive waters to the Yucatan passage, and promote retention in the Antilles zone. Hurricanes are rare during years of high fish catch, in conjunction with cooler SST. Fewer hurricanes offer chances for increased fishing effort in summer. Further work could be done to study the environmental modulation of fish catch per unit effort and dominant species; analysis of how the fish integrate environmental conditions over their lifespan; a comparison of local versus regional nitrification processes; and studies that relate global climatic trends to conditions that influence Caribbean fish catch.

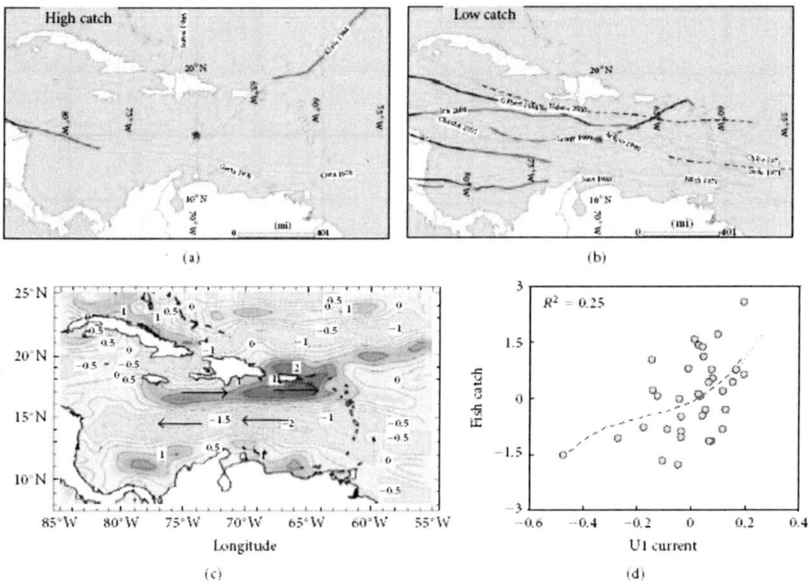

Figure 5: Composite tropical cyclone tracks for (a) high and (b) low-catch years (color = intensity, red > cat.2). (c) First mode of 0–200 m zonal current variability based on EOF analysis, and (d) its time score scatterplot with fish catch (dotted curve is best fit).

REFERENCES

1. A. L. Gordon, "Circulation of the Caribbean Sea," Journal of Geophysical Research, vol. 72, pp. 6207–6223, 1967.

2. S. J. Murphy, H. E. Hurlburt, and J. J. O'Brien, "The connectivity of eddy variability in the Caribbean Sea, the Gulf of Mexico, and the Atlantic Ocean," Journal of Geophysical Research, vol. 104, no. 1, pp. 1431–1453, 1999.

3. C. A. Andrade and E. D. Barton, "Eddy development and motion in the Caribbean Sea," Journal of Geophysical Research, vol. 105, no. 11, pp. 26191–26201, 2000.

4. W. E. Johns, T. L. Townsend, D. M. Fratantoni, and W. D. Wilson, "On the Atlantic inflow to the Caribbean Sea," Deep-Sea Research Part I, vol. 49, no. 2, pp. 211–243, 2002.

5. J. Gyory, A. J. Mariano, and E. H. Ryan, "The Caribbean Current in Ocean Surface Currents," 2005,http://oceancurrents.rsmas.miami.edu/caribbean/caribbean.html.

6. R. L. Molinari, M. Spillane, I. Brooks, D. Atwood, and C. Duckett, "Surface current in the Caribbean Sea as deduced from Lagrangian observations," Journal of Geophysical Research, vol. 86, pp. 6537–6542, 1981.

7. L.-L. Fu and B. Holt, "Some examples of detection of oceanic mesoscale eddies by the Seasat synthetic-aperture radar," Journal of Geophysical Research, vol. 88, no. 3, pp. 1844–1852, 1983.

8. T. H. Kinder, G. W. Heburn, and A. W. Green, "Some aspects of the Caribbean circulation," Marine Geology, vol. 68, no. 1–4, pp. 25–52, 1985.

9. D. M. Fratantoni, "North Atlantic surface circulation during the 1990's observed with satellite-tracked drifters," Journal of Geophysical Research, vol. 106, no. 10, pp. 22067–22093, 2001.

10. D. Roemmich, "Circulation in the Caribbean Sea: a resolved inverse problem," Journal of Geophysical Research, vol. 86, pp. 7993–8005, 1981.

11. A. Hernandez-Guerra and T. M. Joyce, “Water masses and circulation in the surface layers of the Caribbean at 66°W,” Geophysical Research Letters, vol. 27, no. 21, pp. 3497–3500, 2000.

12. J. T. Gunn and D. R. Watts, “On the currents and water masses north of the Antilles/Bahamas Arc,”Journal of Marine Research, vol. 40, no. 1, pp. 1–18, 1982.

13. T. N. Lee, W. E. Johns, R. J. Zantopp, and E. R. Fillenbaum, “Moored observations of western boundary current variability and thermohaline circulation at 26.5∘N in the subtropical North Atlantic,” Journal of Physical Oceanography, vol. 26, no. 6, pp. 962–983, 1996.

14. G. L. Hitchcock, T. N. Lee, P. B. Ortner, S. Cummings, C. Kelble, and E. Williams, “Property fields in a Tortugas Eddy in the southern straits of Florida,” Deep-Sea Research Part I, vol. 52, no. 12, pp. 2195–2213, 2005.

15. W. J. Schmitz and W. S. Richardson, “On the transport of the Florida current,” Deep-Sea Research, vol. 38, supplement, pp. 379–409, 1991.

16. R. C. Wajsowicz, “A modified sverdrup model of the Atlantic and Carribbean circulation,” Journal of Physical Oceanography, vol. 32, no. 3, pp. 973–993, 2002.

17. W. D. Wilson and W. E. Johns, “Velocity structure and transport in the Windward Islands Passages,”Deep-Sea Research Part I, vol. 44, no. 3, pp. 487–520, 1997.

18. A. Alvera-Azcárate, A. Barth, and R. H. Weisberg, “The surface circulation of the Caribbean Sea and the Gulf of Mexico as inferred from satellite altimetry,” Journal of Physical Oceanography, vol. 39, no. 3, pp. 640–657, 2009.

19. J. M. Leis, “The pelagic stage of reef fishes: the larval biology of coral reef fishes,” in The Ecology of Fishes on Coral Reefs, P. F. Sale, Ed., pp. 183–230, Academic Press, San Diego, Calif, USA, 1991.

20. L. W. Botsford, C. L. Moloney, A. Hastings et al., “The influence of spatially and temporally varying oceanographic conditions on meroplanktonic metapopulations,” Deep-Sea Research Part II, vol. 41, no. 1, pp. 107–145, 1994.

21. C. B. Paris and R. K. Cowen, "Direct evidence of a biophysical retention mechanism for coral reef fish larvae," Limnology and Oceanography, vol. 49, no. 6, pp. 1964–1979, 2004.

22. J. D. Neilson and R. I. Perry, "Diel vertical migrations of marine fishes: an obligate or facultative process?" Advances in Marine Biology, vol. 26, pp. 115–168, 1990.

23. H. H. Roberts, P. A. Wilson, and A. Lugo-Fernández, "Biologic and geologic responses to physical processes: examples from modern reef systems of the Caribbean-Atlantic region," Continental Shelf Research, vol. 12, no. 7-8, pp. 809–834, 1992.

24. J. M. Shenker, E. D. Maddox, E. Wishinski, A. Pearl, S. R. Thorrold, and N. Smith, "Onshore transport of settlement-stage Nassau grouper Epinephelus striatus and other fishes in Exuma Sound, Bahamas,"Marine Ecology Progress Series, vol. 98, no. 1-2, pp. 31–43, 1993.

25. R. K. Cowen and L. R. Castro, "Relation of coral reef fish larval distributions to island scale circulation around Barbados, West Indies," Bulletin of Marine Science, vol. 54, no. 1, pp. 228–244, 1994.

26. G. Symonds, K. P. Black, and I. R. Young, "Wave-driven flow over shallow reefs," Journal of Geophysical Research, vol. 100, no. 2, pp. 2639–2648, 1995.

27. E. D. Barton, J. Aristegui, P. Tett et al., "The transition zone of the Canary Current upwelling region,"Progress in Oceanography, vol. 41, no. 4, pp. 455–504, 1998.

28. R. S. Appeldoorn, D. A. Hensley, D. Y. Shapiro, S. Kioroglou, and B. G. Sanderson, "Egg dispersal in a Caribbean coral reef fish," Bulletin of Marine Science, vol. 54, no. 1, pp. 271–280, 1994.

29. M. J. Milicich, "Dynamic coupling of reef fish replenishment and oceanographic processes," Marine Ecology Progress Series, vol. 110, no. 2-3, pp. 135–144, 1994.

30. S. U. Sponaugle and R. K. Cowen, "Early life history traits and recruitment patterns of Caribbean wrasses (Labridae)," Ecological Monographs, vol. 67, no. 2, pp. 177–202, 1997.

31. D. R. Robertson, S. E. Swearer, K. Kaufmann, and E. B. Brothers, "Settlement vs. environmental dynamics in a pelagic-spawning reef fish at Caribbean Panama," Ecological Monographs, vol. 69, no. 2, pp. 195–218, 1999.

32. R. K. Cowen, "Larval dispersal and retention and consequences for population connectivity," in Coral Reef Fishes. Dynamics and Diversity in a Complex Ecosystem, P. F. Sale, Ed., pp. 149–170, Academic Press, San Diego, Calif, USA, 2002.

33. J. L. Munro, V. C. Gaut, R. Thompson, and P. H. Reeson, "The spawning seasons of Caribbean reef fishes,"Journal of Fish Biology, vol. 5, no. 1, pp. 69–84, 1973.

34. P. L. Colin, "Reproduction of the Nassau grouper, Epinephelus striatus (Pisces: Serranidae) and its relationship to environmental conditions," Environmental Biology of Fishes, vol. 34, no. 4, pp. 357–377, 1992.

35. P. J. S. Franks, "Sink or swim: accumulation of biomass on fronts," Marine Ecology Progress Series, vol. 82, pp. 1–12, 1992.

36. W. M. Graham, J. G. Field, and D. C. Potts, "Persistent "upwelling shadows" and their influence on zooplankton distributions," Marine Biology, vol. 114, no. 4, pp. 561–570, 1992.

37. P.S. Lobel and A.R. Robinson, "Reef fishes at sea: ocean currents and the advection of larvae," in The Ecology of Deep and Shallow Coral Reefs, M. L. Reaka, Ed., vol. 11 of NOAA Symposium Series for Undersea, pp. 29–38, Noaa's Undersea Research Program (NURP), Rockville, Md, USA, 1983.

38. P. S. Lobel and A. R. Robinson, "Transport and entrapment of fish larvae by ocean mesoscale eddies and currents in Hawaiian waters," Deep Sea Research Part A, vol. 33, no. 4, pp. 483–500, 1986.

39. S. U. Sponaugle and K. Grorud-Colvert, "Environmental variability, early life-history traits, and survival of new coral reef fish recruits," Integrative and Comparative Biology, vol. 46, no. 5, pp. 623–633, 2006.

40. F. E. Muller-Karger, C. R. McClain, and P. L. Richardson, "The dispersal of the Amazon's water," Nature, vol. 333, no. 6168, pp. 56–59, 1988.

41. J. M. Morrison and O. P. Smith, "Geostrophic transport variability along the Aves Ridge in the eastern Caribbean Sea during 1985-1986," Journal of Geophysical Research, vol. 95, pp. 699–710, 1990.

42. C. B. Grimes and M. J. Kingsford, "How do riverine plumes of different sizes influence fish larvae: do they enhance recruitment?" Marine and Freshwater Research, vol. 47, no. 2, pp. 191–208, 1996.

43. R. J. Allman and C. B. Grimes, "Growth and mortality of little tunny (Euthynnus alletferatus) larvae off the Mississippi River plume and Panama City, Florida," Bulletin of Marine Science, vol. 62, no. 1, pp. 189–197, 1998.

44. A. Scotti and J. Pineda, "Plankton accumulation and transport in propagating nonlinear internal fronts,"Journal of Marine Research, vol. 65, no. 1, pp. 117–145, 2007.

45. A. R. Robinson, Eddies in Marine Science, Springer, New York, NY, USA, 1983.

46. A. Okubo, "The role of diffusion and related physical processes in dispersal and recruitment of marine populations," in The Bio-Physics of Marine Larval Dispersal, P. W. Sammarco and M. L. Heron, Eds., pp. 5–32, American Geophysical Union, Washington, DC, USA, 1994.

47. C. B. Limouzy-Paris, H. C. Graber, D. L. Jones, A. W. Ropke, and W. J. Richards, "Translocation of larval coral reef fishes via sub-mesoscale spin-off eddies from the Florida current," Bulletin of Marine Science, vol. 60, no. 3, pp. 966–983, 1997.

48. S. U. Sponaugle and D. R. Pinkard, "Impact of variable pelagic environments on natural larval growth and recruitment of the reef fish Thalassoma bifasciatum," Journal of Fish Biology, vol. 64, no. 1, pp. 34–54, 2004.

49. C. Yeung, D. L. Jones, M. M. Criales, T. L. Jackson, and W. J. Richards, "Influence of coastal eddies and counter-currents on the influx of spiny lobster, Panulirus argus, postlarvae into Florida Bay," Marine and Freshwater Research, vol. 52, no. 8, pp. 1217–1232, 2001.

50. A. R. Emery, "Eddy formation from an oceanic island: ecological effects," Caribbean Journal of Science, vol. 12, pp. 121–128, 1972.

51. D. Rissik, I. M. Suthers, and C. T. Taggart, "Enhanced zooplankton abundance in the lee of an isolated reef in the south Coral Sea: the role of flow disturbance," Journal of Plankton Research, vol. 19, no. 9, pp. 1347–1368, 1997.

52. G. Basterretxea, E. D. Barton, P. Tett, P. Sangrá, E. Navarro-Perez, and J. Arístegui, "Eddy and deep chlorophyl maximum response to wind-shear in the lee of Gran Canaria," Deep-Sea Research Part I, vol. 49, no. 6, pp. 1087–1101, 2002.

53. R. Froese and D. Pauly, Eds., "FishBase," World Wide Web electronic publication, 2010,http://www.fishbase.org/.

54. FAO (Food and Agriculture Organization), "Marine fishery resources of the Antilles: Lesser Antilles, Puerto Rico and Hispaniola," Tech. Rep. 326, FAO Fisheries, Jamaica, Cuba, 1993.

55. R. Mahon and H. A. Oxenford, "Precautionary assessment and management of dolphinfish in the Caribbean," Scientia Marina, vol. 63, no. 3-4, pp. 429–438, 1999.

56. M. J. Paddack, J. D. Reynolds, C. Aguilar et al., "Recent region-wide declines in Caribbean reef fish abundance," Current Biology, vol. 19, no. 7, pp. 590–595, 2009.

57. E. Kalnay, M. Kanamitsu, R. Kistler et al., "The NCEP/NCAR 40-year reanalysis project," Bulletin of the American Meteorological Society, vol. 77, no. 3, pp. 437–471, 1996.

58. J. Sheffield, G. Goteti, and E. F. Wood, "Development of a 50-year high-resolution global dataset of meteorological forcings for land surface modeling," Journal of Climate, vol. 19, no. 13, pp. 3088–3111, 2006.

59. J. A. Carton and B. S. Giese, "A reanalysis of ocean climate using simple ocean data assimilation (SODA),"Monthly Weather Review, vol. 136, no. 8, pp. 2999–3017, 2008.

60. W. W. Gregg, "Assimilation of SeaWiFS ocean chlorophyll data into a three-dimensional global ocean model," Journal of Marine Systems, vol. 69, no. 3-4, pp. 205–225, 2008.

61. E. Wolanski and W. M. Hamner, "Topographically controlled fronts in the ocean and their biological influence," Science, vol. 241, no. 4862, pp. 177–181, 1988.

62. F. E. Werner, J. A. Quinlan, B. O. Blanton, and R. A. Luettich Jr., "The role of hydrodynamics in explaining variability in fish populations," Journal of Sea Research, vol. 37, no. 3-4, pp. 195–212, 1997.

63. M. R. Jury, "A quasi-decadal cycle in Caribbean climate," Journal of Geophysical Research, vol. 114, no. 13, Article ID D13102, 8 pages, 2009.

64. M. R. Jury and D. B. Enfield, "Environmental patterns associated with active and inactive Caribbean hurricane seasons," Journal of Climate, vol. 23, no. 8, pp. 2146–2160, 2010.

8

TROPHIC RELATIONSHIPS AMONG FISH ASSEMBLAGES ON A MUDFLAT WITHIN A BRAZILIAN MARINE PROTECTED AREA

Dafne Marcelle de Almeida Ramos Campos[1] , Adna Ferreira da Silva[1] , Natalice dos Santos Sales[1] , Ronnie Enderson Mariano Carvalho Cunha Oliveira[1] , Andre Luiz Machado Pessanha[1] *

[1]Universidade Estadual da Paraíba. (Rua Juvência Arruda, Bodocongó, Campina Grande - PB)

ABSTRACT

The present study deals with the temporal variations in diet and the trophic guilds of dominant fish species on a tidal mudflat during the dry and rainy seasons. We sought to classify the diet composition of 17 species in the Mamanguape river estuary, northeastern Brazil, identifying the dominant food components and evaluating the effects of seasonality on the guild organization. Diets varied little between species and seasons, though they seemed to be more heterogeneous during the rainy season. Five primary feeding guilds were identified, in accordance with the importance of prey in the diets: (1) Detritivore, (2) Zooplanktivore, (3) Zoobenthivore-epifaune, (4) Zoobenthivore-infaune, and (5) Piscivore. Most fishes fed on a diverse range of food items but relied heavily on zooplankton prey. Several fish species showed a tendency to a specialised diet, with almost all species showing some degree of opportunistic feeding. A high degree of diet overlap was found among some species; however, the presence of exploitative competition could not be determined.

KEYWORDS

Diet; Feeding ecology; Fish community; Diet overlap; Estuary

1. INTRODUCTION

Estuaries are recognised for encompassing a variety of habitats, including mangrove swamps, algal banks, and mudflats (TSE et al., 2008), with vegetated habitats exhibiting greater fish abundance and species richness than non-vegetated habitats (LUGENDO et al., 2006). This pattern is the result of the greater structural complexities and resources that develop on a vegetated substrate, providing greater protection from predators and increased food availability (WANG et al., 2009). This structural complexity increases the availability of consolidated substrates for the diatom colonisation of micro/macro algae that supports great fish diversity (HINDELL; JENKINS, 2004).

However, as non-vegetated habitats, tidal mudflats have been identified as important environments for supporting the estuarine life cycle and are considered to be key habitats for the estuarine food chain because of their high primary productivity when compared with adjacent areas (FRANÇA et al., 2008, 2009). Mudflats are most commonly located at the mouth of estuaries and serve multiple roles, such as transit routes and/or habitats for some fish species in tropical and subtropical estuaries, particularly for juveniles and school-forming species (HINDELL; JENKINS, 2004; BARLLETA; BLABER, 2007), while the use of structured areas, such as mangrove swamps, has been suggested for small specimens, such as the Gobiidae (HINDELL; JENKINS, 2004).

The loss of these intertidal areas, due to the constant demands of human activities in coastal areas, is a global- scale problem (ROPER et al., 1988; BOER; PRINS, 2002; MARTINHO et al., 2007; TSE et al., 2008; FRANÇA et al., 2012). On tidal mudflats, fish find refuge near the coast; thus, these habitats are considered to be additional recruitment areas that are essential for the life cycle of these fish species. However, contradictions exist, creating the need to study tidal mudflats independently to better understand how resources are used by different fish species, especially by juveniles.

In northeastern Brazil, river runoff is strongly influenced by seasonal rainfall. Interannual and seasonal variability in river flows can cause changes in the circulation and physicochemical conditions of the estuary. Changes in salinity can affect several variables, from the submerged vegetation in tidal channels to the diversity of benthic organisms and fish. Salinity affects fish distribution in estuaries through its influence on food availability (VIVIER et al., 2010). Few studies have incorporated temporal heterogeneity when examining trophic structures, despite its apparent relevance for the structure and function of fish assemblages (WILSON; SHEAVES, 2001; HARRINSON; WHITFIELD, 2012).

Trophic ecology studies seek to identify the feeding habits of species through the analysis of the major items consumed. Knowledge of the diets of species is one of the basic requirements for a closer examination of the relationships among organisms in a given ecosystem, as very close relationships exist between the quantity, quality, and availability of food and the distribution and abundance of consumer organisms. The structures of fish assemblages that use the shallow areas of estuaries are strongly influenced by trophic relationships. Knowledge of the trophic web allows one to describe energy flows in ecosystems and ecological relationships among organisms (PAIVA et al., 2008; DANTAS et al., 2013). Thus, it is important to understand the trophic relationships between fish and their potential prey to understand the relationships among fish in a community.

Marine Protected Areas (MPAs) are increasingly seen as important tools for the mangement of coastal ecosystems and fisheries, where significant effects at the fish assemblage level is marked in the total abundance, species richness and diversity of fishes in these areas (CLAUDET et al., 2006). Therefore, the present study aimed to characterise the trophic ecology of the major fish species caught on a tropical tidal mudflat to accomplish the following: (1) describe and examine the diets of species during different phases of the hydrological regime and (2) define the trophic guilds within the fish community of a tidal mudflat located in a tropical estuary with little anthropogenic influence.

2. MATERIAL AND METHODS

2.1 Study area

The Mamanguape River estuary is located on the north coast of the Paraíba state and extends for 25 km in an east-west direction and for 5 km in a north-south direction. It is part of the Environmental Protection

Area (Área de Proteção Ambiental - APA) of Barra de Mamanguape (Figure 1). Its primary purpose is to protect marine biodiversity, to favor social and economic activities linked to the sea, especially fisheries, and to promote public education and scientific research. The regional climate is classified by Köppen as As-type (hot and humid). The rainy season begins in February and lasts until July, with maximum rainfall occurring from April to June, whereas the dry season occurs in the period from August to January, the lowest rainfall occurring between October and December (PEREIRA; ALVES, 2006). The average rainfall recorded in the area is between 1,750 and 2,000 mm annually, and the average temperature is approximately 24-26ºC. There is a well-preserved mangrove swamp in the area, composed of *Avicennia germinans, Avicennia schaweriana, Conocarpus erectus, Laguncularia racemosa* and *Rhizophora mangle*, which extends for 600 ha around the primary channel and tidal creek, in addition to Atlantic Forest remnants (ROCHA et al., 2008). Endangered species, such as the seahorse, *Hippocampus reidi*, and the West Indian manatee, *Trichechus manatus*, are also found in this estuary (MOURÃO; NORDI, 2003; CASTRO et al., 2008).

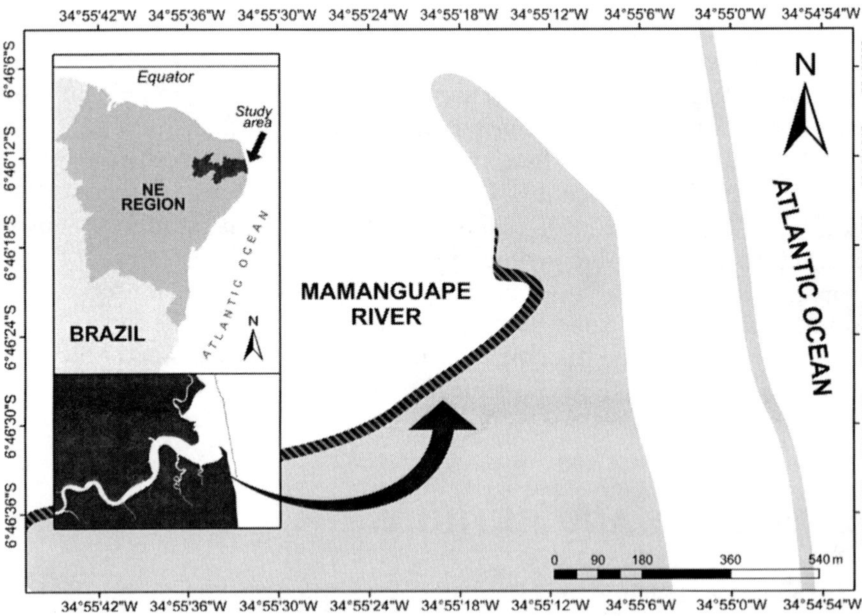

Figure 1 Map of the study area. The hatched area indicates the tidal mudflat studied in the Mamanguape River Estuary, Brazil.

The Curva do Pontal beach (6º46'27'' S; 34º55'20'' W), located 2.3 km upstream from the estuary mouth, is 1,200 m long and has very calm waters because of the diminished influence of waves (Figure 1). The beach is greatly influenced by the entrance of ocean waters, where marine sediments are regularly exposed and submerged by tidal action, and has a maximum depth of 4 m. The tidal mudflat examined is a non-vegetated area with a gentle slope and fine muddy sediment in the intertidal zone; in the subtidal zone, seagrass, sessile invertebrates, macroalgae, mangrove leaves, and fallen branches form the benthic cover (XAVIER et al., 2012).

2.2 Sampling programme

Samples were collected on monthly daytime excursions during spring low tides, in both the dry season (October 2010 to January 2011) and the rainy season (March to June 2011), to sample the ichthyofauna of the tidal mudflat more completely and efficiently during the local hydrological regime.

The fish were sampled using a beach seine (10.0 x 1.5 m; 8 mm mesh size). The seine hauls were 30 m long, parallel and close to the shore, and were taken out to a depth of approximately 1.5 m. The sampling unit was standardised with 5 replicates in an effort to capture individuals that use the area for feeding, 5 m apart to minimize the influence on the following haul. The collected fish were fixed in 10% formalin for later identification in the laboratory, according to FIGUEIREDO and MENEZES (1978, 1980, 2000), MENEZES and FIGUEIREDO (1980, 1985), and ARAÚJO et al. (2004). The total length (TL, mm) and body weight (g) were measured for each individual. Temperature, salinity, and water transparency were measured, using a thermometer, an optical refractometer, and a Secchi disc, respectively.

2.3 Data analysis

Data from 17 predator species were included in the current analysis. We include numerically dominant, commercially valuable, and ecologically important species in the Mamanguape river estuary (XAVIER et al., 2012;OLIVEIRA; PESSANHA, 2014). To remove the stomach of each captured specimen, an incision was made in the abdominal region, from the anus toward the head. The stomachs were opened, and their contents arranged in a Petri dish for viewing. Content analysis was performed using a stereoscopic microscope, and food items were identified to the lowest possible taxonomic level and subsequently grouped into the following taxonomic categories: Plant Material (plant and algae debris); Zooplankton (Decapoda larvae, Calanoida and Cyclopoida copepods, invertebrates, and fish eggs); Insects; Polychaeta; Fish (scales and fish); Mollusca (Bivalvia, Gastropoda, and Scaphopoda); Decapoda (crabs and shrimp); Benthic or Epibenthic Crustaceans (Isopoda, Tanaidacea, Ostracoda, *Caprella* amphipods, Mysidacea, Gammaridea amphipods, Harpacticoida copepods, and Cirripedia); Infaune (Foraminifera, Sipuncula, Nematoda, and Trematoda); and Phytobenthos (diatoms).

To analyse each diet, the frequency of occurrence (%F), the percentage number (%N), and the volume (%V) of different food items were calculated (HYSLOP, 1980), and the Index of Relative Importance (IRI) was subsequently applied (PINKAS et al., 1971). Although the IRI may not be a good index, as pointed out byMACDONALD and GREEN (1983), it was used in this paper because IRI has often been used in other studies on stomach contents and it facilitates comparison with the results of other studies (BARRY et al., 1996). For items that cannot be counted, a value of 0.1 was given for their number (%N) when they were present in the diet, to offset distortions in the index (ABDURAHIMAN et al., 2010).

A non-hierarchical cluster analysis, based on the UPGAM (Unweighted Pair-Group Average) linkage method, was used to construct a dendrogram to identify guilds for each phase of the hydrological regime. These data were used to build a similarity matrix using the correlation coefficient. A cophenetic correlation analysis was employed to increase the reliability of conclusions drawn from the interpretation of the dendrogram. We adopted rc = 0.80 as the fidelity criterion for this analysis (ROMESBURG, 1984; KOPP et al., 2007). The trophic guilds were defined according to ELLIOTT et al. (2007).

Non-metric multidimensional scaling (nMDS) was applied to obtain a graphical representation of the trophic groups, using the volumetric measurement of food items. The Bray-Curtis coefficient was calculated for each fish species after a fourth-root transformation. For easy interpretation, the items were coded and grouped according to the taxonomic categories previously described. An ANOSIM (analysis of similarity) was used to test for differences in species between the phases of the hydrological regime and among the guilds. Prior to ordination (nMDS), the volumetric dietary data of each category for the diets of each individual of each fish species were treated in different ways according to the purpose of the analysis.

The trophic niche breadth (diet breadth) for each period of the hydrologic cycle was calculated using Levins' standardised index (KREBS,1989), which ranges from 0, where a species consumed only 1 type of prey, to 1, where a species similarly consumed various types of prey. To standardise the measure of each trophic niche we applied HURLBERT (1978). To calculate niche overlap, PIANKAS index (KREBS, 1989) was used, which measures the extent to which 2 species share a number of resources on a scale ranging from 0 to 1. A score of 0 indicates that the 2 species are completely dissimilar, while a score of 1 indicates complete overlap.

To explain trophic relationships, dietary data were summarised into a trophic diagram. Links between fish and their prey were constructed by incorporating the contribution of the volume of different food items in the diet of each fish species during the different phases of the hydrological regime (rainy and dry). The volumetric dietary data is the better metric for quantifying the relative importance of different food items (HAMMERSCHLAG et al., 2010).

3. RESULTS

The temperature, salinity and transparency values were lowest in the rainy season and highest in the dry season, following a typical pattern for this tropical region (Figure 2). The temperature varied between 26ºC and 34ºC (dry season: = 29.6, standard error (SE) = 0.49; rainy season: = 29.5, SE = 0.32); salinity ranged between 10 and 42 (dry season: = 33.3, SE = 1.03; rainy season: = 27.5, SE = 1.86); and transparency ranged between 10 and 80 cm (dry season: = 43.3, SE = 2.58; rainy season: = 40.6, SE = 4.20).

The stomach contents of the total 1,867 individuals (representing more than 80% of the numerical abundance of fish on the tidal mudflat caught during both periods of the hydrological regime were analysed (Table 1). The diets of the species studied were composed of 35 prey items, the Crustacea group being the most diverse, primarily because of zooplankton and small benthic crustaceans (Tables 2 and 3). The Zooplanktivore guild was the most species-rich, while the

Zoobenthivore-infaune and Insectivore were the most species-poor, with the latter guild present only during the rainy season. An analysis of the trophic guilds did not reveal any differences between the hydrological regime phases on the tidal flat studied (Figure 3). In the dry season, the cluster formed (cophenetic coefficient rc = 0.89) comprised 5 different trophic guilds separated at the 40% similarity level: the first group consisted of 4 species of Zoobenthivore-epifaune feeding predominantly on Bivalvia and Gastropoda; the second group, containing 8 species, formed the Zooplanktivore guild; the third group, containing 2 species, formed the Piscivore guild; the fourth group consisted of only 1 species of the Zoobenthivore-infaune guild feeding predominantly on Polychaeta; and the fifth group consisted of 2 species that had diets based on organisms near the substrate (Detritivore), such as diatoms and foraminifera.

In the rainy season, 6 trophic guilds were formed via clustering (cophenetic coefficient rc = 0.92) separated at the 40% similarity level: the first and second groups were each composed of only 1 species that fed on Insects and Polychaeta (Zoobenthivore-infaune), respectively; the third group consisted of 2 species of Mollusca eaters (Zoobenthivore-epifaune); the fourth group was formed by 6 species consuming zooplanktivorous organisms; the fifth group consisted of 4 species of fish consumers; and the sixth group was formed by 2 species that were consumers of organisms near the substrate, such as Infaune and Phytobenthos (diatoms) (Figure 4).

The dendrogram analysis revealed that most species remained within the same trophic guilds during both phases of the hydrological regime (Figure 3 and 4). However, 5 species changed guilds: during the dry season,*A. brasiliensis* and *H. unifasciatus* were categorised as Zoobenthivore- epifaune, but, in the rainy season, they moved to the Zooplanktivore and Insect eater guilds, respectively; *B. soporator*, which was part of the Zooplanktivore guild in the dry season, was categorised as a Zoobenthivore- epifaune in the rainy season; and*C. macrops* and *O. saurus* were in the Zooplanktivore guild during the dry season but were categorised as Piscivore during the rainy season.

The nMDS showed no distinction between hydrological periods, a result corroborated by the ANOSIM value (global R = 0.067, p = 0.122) (Figure 5). However, when the samples were coded by trophic guild, a very clear separation was evident (global R = 0.404, p = 0.001): the Detritivore group were plotted in middle right, Piscivore and Zoobenthivore-infaune in the extreme lower left, Zoobenthivore-epifaune through middle left to upper centre, while the dominant group, the Zooplanktivore guild, was distributed throughout the diagram (Figure 5).

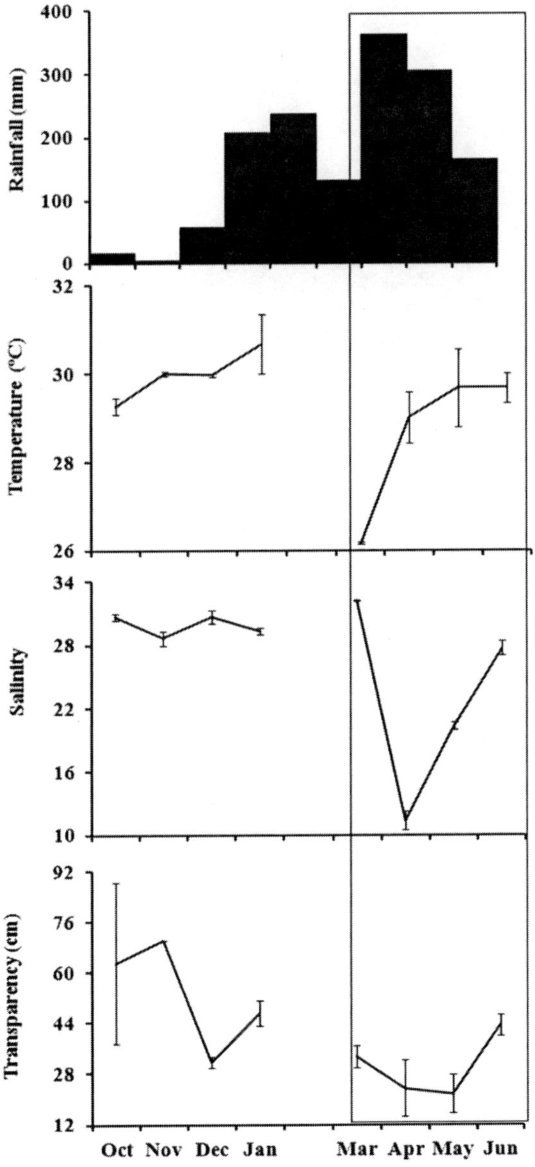

Figure 2 Monthly total rainfall and mean (standard error) temperature, salinity, and transparency in the Mamanguape River Estuary, Brazil. Demarcated area indicating rainy season.

Table 1 Number of individuals captured on the tidal mudflat during the dry and rainy seasons in a tropical estuary in Brazil

Families	Species	Dry season					Rainy season				
		N	B	F	TL (mean ± SD)	TL (range)	N	B	F	TL (mean ± SD)	TL (range)
Engraulidae	*Anchovia clupeoides*	47	89.8	14.8	106 ± 6.7	95 - 134	20	84.6	12.0	81 ± 3.6	75 - 87
	Lycengraulis grossidens	57	1704.1	40.0	58 ± 22.2	20 - 214	328	174.9	24.0	42 ± 12.3	25 - 143
Clupeidae	*Harengula clupeola*	08	78.1	14.2	75 ± 11.1	65 - 95	08	35.4	10.3	65 ± 5.3	60 -76
	Rhinosardinia bahiensis	155	8143.2	37.1	74 ± 8.0	60 - 106	119	350.0	16.0	73 ± 6.4	54 - 88
Hemiramphidae	*Hyphorhamphus unifasciatus*	70	120.1	37.2	137 ± 38.7	60 - 183	26	198.1	36.0	154 ± 28.5	81 - 202
Atherinopsidae	*Atherinella brasiliensis*	191	412.7	45.7	95 ± 19.9	27 -130	258	414.8	60.0	42 ± 32.4	17 -120
Mugilidae	*Mugil liza*	45	43.4	20.0	29 ± 4.7	20 - 58	208	27.1	44.0	31 ± 4.1	17 - 52
Gerreidae	*Eucinostomus argenteus*	20	79.0	28.5	63 ± 22.7	16 - 105	27	52.4	20.0	18 ± 13.7	10 - 100
	Eucinostomus melanopterus	3	0.64	2.9	54 ± 11.9	47 - 68	31	7.0	24.0	23 ± 12.7	15 - 50
Carangidae	*Caranx latus*	18	41.2	28.6	61 ± 14.5	42 - 104	29	148.6	28.0	68 ± 11.5	26 - 95
	Oligoplites saurus	08	5.2	5.7	63 ± 25.0	31 - 103	06	6.5	20.0	62 ± 9.1	52 - 76
Gobiidae	*Bathigobius soporator*	08	49.5	20.0	88 ± 15.8	70 - 122	04	45.9	20.0	74 ± 21.6	37 - 104
Paralichthyidae	*Citharichthys spilopterus*	16	43.7	8.6	82 ± 11.2	67 - 101	07	37.9	20.0	93 ± 37.1	14 - 143
	Citharichthys macrops	33	50.9	20.0	72 ± 28.8	22 - 140	08	26.3	24.0	53 ± 20.9	18 - 130
	Achirus lineatus	19	54.3	31.4	56 ± 46.1	15 - 114	03	43.6	8.0	45 ± 10.8	15 - 62
Tetraodontidae	*Sphoeroides testudineus*	51	698.9	45.7	125 ± 48.9	21 - 197	27	1083.7	48.0	104 ± 54.5	12 - 236
	Sphoeroides greeley	03	62.7	14.3	68 ± 17.3	39 - 81	06	75.4	20.0	52 ± 18.7	23 - 74

N: number of individuals; B: Biomass (in grams); F: frequency of occurrence; TL: total length, in mm. (mean ± SD - standard deviation).

Table 2 Volumetric frequency of food items in the diets of 17 fish species collected on the tidal mudflat during the dry season in a tropical estuary in Brazil. The number of fish analysed (N) is in parentheses. The abbreviation used for each species is the first letter of the taxa.

Items	Lg N = 111	Ac N = 47	Rb N = 155	Hc N = 8	Hu N = 70	Ab N = 191	Ml N = 45	Os N = 8	Cl N = 18	Ea N = 20	Em N = 3	Bs N = 8	Al N = 19	Cs N = 16	Cm N = 33	Sr N = 51	Sg N = 3
Diatoms	1.67	0.32	0.83			0.12	7.84							0.02	0.01		
Foraminifera	0.27	3.88	0.90		0.11	0.21	76.19		0.14	0.27					0.10	< 0.01	
Trematoda	11.56			1.80					0.28							0.19	
Nematoda	0.13	9.06	1.88	0.90	0.17	0.10			0.28	1.24	0.65	0.71	0.48		0.30	0.07	0.54
Sipuncula		3.23				0.01	2.38					0.11					
Gastropoda	4.04	8.08	11.17		13.23	63.65	2.38			1.93					0.10	1.48	13.58
Bivalvia	2.09	1.94	7.14	1.80	1.41	0.53				1.10	0.65					67.66	54.34
Scaphopoda						0.02											
Polychaeta Errant									0.43	16.27			57.83		0.30	0.07	8.15
Polychaeta Sedentary	2.50	0.97	0.24					1.08		47.30			15.90		0.60	0.09	
Brachyura	1.95	0.97	0.32	2.70		0.36		5.43	14.57			1.89		3.64	3.10	23.35	9.51
Dendobranchiata								5.43	30.31					1.56	2.70	0.07	
Decapod larvae	20.76		12.56		0.17	0.06		18.47	0.28	0.27						0.03	
Amphipod Caprellidae	0.43					< 0.01				0.27						0.01	
Amphipod Gammaridae		0.64													3.51		
Mysida	4.87								0.14								
Copepoda n.i.	0.27		1.97		1.05	0.06	2.38				0.10						
Cyclopoida	15.88	0.97	39.36	58.55	0.41	< 0.01		1.08	0.14	0.68			0.48	0.26	7.42	0.01	
Calanoida	15.32	1.94	11.91	22.52	35.44	1.48		3.26	3.03	20.27			11.56		15.04		
Isopoda	0.27		0.16			0.04				8.13							
Tanaidacea			0.08												3.10		
Ostracoda	0.41	12.94	8.37	3.60	1.23	0.83		1.08	0.87	1.24			1.44		0.10	0.02	
Cirripedia																0.32	6.79
Insecta ni					0.41	0.01	2.38			0.13							
Coleoptera					0.76	0.04											
Diptera					1.70	< 0.01										0.39	
Hymenoptera					20.92	0.31											
Simuliidae pupae					0.17												
Araneae						0.47											
Fish eggs	0.13	0.32	1.15	3.60	0.05						79.87						
Fish	17.14				0.05			4.34	49.22			0.94	12.04	93.73	63.27	0.38	
Scale	0.91	0.64	0.41	1.80	0.11	0.01	2.33	30.43	0.14	0.27		1.89			0.10	0.03	
Invertebrate eggs			0.65		6.29	1.11	2.38	1.08				39.69			0.20		5.97
Algae		0.64			2.11	< 0.01	0.47					54.74					
Plant Material	0.1	53.39	0.83	2.70	2.35	< 0.01		6.52		0.13	14.28					0.35	1.08

Table 3 Volumetric frequency of food items in the diets of 17 fish species collected in the tidal mudflat during the rainy season in a tropical estuary in Brazil. The number of fish analysed (N) is in parentheses. The abbreviation used for each species is the first letter of the taxa

Items	Lg N = 144	Ac N = 20	Rb N = 119	Hc N = 14	Hu N = 26	Ab N = 258	Ml N = 208	Os N = 6	Cl N = 19	Ea N = 27	Em N = 31	Bs N = 4	Al N = 3	Cs N = 7	Cm N = 8	St N = 27	Sg N = 6
Diatoms	0.08	5.30	0.10	2.67	0.45	0.10	13.04			51.35	8.12						0.03
Foraminifera	0.33	24.48	0.23	0.44		0.06	36.58									0.05	0.65
Trematoda	4.96			2.67					0.02							0.02	0.32
Nematoda	0.33		0.15	0.89	0.06	0.41	2.43		0.02		1.48			0.66			
Sipuncula			0.64	0.66													9.19
Gastropoda	7.28	4.08	1.66	0.44	0.12	1.87	6.09		0.04							7.55	84.04
Bivalvia	3.80	4.08	0.56	0.44	0.06	2.25	1.21				0.74			0.22		29.69	0.32
Scaphopoda																	
Polychaeta Errant									1.21			2.92	60.97		0.58		
Polychaeta Sedentary	0.41		0.03							1.00	12.63						
Brachyura	10.09		0.35		0.06	2.81			2.93			66.34	30.40	1.32	3.50	17.56	1.96
Dendobranchiata				0.22		6.94		1.33	47.88					1.10			0.65
Decapod larvae	2.06	8.16	11.00	0.22	1.03	27.24	3.65	0.66	0.09								
Amphipod Caprellidae	0.75		0.26						0.02								
Amphipod Gammaridae	0.08		0.03														
Mysida	0.08														0.29		
Copepoda ni			1.99			1.00					0.74						
Cyclopoida	32.93	4.08	39.81	59.15	1.86	5.45	2.43	4.46	0.14	5.40	40.29				2.35	0.05	
Calanoida	21.09	12.24	40.26	22.09	2.18	15.91		0.31	0.11	43.24	26.74			0.44	1.46	0.05	0.32
Isopoda	1.15		0.26	1.11		0.06	1.21		0.02								
Tanaidacea	0.17																
Ostracoda	1.07	8.16	0.94	4.24	0.38	1.25	7.31		0.02						0.29	0.06	1.31
Cirripedia																14.90	
Insecta ni						0.83			0.71								
Coleoptera						0.31											
Diptera					5.21	5.07		0.08									
Hymenoptera				0.22	84.14	20.64		0.89								0.21	
Simuliidae pupae					<0.01												
Ceratopogonidae				0.22													
Araneae						0.06											
Fish eggs	0.08		0.30	0.22	0.12	5.48											
Fish	11.98							91.61	43.25			17.07	1.21	96.26	86.52	29.80	
Scale			0.30	1.33		1.00	3.65	0.40	0.04			0.48			0.29	<0.01	
Invertebrate eggs	0.16		0.45		0.06	1.11		0.22	3.36			12.19			3.80		
Algae			0.03		0.01												
Plant Material	0.41	28.57	0.52	2.67	3.28		1.21			0.74		29.32	7.31		0.87		0.98

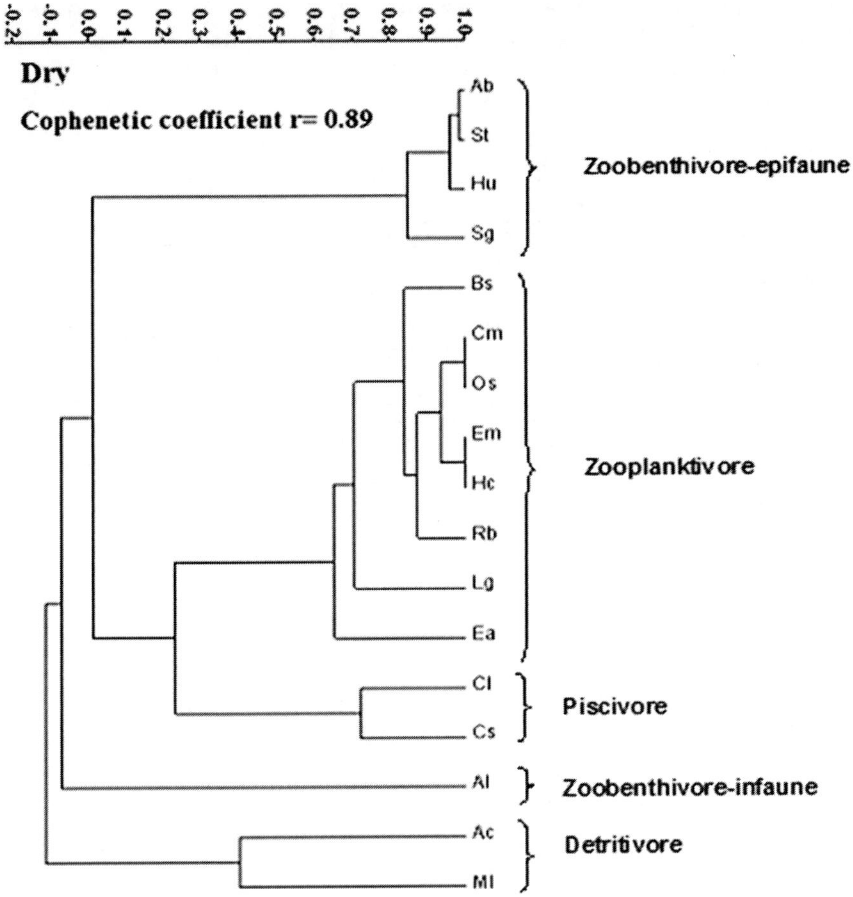

Figure 3 Dendrogram of the guild structure of the fish assemblage on the tropical tidal mud flat during the dry period, determined using correlation similarities of the Index of Relative Importance. The abbreviation used for each species is the first letter of the taxa.

Overall, the niche breadth values were higher during the rainy season. The species *A. lineatus, E. argenteus*, and *A. clupeoides* had the highest niche breadth values during the rainy season, and *O. saurus, L. grossidens*, and *S. greeleyi* had the highest values in the dry season. The flatfish *C. spilopterus* showed the lowest values during the dry season whereas *O. saurus* during the rainy season (Figure 6). Interspecific overlap values varied according to the period analysed, with only one comparison (Rb x Hc) remaining significant during both phases of the hydrological regime; the largest number of overlaps was observed among individuals within the same trophic guilds, primarily in the Zooplanktivore and Zoobenthivore-epifaune guilds (Table 4).

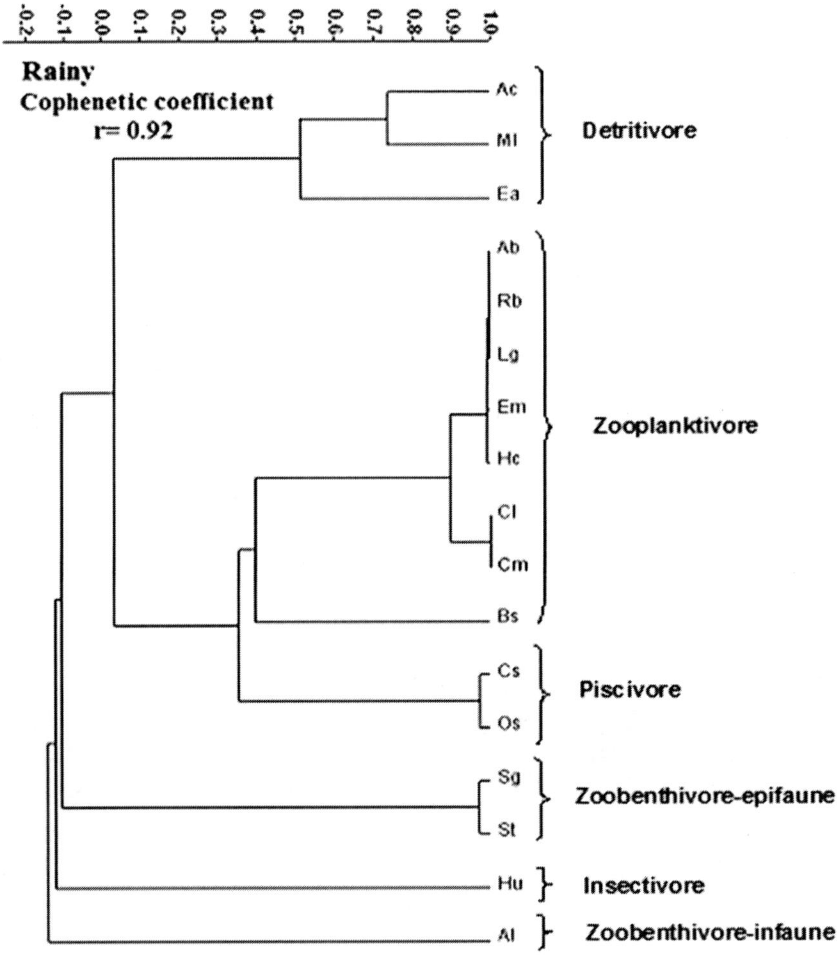

Figure 4 Dendrogram of the guild structure of the fish assemblage on the tropical tidal mud flat during the rainy season, determined using correlation similarities of the Index of Relative Importance. The abbreviation used for each species is the first letter of the taxa.

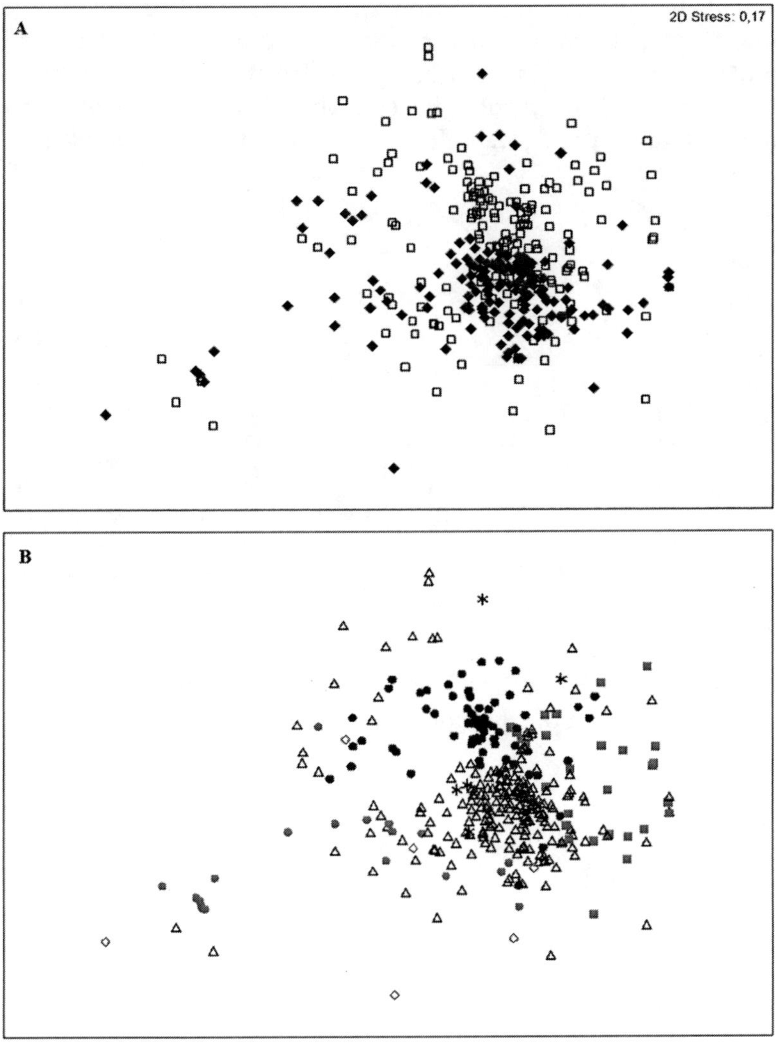

Figure 5 Nonmetric multidimensional scaling (nMDS) ordination of the dietary similarities for 17 species collected on the tropical tidal mud flat: (A) during the dry (black diamond) and rainy (square) seasons; (B) Trophic guilds are: Gray square: Detritivore; triangle: Zooplanktivore; gray circle: Piscivore; black circle: Zoobenthivore-epifaune; asterisk: Insectivore; diamond: Zoobenthivore-infaune.

The trophic organisations differed during different phases of the hydrological regime, with a larger number of trophic links being observed during the rainy season when compared to the dry season (Figure 7). A pattern was also observed in the values of trophic links, with higher values occurring during the dry season when compared to the rainy season. During both phases of the hydrological regime, a larger number of links with low values was indicative of trophic resource partitioning

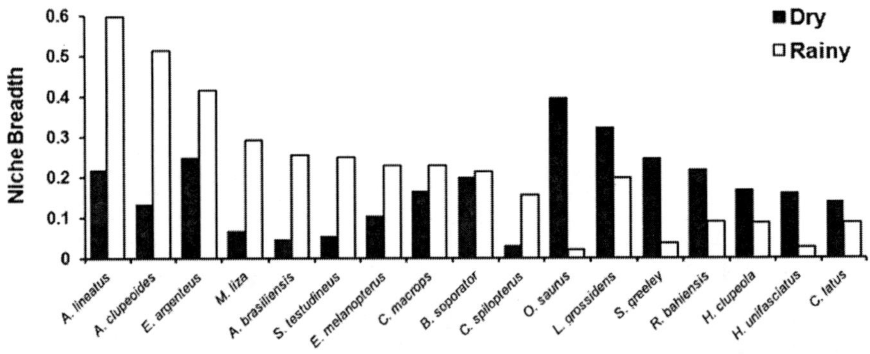

Figure 6 Dietary breadths of the 17 species collected in the tropical tidal mud flat during the dry and rainy seasons.

4. DISCUSSION

The trophic guilds described for the tidal mudflat of the Mamanguape River estuary did not vary between hydrological regime periods; for both periods, the Zooplanktivore was the most species-rich, indicating the high abundance of zooplankton in the study area. In general, the results of the present study were similar to those of other studies of estuarine environments, corroborating the hypothesis that, although fish species composition may differ considerably among estuaries, the basic trophic structure within an estuary is usually very similar (ELLIOTT; DEWAILLY, 1995; BARLETTA; BLABER, 2007). Studies of South African estuarine ichthyofauna (HARRINSON; WHITFIELD, 2012) indicate that Zooplanktivore species were a dominant component in subtropical estuaries, probably due to river flow and nutrient supply within this region, whereas another study at that tropical bay also found Calanoida as the primary food resource for dominant fishes species (HAJISAME; IBRAHIM, 2008). Comparative studies indicate that, in environments devoid of vegetation (non- vegetated), carnivorous and omnivorous species consume pelagic prey present in the water column (ANGEL; OJEDA, 2001). In addition, tidal mudflats are recognised as areas that are protected from wind and wave action, having calm waters that retain holo- and meroplankton (ARCHAMBAULT et al., 1998), which, in the present study, stood out as consumed prey. Although the number of guilds was constant, their compositions changed, as observed in the Zooplanktivore and Zoobenthivore-epifaune guilds. The increase, decrease, or replacement of species within a guild is a result to be expected in tropical regions (BARLETTA; BLABER, 2007). For example, BALDÓ and DRAKE (2002) suggested that the mysids (hyperbenthic) replace copepods (zooplanktonic) progressively in the diet of juvenile fish species as they grow in the tropical estuary. DEGRÉ et al. (2006) found that the grazing fish, which kept a roughly balanced diet between microphytobenthos and detritus, changed their exclusively herbivorous diet during summer to prey on nematofauna during winter on a mudflat This fluidity among guilds exists because ecological interactions among species are intensified at certain periods, with the less competitive species or species more susceptible to predation being excluded or replaced by species with greater adaptability to environmental pressures (ARAÚJO et al., 2006). Although ontogenetic changes were not specifi examined in the present study, some authors have asserted that certain species may segregate into different guilds during growth (LEGUERRIERA et al., 2007;ABDURAHIMAN et al., 2010).

Autochthonous prey items, represented primarily by zooplankton and, in particular, by copepods, are of fundamental importance for juveniles. The dominance and wide range of small organisms, which confer immediate availability and ease of capture by fish, are striking features in estuarine areas (INOUE et al., 2005;KELLNREITNER et al., 2012). The wide availability of these prey items is a good indicator of the value of tidal mudflats as nursery grounds for biotopes, as fish choose habitats that provide high abundance and diversity of prey. Prey density is important in determining the survival of fish larvae and juveniles, and growth may be affected when prey density becomes critical (BUSKEY et al., 1993). In estuarine ecosystems, zooplankton are considered to be the basis of the pelagic chain, connecting the input of matter transported by rivers and macrofauna (LOBRY et al., 2008), in addition to being used by pelagic species of the families Engraulidae (*A. clupeoides* and *L. grossidens*) and Clupeidae (*R. bahiensis* and *H. clupeola*), which were highly abundant in this tropical estuary. The clupeiformes were important consumers of this prey in the Guadalquivir Estuary (BALDÓ; DRAKE, 2002) and in habitats of the lower reaches of the Caeté Estuary (BARLETTA; BLABER, 2007).

Table 4 Pianka overlap index between the diets of 17 species collected in the tropical tidal mud flat during the dry (lower half of the matrix) and rainy period (upper half of the matrix). Values > 0.6 are in bold. The abbreviation used for each species is the first letter of the taxa

Dry/Rainy	Al	Ac	Ab	Bs	Cl	Cs	Cm	Ea	Em	Hc	Hu	Ml	Os	Rb	Sg	St	Lg
Al	-	0.02	0.00	0.48	0.29	0.37	0.14	0.00	0.09	0.00	0.01	0.01	0.40	0.00	0.00	0.39	0.16
Ac	0.00	-	0.28	0.09	0.00	0.00	0.02	0.29	0.24	0.24	0.04	0.62	0.00	0.31	0.11	0.07	0.19
Ab	0.00	0.13	-	0.02	0.00	0.00	0.02	0.27	0.30	0.27	0.54	0.08	0.02	0.50	0.05	0.05	0.27
Bs	0.00	0.01	0.31	-	0.21	0.23	0.14	0.00	0.05	0.00	0.00	0.00	0.24	0.00	0.00	0.49	0.26
Cl	0.43	0.01	0.00	0.01	-	**0.61**	0.19	0.01	0.15	0.00	0.00	0.00	**0.66**	0.00	0.00	0.42	0.12
Cs	0.53	0.00	0.00	0.00	0.01	-	0.24	0.00	0.20	0.00	0.00	0.00	**0.89**	0.00	0.00	0.55	0.15
Cm	0.25	0.01	0.00	0.01	**0.76**	0.37	-	0.02	0.11	0.05	0.00	0.00	0.27	0.05	0.00	0.19	0.15
Ea	0.02	0.04	0.05	0.00	0.02	0.00	0.13	-	0.46	0.33	0.02	0.23	0.00	0.50	0.00	0.00	0.32
Em	0.02	0.17	0.00	0.00	0.00	0.00	0.00	0.00	-	**0.95**	0.03	0.08	0.26	**0.91**	0.00	0.14	0.27
Hc	0.04	0.08	0.01	0.00	0.03	0.00	0.24	0.17	0.06	-	0.03	0.08	0.04	**0.88**	0.01	0.00	0.18
Hu	0.03	0.12	0.29	0.00	0.04	0.00	0.23	0.36	0.01	0.29	-	0.08	0.01	0.03	0.00	0.00	0.01
Ml	0.00	0.07	0.03	0.00	0.00	0.00	0.00	0.00	0.00	0.00	0.01	-	0.00	0.06	0.15	0.04	0.11
Os	0.00	0.16	0.02	0.00	0.12	0.00	0.12	0.04	0.02	0.06	0.07	0.02	-	0.03	0.00	**0.61**	0.16
Rb	0.04	0.11	0.24	0.00	0.02	0.00	0.20	0.15	0.02	**0.90**	0.29	0.03	0.17	-	0.03	0.01	0.36
Sg	0.00	0.08	0.22	0.00	0.00	0.00	0.00	0.08	0.10	0.03	0.09	0.00	0.00	0.20	-	0.15	0.17
St	0.00	0.04	0.02	0.00	0.05	0.00	0.01	0.02	0.00	0.03	0.03	0.00	0.02	0.15	**0.93**	-	0.26
Lg	0.03	0.56	0.11	0.00	0.41	0.00	0.57	0.22	0.10	0.56	0.38	0.01	0.20	0.51	0.08	0.07	-

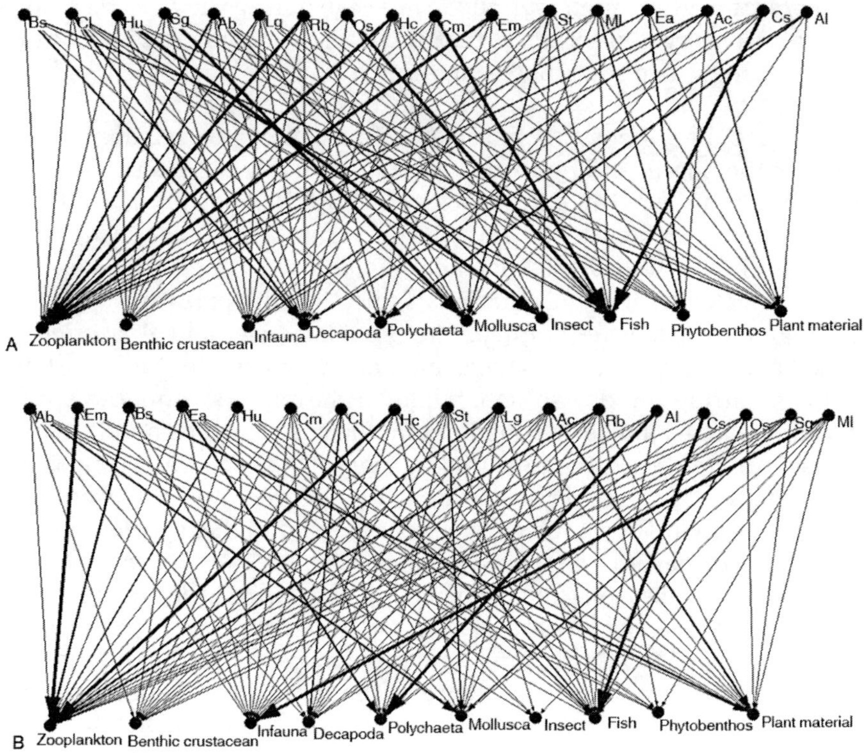

Figure 7 Trophic model of the fishes collected from the tidal mudflat (Brazil) and their prey in the (A) dry and (B) rainy seasons. The fishes (top) are arranged from left to right in order of increasing population based on the dietary niche breadth. The widths of the linking lines correspond to the contributions by volume of different food items in the fish diets.

The stability of biological communities partially depends on the availability of food items and on the stability of the trophic interactions among species. Changes in the proportion of available allochthonous and autochthonous resources and in the primary productivity of an estuarine system are common when comparing dry and rainy seasons, and the presence of some trophic guilds in one season but not the other may be the result of changes in prey availability (LOBRY et al., 2008). The input of allochthonous material during the rainy season and its importance to the tidal mudflat is evidenced by the formation of the Insectivorous guild (represented by *H. unifasciatus*). The uniform presence of some species in the same guild, such as *A. lineatus, C.*

spilopterus, and*S. testudineus* and *S. greeleyi*, during the periods studied was also observed, indicating that the environmental stability of tidal mudflats may have contributed to this result.

A large number of possible interconnections among prey and predators reflects the degree of opportunism, which is likely to be a key factor in promoting the survival of the species (DOLBETH et al. 2008), in addition to generalism, which tends to reduce overlap among estuarine fishes (GARRISON; LINK, 2000). Diet analysis showed changes in key prey types between hydrological periods, indicating an opportunistic trophic strategy for the species studied. The Brazilian silverside, *A. brasiliensis*, consumed Gastropoda, Bivalvia, and algae, linked to the benthonic domain, during the dry season, while, in the rainy season, the species consumed items linked to zooplankton, such as copepods and Decapoda larvae; *H. unisfasciatus* consumed primarily algae, Gastropoda, and Hymenoptera in the dry season, while, in the rainy season, only Hymenoptera had a significant value in the diet of this species; and, for the Clupeiformes, *L. grossidens, A. clupeoides*, and *R. bahiensis*, the diet during the dry season consisted mostly of zooplankton or epibenthic organisms, while, in the rainy season, their diet was based primarily on zooplankton. In other shallow ecosystems along the Brazilian coast, both in subtropical and tropical areas, the reported dietary items have been very similar to those found in the present study (BORTOLUZZI et al., 2006; ROCHA et al., 2008; VASCONCELOS-FILHO et al., 2009; CONTENTE et al., 2011).

When several teleosts are found in the same environment and are characterised by having wide a range of ecological niches and high trophic flexibility, the potential for interspecific competition is reduced through resource partitioning (ROSS, 1986). In the estuary studied, teleosts took advantage of temporal variations that affect the density of prey and/or the abundance of resources. Trophic flexibility leads to high diet overlap among species, as there is no need to partition the available resources, especially among species of the same guild, as was observed in some comparisons between the Zooplanktivore and Zoobenthivore-epifaune. Despite significant overlap values, there was no indication of competition for resources because, when resources are abundant, competition is absent (SCHONER, 1982), and temporal segregation is a strategy that minimises direct confrontation in the use of resources. When resource partitioning becomes important in the organisation of fish assemblages, species segregate along the resource dimension and demonstrate decreased niche breadth to maintain the minimum level of niche separation (LABROPOULOU; ELEFTHERIOU, 1997).

Niche breadth was higher during one season of the hydrological regime than the other, depending on the species group analysed. Some Zooplanktivores and Piscivores had lower niche breadth values during the rainy season. During the rainy season, system productivity is greater because of increased organic matter input, resulting in greater prey abundance. Increased prey abundance should result in lower niche breadth values (WOOTON, 1990), as was observed for *H. clupeola, S. greeleyi, R. bahiensis*, and *O. saurus*. During the dry season, predators become less selective and niche breadth increases, as was observed for *M. liza, E. argenteus* and *A. clupeoides*. Resident estuarine species, such as *A. brasiliensis, S. testudineus, A. linetaus*, and *C. spilopterus*, had higher niche breadth values in the rainy season, when dominant prey were abundant, but tended to specialise in the dry season, when organisms were less abundant. TSE et al. (2008) suggested that the foraging efficiency of juvenile fishes might be responsible for differences in diet breadth among juvenile fishes in a mudflat habitat.

Tropical environments have great fish diversity, a phenomenon that can be explained by the existence of trophic resource partitioning, which allows for the coexistence of species in shallow coastal ecosystems. By analysing trophic links in the dry and rainy seasons, a wide variety of items consumed by all species was observed, indicating high complexity and connectivity, as well as partitioning of the same food item by several species. Trophic web models are frequently used to describe relationships among individuals in estuaries because they summarise the vast complexity of the data (LUCZKOVICH et al., 2002). In the present study, most of the arrows in the trophic webs represented low percentage volume values, indicating the low consumption of certain food items, which allows more individuals to consume the items; that is, there was a greater partitioning of the resources available on this tidal mudflat. The large number of possible interconnections among prey and predators reflects the degree of opportunism, which is most likely a key aspect that enables success in estuarine environments (DOLBETH et al., 2008).

ACKNOWLEDGEMENTS

We thank biologists Nathalia Italiano Medeiros, Gabriela Guerra Araújo Abrantes, Bianca Bezerra Fialho Soares and Renato Pereira Dantas for helping in the field work and SISBIO for the permission to carry out the research in the protected area (24557). This study was partially supported by the National System of Research on Biodiversity (SISBIOTA/BRASIL) and CNPq - Brazilian National Agency for Scientific and Technological Development (Proc. 563202/2010-6 and Proc. 477663/2011-7).

REFERENCES

1. ABDURAHIMAN, K. P.; NAYAK, T. H.; ZACHARIA, P.U.; MOHAMED, K. S. Trophic organisation and predatorprey interactions among commercially exploited demersal finfishes in the coastal Waters of the southeastern Arabian sea. Estuar. Coast. Shelf Sci., v. 87, n. 4, p. 601-610, 2010.

2. ANGEL, A.; OJEDA, F. P. Structure and trophic organization of subtidal fish assemblages on the northern Chilean coast: the effect of habitat complexity. Mar. Ecol. Prog. Ser., v. 217, p. 81-91, 2001.

3. ARAÚJO, F. G.; GUIMARÃES, F. J. C.; COSTA, R. C. Environmental influences on distribution of four Sciaenidae species (Actinopterygii, Perciformes) in a tropical bay at Southeastern Brazil. Rev. Bras. Zool., v. 23, n. 2, p. 497-508, 2006.

4. ARAÚJO, M. E.; TEIXEIRA, J. M. C.; OLIVEIRA, A. M. E. Peixes estuarinos do nordeste brasileiro: Guia Ilustrado. Recife: Universitária UFPE e EFC, 2004. 260 p.

5. ARCHAMBAULT, P.; ROFF, J. C.; BOURGET, E.; BANG, B.; INGRAM, G. R. Nearshore abundance of zooplankton in relation to shoreline configuration and mechanisms involved. J. Plankton Res., v. 20, n. 4, p. 671-690, 1998.

6. BALDÓ, F.; DRAKE, P. A multivariate approach to the feeding habits of small fishes in the Guadalquivir Estuary. J. Fish Biol., v. 61, p. 21-32, 2002.

7. BARLLETA, M.; BLABER, S. J. M. Comparison of fish assemblages and guilds in tropical habitats of the Embley (Indo-West Pacific) and Caeté (Western Atlantic) estuaries. Bull. Mar. Sci., v. 80, n. 3, p. 647-680, 2007.

8. BARRY, J. M.; YOKLAVICH, M. M.; CAILLIET, G. M.; AMBROSE, D. A.; ANTRIM, B. S. Trophic ecology of dominant fishes in Elkhorn Slough, California, 1974-1980. Estuaries, v. 19, n. 1, p. 115-138, 1996.

9. BOER, W. F.; PRINS, H. H. T. The community structure of a tropical intertidal mudflat under human exploitation. ICES J. Mar. Sci., v. 59, n. 6, p. 1237-1247, 2002.

10. BORTOLUZZI, T.; ASCHENBRENNER, A. C.; SILVEIRA, C. R.; ROOS, D. C.; LEPKOSKI, E. D.; MARTINS, J. A.; GOULART, M. G.; QUEROL, E.; QUEROL, M. V. Hábito alimentar da Sardinha Prata, Lycengraulis grossidens (Spix & Agassiz, 1829), (Pisces, Engraulidae), Rio Uruguai Médio, Sudoeste do Rio Grande do Sul, Brasil. Biodivers. Pampeana, v. 4, p. 11-23, 2006.

11. BUSKEY, E. J.; COUTER, C.; STROM, S. Locomotory patterns of microzooplankton: potential effects on food selectivity of larval fish. Bull. Mar. Sci., v. 53, n. 1, p. 29-43, 1993.

12. CASTRO, A. L. C.; DINIZ, A. F.; MARTINS, I. Z.; VENDEL, A. L.; OLIVEIRA, T. P. R.; ROSA, I. M. L. Assessing diet composition of seahorses in the wild using a non-destructive method: Hippocampus reidi (Teleostei: Syngnathidae) as a study-case. Neotrop. Ichthyol., v. 6, n. 4, p. 637-644, 2008.

13. CLAUDET, J.; PELLETIER, D.; JOUVENELC, J. Y.; BACHETD, F.; GALZIN, R. Assessing the effects of marine protected area (MPA) on a reef fish assemblage in a northwestern Mediterranean marine reserve: identifying community-based indicators. Biol. Conserv., v. 130, n. 6, p. 349-369, 2006.

14. CONTENTE, R. F.; STEFANONI, M. F.; SPACH, H. L. Feeding ecology of the Brazilian silversideAtherinellabrasiliensis (Atherinopsidae) in a sub-tropical estuarine ecosystem. J. Mar. Biol. Assoc. U. K., v. 91, n. 6, p. 1197-1205, 2011.

15. DANTAS, D. V.; BARLETTA, M.; RAMOS, J. A. A.; LIMA, A. R. A.; COSTA, M. F. Seasonal diet shifts and overlap between two sympatric catfishes in an estuarine nursery. Estuaries, v. 36, n. 2, p. 237-256, 2013.

16. DEGRÉ, D.; LEGUERRIER, D.; DU CHATELET, E. A.; RZEZNIK, J.; AUGUET, J.; DUPUY, C.; MARQUIS, E.; FICHET, D.; STRUSKI, C.; JOYEUX, E.; SAURIAU, P.; NIQUIL, N. Comparative analysis of the food webs of two intertidal mudflats during two seasons using inverse modelling: Aiguillon Cove and Brouage Mudflat, France. Estuar. Coast. Shelf Sci., v. 69, n. 1/2, p. 107-124, 2006.

17. DOLBETH, M.; MARTINHO, F.; LEITÃO, R.; CABRAL, H.; PARDAL, M. A. Feeding patterns of the dominant benthic and demersal fish

community in a temperate estuary. J. Fish Biol., v. 72, n. 10, p. 2500-2517, 2008.

18. ELLIOTT, M.; DEWAILLY, F. The structure and components of European estuarine fish assemblages. Neth. J. Aquatic. Ecol., v. 29, n. 3/4, p. 397-417, 1995.

19. ELLIOTT, M.; WHITFIELD, A. K.; POTTER, I. C.; BLABER, S. J. M.; CYRUS, D. P.; NORDLI, E. F. G.; HARRISON, T. D. The guild approach to categorizing estuarine fi assemblages: a global review. Fish Fish., v. 8, p. 241-268, 2007.

20. FIGUEIREDO, J. L.; MENEZES, N. A. Manual de peixes marinhos do sudeste do Brasil. II. Teleostei (1). São Paulo: Museu de Zoologia da Universidade de São Paulo, 1978. 110 p.

21. FIGUEIREDO, J. L.; MENEZES, N. A. Manual de peixes marinhos do sudeste do Brasil. III. Teleostei (2). São Paulo: Museu de Zoologia da Universidade de São Paulo, 1980. 90 p.

22. FIGUEIREDO, J. L.; MENEZES, N. A. Manual de peixes marinhos do sudeste do Brasil. VI. Teleostei (5). São Paulo: Museu de Zoologia da Universidade de São Paulo, 2000. 116 p.

23. FRANÇA, S.; PARDAL, M. A.; CABRAL, H. N. Mudflat nekton assemblages in the Tagus estuary (Portugal): distribution and feeding patterns. Sci. Mar., v. 73, n. 2, p. 591-602, 2008.

24. FRANÇA, S.; VINAGRE, C.; PARDAL, M. A.; CABRAL, H. N. Spatial and temporal patterns of benthic invertebrates in the Tagus estuary, Portugal: comparison between subtidal and an intertidal mudflat. Sci. Mar., v. 73, n. 2, p. 307-318, 2009.

25. FRANÇA, S.; VASCONCELOS, R. P.; FONSECA, V. F.; TANNER, S. E.; REIS-SANTOS, P.; COSTA, M. J.; CABRAL, H. N. Predicting fish community properties within estuaries: influence of habitat type and other environmental features. Estuar. Coast. Shelf Sci., v. 107, p. 22-31, 2012.

26. GARRISON, L. P.; LINK, J. S. Dietary guild structure of the fish community in the Northeast United States continental shelf ecosystem. Mar. Ecol. Prog. Ser., v. 202, p. 231-240, 2000.

27. HAJISAMAE, S.; IBRAHIM, S. Seasonal and spatial variations of fish trophic guilds in a shallow, semi-enclosed tropical estuarine bay. Environ. Biol. Fish., v. 82, n. 3, p. 251-264, 2008.

28. HAMMERSCHLAG, N.; OVANDO, D.; SERAFY, J. E. Seasonal diet and feeding habits of juvenile fishes foraging along a subtropical marine ecotone. Aquat. Biol., v. 9, n. 3, p. 279-290, 2010.

29. HARRISON, T. D; WHITFIELD, A. K. Fish trophic structure in estuaries, with particular emphasis on estuarine typology and zoogeography. J. Fish Biol., v. 81, n. 6, p. 2005-2029, 2012.

30. HINDELL, J. S.; JENKINS, G. P. Spatial and temporal variability in the assemblage structure of fishes associated with mangroves (Avicennia marina) and intertidal mudflats in temperate Australian embayments. Mar. Biol., v. 144, p. 85-395, 2004.

31. HURLBERT, S. H. The measurement of niche overlap and some relatives. Ecology, v. 59, n. 1, p. 67-77, 1978.

32. HYSLOP, E. J. Stomach contents analysis - a review of methods and their application. J. Fish Biol., v. 17, n. 4, p. 411-429, 1980.

33. INOUE, T.; SUDA, Y.; SANO, M. Food habits in the surf zone of a sandy beach at Sanrimatsubara, Fukuoka Prefecture, Japan. Ichthyol. Res., v. 52, n. 1, p. 9-14, 2005.

34. KELLNREITNER, F.; POCKBERGER, M.; ASMUS, H. Seasonal variation of assemblage and feeding guild structure of fish species in a boreal tidal basin. Estuar. Coast. Shelf Sci., v. 108, p. 97-108, 2012.

35. KOPP, M. M.; DE SOUZA, V. Q.; COIMBRA, J. L. M.; LUZ, V. K.; MARINI, N.; OLIVEIRA, A. C. Melhoria da correlação cofenética pela exclusão de unidades experimentais na construção de dendogramas. Rev. Fac. Zoo. Vet. Agr., v. 14, n. 2, p. 46-53, 2007.

36. KREBS, C. J. Ecological methodology. New York: Harper-Collins Publ., 1989. 624 p.

37. LABROPOULOU, M.; ELEFTHERIOU, A. The foraging ecology of two pairs of congeneric demersal fish species: importance of morphological characteristics in prey selection. J. Fish Biol., v. 50, n. 2, p. 324-340, 1997.

38. LEGUERRIERA, D.; DEGRÉA, D.; NIQUILA, N. Network analysis and inter-ecosystem comparison of two intertidal mudflat food webs (Brouage Mudflat and Aiguillon Cove, SW France). Estuar. Coast. Shelf Sci., v. 74, n. 3, p. 403-418, 2007.

39. LOBRY, J.; DAVID, V.; PASQUAD, S.; LEPAGE, M.; SAUTOUR, B.; ROCHARD, E. Diversity and stability of an estuarine trophic network. Mar. Ecol. Prog. Ser., v. 358, p. 13-25, 2008.

40. LUCZKOVICH, J. J.; WARD, G. P.; JOHNSON, J. C.; CHRISTIAN, R. R.; BAIRD, D.; NECKLES, H.; RIZZO, W. M. Determining the trophic guilds of fishes and macroinvertebrates in a seagrass food web. Estuaries, v. 25, n. 6A, p. 1143-1163, 2002.

41. LUGENDO, B. R.; NAGELKERKEN, I.; VELDE, G. V.; MGAYA, Y. D. The importance of mangroves, mud and sand flats, and sea-grass beds feeding areas for juvenile fishes in Chwaka Bay, Zanzibar: gut content and stable isotope analyses. J. Fish Biol., v. 69, n. 6, p.1639-1661, 2006.

42. MACDONALD, J. S.; GREEN, R. H. Redundancy of variables used to describe importance of prey species in fish diets. Can. J. Fish. Aquat. Sci., v. 40, p. 635-637, 1983.

43. MARTINHO, F.; LEITÃO, R.; NETO, J. M.; CABRAL, H. N.; MARQUES, J. C.; PARDAL, M. A. The use of nursery areas by juvenile fish in a temperate estuary, Portugal. Hydrobiologia, v. 587, n. 1, p. 281-290, 2007.

44. MENEZES, N. A.; FIGUEIREDO, J. L. Manual de peixes marinhos do sudeste do Brasil. IV. Teleostei (3). São Paulo: Museu de Zoologia da Universidade de São Paulo, 1980. 96 p.

45. MENEZES, N. A.; FIGUEIREDO, J. L. Manual de peixes marinhos do sudeste do Brasil. V. Teleostei (4). São Paulo: Museu de Zoologia da Universidade de São Paulo, 1985. 105 p.

46. MOURÃO, J. S.; NORDI, N. Etnoecologia de pescadores artesanais do estuário do Rio Mamanguape, Paraíba, Brasil. Bol. Inst. Pesca, v. 29, n. 1, p. 9-17, 2003.

47. OLIVEIRA, R. E. M. C. C.; PESSANHA, A. L. M. Fish assemblage along a morphodynamic continuum on three tropical beaches. Neotrop. Ichthyol., v. 12, n. 1, p. 165-175, 2014.

48. PAIVA, A. C. G.; CHAVES, P. T. C.; ARAÚJO, M. E. Estrutura e organização trófi da ictiofauna de águas rasas em um estuário tropical. Rev. Bras. Zool., v. 25, n. 4, p. 647-661, 2008.

49. PEREIRA, M. S.; ALVES, R. R. N. Composição Florística de um remanescente de Mata Atlântica na Área de Proteção Ambiental Barra do Rio Mamanguape, Paraíba, Brasil. Rev. Biol. Ciênc. Terra, v. 6, n. 1, p. 357-366, 2006.

50. PINKAS, L.; OLIPHONT, M. S.; IVERSON, I. L. K. Food habits of albacore, bluefin tuna, and bonito in California waters. California

Fish Games, California Sacramento: State of California. Fish Bul., v. 152, p. 105, 1971.

51. ROCHA, M. S. P.; MOURÃO, J. S.; SOUTO, W. M. S.; BARBOZA, R. R. D.; ALVES, R. R. N. O uso dos recursos pesqueiros no estuário do Rio Mamanguape, estado da Paraíba, Brazil. Interciência, v. 33, n. 12, p. 903-909, 2008.

52. ROMESBURG, H. C. Cluster analysis for researchers. Morrisville: Lulu Press, 1984. 340 p.

53. ROPER, D. S.; THRUSH, S. F.; SMITH, D. G. The influence of runoff on intertidal mudflat benthic communities. Mar. Environ. Res., v. 26, n. 1, p. 1-18, 1988.

54. ROSS, S. T. Resource partitioning in fish assemblages: a review of field studies. Copeia, v. 1986, p. 352-388, 1986.

55. SCHOENER, T. W. The controversy over interspecific competition. Am. Nat., v. 70, p. 586-595, 1982.

56. TSE, P.; NIP, T. H. M.; WONGCK, C. K. Nursery function of mangrove: A comparison with mudflat in terms of fish species composition and fish diet. Estuar. Coast. Shelf Sci., v. 80, n. 2, p. 235-242, 2008.

57. VASCONCELOS-FILHO, A. L.; NEUMANN-LEITÃO, S.; ESKINAZI-LEÇA, E.; OLIVEIRA, A. M. E.; PORTO-NETO, F. F. Hábitos alimentares de consumidores primários da ictiofauna do Sistema Estuarino de Itamaracá (Pernambuco - Brasil). Rev. Bras. Eng. Pesca, v. 4, n. 1, p. 21-31, 2009.

58. VIVIER, L.; CYRUS, D. P.; JERLING, H. L. Fish community structure of the St Lucia estuarine system under prolonged drought conditions and its potential for recovery after mouth breaching. Estuar. Coast. Shelf Sci., v. 86, n. 4, p. 568-579, 2010.

59. WANG, M.; HUANG, Z.; SHI, F.; WANG, W. Are vegetated areas of mangroves attractive to juvenile and small fish? The case of Dongzhaigang Bay, Hainan Island, China. Estuar. Coast. Shelf Sci., v. 85, n. 2, p. 208-216, 2009.

60. WILSON, J. P.; SHEAVES, M. Short-term temporal variations in taxonomic composition and trophic structure of a tropical estuarine fish assemblage. Mar. Biol., v. 139, n. 4, p. 787-796, 2001.

61. WOOTTON, R. J. Ecology of Teleost Fishes (Fish and Fisheries Series). Boston: Kluwer Academic Publishers, 1990. 386 p.

62. XAVIER, J. H. A.; CORDEIRO, C. A. M. M.; TENÓRIO, G. D.; DINIZ, A. F.; PAULO-JÚNIOR, E. P. N.; ROSA, R. S.; ROSA, I. L. Fish assemblage of the Mamanguape Environmental Protection Area, NE Brazil: abundance, composition and microhabitat availability along the mangrove-reef gradient. Neotrop. Ichthyol., v. 10, n. 1, p. 109-122, 2012.

9

INFLUENCE OF EL NIÑO WIND STRESS ANOMALIES ON SOUTH BRAZIL BIGHT OCEAN VOLUME TRANSPORTS

Luiz Paulo de Freitas Assad,1 Carina Stefoni Böck,1 Rogerio Neder Candella,2 and Luiz Landau1

1Universidade Federal do Rio de Janeiro, COPPE, Programa de Engenharia Civil, Laboratório de Métodos Computacionais em Engenharia, Avenida Athos da Silveira Ramos 149, Cidade Universitária, Ilha do Fundão, 21941-909 Rio de Janeiro, RJ, Brazil

2Instituto de Estudos do Mar Almirante Paulo Moreira, Divisão de Oceanografia Física, Rua Kioto 253, Arraial do Cabo, 28930-000 Rio de Janeiro, RJ, Brazil

ABSTRACT

The knowledge of wind stress variability could represent an important contribution to understand the variability over upper layer ocean volume transports. The South Brazilian Bight (SBB) circulation had been studied by numerous researchers who predominantly attempted to estimate its meridional volume transport. The main objective and contribution of this study is to identify and quantify possible interannual variability in the ocean volume transport in the SBB induced by the sea surface wind stress field. A low resolution ocean global circulation model was implemented to investigate the volume transport variability. The results obtained indicate the occurrence of interannual variability in meridional ocean volume transports along three different zonal sections. These results also indicate the influence of a wind driven large-scale atmospheric process that alters locally the

SBB and near-offshore region wind stress field and consequently causes interannual variability in the upper layer ocean volume transports. A strengthening of the southward flow in 25°S and 30°S was observed. The deep layer ocean volume transport in the three monitored sections indicates a potential dominance of other remote ocean processes. A small time lag between the integrated meridional volume transports changes in each monitored zonal section was observed.

1. INTRODUCTION

Very little is known about the influence of the wind stress field over the upper ocean volume transport in the South Brazil Bight (SBB) and the interannual variability associated with them.

One of the most important oceanographic features of the SBB is the Brazil Current (BC). This current flows along most part of the Brazilian tropical and subtropical continental shelf and the adjacent offshore region, and it is one of the major currents associated with the South Atlantic subtropical gyre (SASG) ocean circulation system. The BC has been studied by numerous researchers who predominantly attempted to estimate its meridional volume transport based on hydrographic datasets acquired at different latitudes [1, 2]. Only a few studies in the scientific literature treat specifically about the interannual variability of the BC volume transport even though many studies describing general dynamic aspects in different variability scales for the whole South Atlantic basin had been published such as Schott et al. [3], Stramma et al. [4], Lumpkin and Garzoli [5] and Veleda et al. [6]. Goni et al. [7], using satellite-derived sea height anomaly and sea surface temperature (SST) observations from 1993 to 2008, observed low-frequency variability of the BC front. The authors were not able to observe statistically significant changes in BC geostrophic transport, suggesting that the low-frequency changes in the BC front are governed by different mechanisms compared to the seasonal variability of the surface currents. The authors suggested the analysis of longer records, together with comprehensive numerical experiments using state-of-the-art ocean models, to determine the origin of these variabilities.

The region where the BC is located presents a complex vertical extension composed of different water masses that, in some regions, flow in opposite directions. The BC is important for understanding the South American and especially the Brazilian climate because of its important contribution as a source of sea surface temperature variability in the region. Cataldi et al. [8], using an atmospheric general circulation model (AGCM) forced with synthetic SST anomaly fields from observed data in BMC region, observed a strong correlation between positive SST anomalies near the Brazil Malvinas confluence (BMC) region and the hydrometeorological regime over the south and southeast Brazilian regions. This motivates the need to better understand how the upper layer transports would vary under the influence of interannual wind stress variability.

The ENSO-driven, large-scale atmospheric teleconnections alter the near-surface air temperature, humidity, and wind, in addition to the distribution of clouds far from the Equatorial Pacific. The resulting variations in the surface heat, momentum, and freshwater fluxes can induce changes in the SST, salinity, mixed layer depth (MLD), and ocean currents [9]. Vivier et al. [10], studying time series of current meter data located in the BMC region, pressure sensor data sets located in the Drake Passage and a five year Malvinas Current (MC) transport series, observed an important annual variability signal. The authors suggested that the main source of BMC front movements is the BC. Lentini et al. [11], studying sea surface temperature variability in the southwestern Atlantic region from thirteen years remote sensing data set, observed a northward (southward) advection of cold (warm) anomalies (±1°C) during and immediately after warm (cold) phases of the ENSO and noted that the larger amplitudes of these anomalies were situated at the La Plata River and the Patos Lagoon offshore regions.

The South Atlantic basin large scale atmospheric circulation could be influenced by remote ocean-atmosphere processes. Torres Jr. [12], using an AGCM to force an Oceanic General Circulation Model (OGCM), demonstrated that on seasonal and interannual scales, the SASH is sensitive to the occurrence of sea surface anomalies in the Equatorial Pacific. The author also observed changes in Hadley and Walker circulations that influenced the SASH behavior and consequently increased the BC and MC transports.

Assad et al. [13] studied the influence of an ENSO-type global wind stress field that was represented by a monthly mean global wind stress field where it was introduced anomaly signals related with 1982-1983 ENSO event. The results revealed the occurrence of positive SST anomalies, near the BMC region associated with upper ocean velocity anomalies. The occurrence of positive anomalies was always associated with an enhanced BC meridional surface volume transport, together with a decrease of the ACC zonal volume transport, vertically integrated across the Drake Passage, and a correspondent decrease in the upper layer MC volume transport. It was possible to see negative upper layer temperature anomaly predominance just before the peak ocean kinetic energy month and positive anomaly predominance immediately following this month.

Most of the knowledge that exists about the seasonality of BC volume transport was obtained by the application of the geostrophic method using in situ temperature and salinity data collected along the Brazilian coast and the adjacent offshore region [2]. This method effectively determines numerous dynamic ocean properties and has historical importance in understanding features of the global ocean circulation. However, this method has several intrinsic problems in determining the zero-motion level, which is fundamental for its application. This fact explains the discrepancies in volume transport estimates for the same region, which are commonly found in the literature. This problem is minimized when volume transport estimates are obtained by a numerical deterministic ocean model, which allows for the direct identification of the zero-motion level in the water column.

The main objective and contribution of this study is to identify and quantify possible interannual variability in the ocean volume transport in the SBB in the region confined between 23°S and 30°S, in response to a transient three year global wind stress field. The results obtained clearly indicate the occurrence of interannual variability in BC volume transport when the three-dimensional field is analyzed. These results also indicate the influence of a wind-driven, large-scale atmospheric teleconnection process that alters the SBB and near-offshore region wind stress field and consequently causes interannual variability in the upper layer volume transports.

2. STUDY AREA

The SBB is defined as the region of the southwestern Brazilian continental shelf extending from Cabo de São Tomé (22°S–41°W) to Cabo de Santa Marta (28.63°S–48.75°W) [14].

The BC is characterized by the vertical stacking of four water masses in the first 3000 meters: Tropical Water (TW), South Atlantic Central Water (SACW), Antarctic Intermediate Water (AIW) and North Atlantic Deep Water (NADW) [2]. According to Peterson and Stramma [1], the

BC is formed just a few degrees south of the equator in the South Equatorial Current (SEC) bifurcation region. In fact, the southern branch of the SEC (the SSEC) forms the BC. Near this region, the BC flow is shallow and transports warm and salty water, which is represented by the TW in the upper 150 meters of depth, with an approximate volume transport of 4 Sv. The surface portion of this current reaches the South American coast at approximately 15°S [15].

Notably, most of the water volume transported by the SSEC turns northward, feeding the North Brazil Current (NBC) [16]. This fact explains the occurrence of a weak southward volume transport in this region. Between 15°S and 20°S, the BC flow remains weak, with values less than or equal to 11 Sv. Miranda and Castro [17] estimated a value of 6.5 Sv for a section at 19°S. On the offshore region near the city of Cabo Frio at 23°S, several measures show that the volume transport reaches 11 Sv, with more than half of the flow confined to the first 200 meters [1]. South of 24°S, the BC increases its transport. This fact is associated with an anticyclonic recirculation cell that is observed in hydrographic measures [1]. Several estimates of this recirculation cell obtained values of 7.5 Sv for the first 800 meters, which causes an increase in the volume transport at 33°S of approximately 18 Sv. This circulation is associated with the SACW that enters the subtropical ocean gyre after its formation at the BMC region. When the SACW reaches the South American coast, it bifurcates at approximately 20°S, where part of its volume flows to the north and the other part flows to the south [18]. Stramma and England [18] synthesizes the circulation associated with the western boundary current system flowing along the Brazilian coast.

Stramma and England [18] presented schematic circulation diagrams based on observational data sets from different depths of the South Atlantic Ocean basin. The authors suggested an anticyclonic recirculation cell directly influenced by the SASG, where the SACW flows and, near the Brazil coast, feeds up into the BC flow between 20°S and 25°S at depth ranging between 100 and 500 meters. Below the SACW level, the authors observed the AIW that also follows the SASG and also feeds up into the BC flow farther south near 28°S in depths ranging between 500 and 1000 meters.

3. METHODOLOGY

The methodology of this work consists on estimating the BC system region volume ocean transports estimated along three zonal sections (furthered described) using the Modular Ocean Model 4.0 (MOM). MOM was developed at the Geophysical Fluid Dynamics Laboratory (GFDL) and has been largely used by the scientific community for global climatic research investigations and forecasts [19].

The numerical grid used had space resolution capable to resolve the main low frequency ocean processes presented in the whole South Atlantic Ocean Basin. In this grid, the world ocean is represented by a discrete integration domain with a horizontal resolution of 1° longitude × 1° latitude but increasing to 1/3° of latitudinal resolution within the 10°N–10°S equatorial band, with 360 longitudinal points and 200 latitudinal points composing the horizontal grid. Figure 1 shows the study area with the zonal cross-sections used for the volume transport analysis. The vertical coordinate layers vary in thickness to accommodate higher resolution near the surface of the ocean. There are 50 vertical layers, with 22 layers confined within the first 220 meters. Regions with depths smaller than 40 meters were not considered in the model calculations.

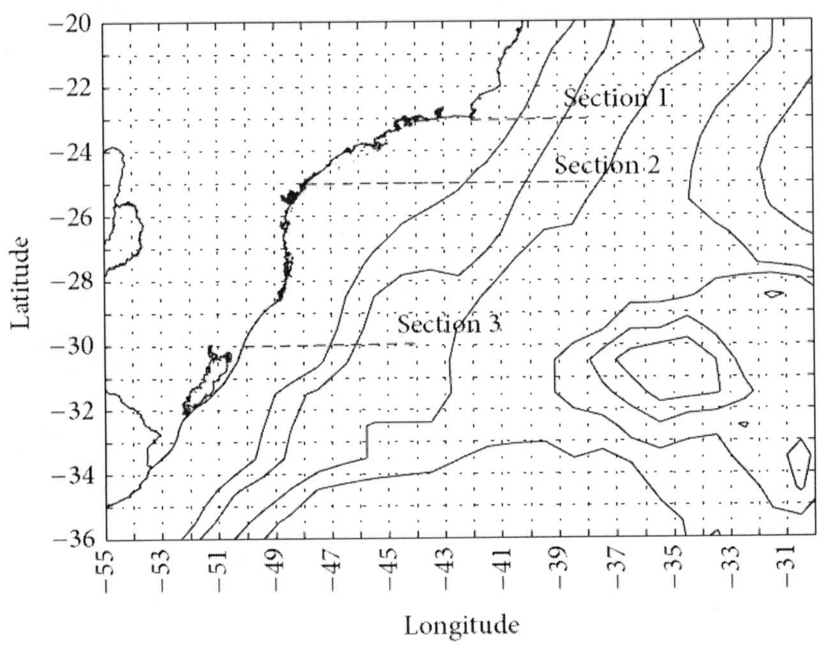

Figure 1: The study area comprising the SBB with the position of the three monitored zonal sections (1, 2 and 3). Isobathymetric lines are also presented.

Two experiments, named the control and perturbed experiments, were performed. The control experiment reproduced the mean seasonal cycle behavior of BC volume transport along the zonal sections (Figure 1). In the perturbed experiment, ocean volume transport fields were produced by forcing the model with an anomalous sea surface wind stress field. The two experiments are detailed in the following sections.

3.1. Control Experiment

Cyclic boundary conditions were imposed on the model until it reached the desired climate, or stable state [20]. The data used as sea surface boundary conditions were taken from the climatologic data set of the Ocean Model Intercomparison Project (OMIP) [21]. OMIP data set was produced to generate robust boundary conditions to be used on oceanic numerical model experiments. The atmospheric variables used to force the model were the meridional and zonal wind stress components, shortwave and net longwave radiation, sensible heat flux, specific humidity flux, and frozen and liquid precipitation. The OMIP product consisted of daily means of an annual cycle, except for the specific humidity and sensible heat flux data, which were represented by the monthly means of an annual cycle.

The experiment used, as its initial condition, the ocean data (sea surface elevation, temperature, salinity and velocity fields) obtained from the ocean data assimilation for seasonal-to-interannual prediction (ODASI) experiment conducted by GFDL [22]. The ODASI experiment also used the MOM 4.0 code and generated 40 years of monthly average ocean data between 1963 and 2003. An assimilation ocean data system was employed during the model integration. The assimilation was conducted only between the years 1993 and 2001. ODASI data was chosen because of its capacity to reproduce the most important features of an ocean in thermodynamic and dynamic equilibrium and a reasonable "ocean climate" state. As the main idea of this work is to investigate the interannual variability of the BC system ocean volume transports exclusively due to wind stress field influence, the mass and velocity fields from January 1985 were chosen as initial boundary conditions. The choice of these fields was justified to ensure the absence of any strong global climate variability processes over such as ENSO over the ocean model prognostic fields.

3.2. Perturbed Experiment

The main goal of this experiment was to identify interannual variability in BC volume transport induced by a three-year transient wind stress field. The perturbed experiment used September monthly mean of dynamic and thermodynamic fields generated by the control experiment as initial conditions, but with only anomalous wind stress boundary conditions. These anomalous boundary conditions were determined by the monthly mean of global wind stress fields generated by an AGCM. The AGCM was integrated for three years, and it used Pacific Equatorial SST synthetic fields as boundary conditions based on the 1982-1983 ENSO event [12, 13, 23]. The choice of a three year period wind stress field was justified because it presented the same duration of the Equatorial Pacific SST anomaly pulse observed for the 1982-1983 ENSO event.

Those anomalous contour conditions were obtained through the monthly mean of global wind stress fields resulting from the difference between the results of two experiments which were carried out using the National Center of Atmospheric Research (NCAR) atmospheric general circulation model Community Climate Model (CCM 3.0), the climatologic (control experiment) and another one with a perturbed condition originated from a SST analytical anomaly, overlapping the field of the mean SST, both experiments being integrated for three years.

The synthetic SST fields were chosen because they allow an independent evaluation of the SST influence on the atmospheric dynamics of the studied region without taking into consideration other perturbations in the oceans, in the cryosphere or in the internal atmosphere structure itself [8].

The SST anomalies synthetic fields were obtained fitting the combination of two Gaussian distributions in space (1) and one in time (2), according to Torres Jr. [12] suggested methodology. Consider

$$f(x) = \frac{1}{\sigma_{\varphi}\sqrt{2\pi}} \exp\left[\frac{(y-\eta_{\varphi})^2}{2\sigma_{\varphi}^2}\right] \frac{1}{\sigma_{\lambda}\sqrt{2\pi}} \exp\left[\frac{(x-\eta_{\lambda})^2}{2\sigma_{\lambda}^2}\right]. \tag{1}$$

In which is Gaussian space distribution function; is longitude index; is latitude index; is latitude displacement; is latitude variance; is longitude displacement; is longitude variance. Consider

$$g(x) = \frac{1}{\sigma_\tau \sqrt{2\pi}} \exp\left[\frac{(t-\eta_\tau)^2}{2\sigma_\tau^2}\right]. \tag{2}$$

In which is Gaussian time distribution function; is time index; is time displacement; is time variance or pulse width.

The adjustment was done based on the SST NASA/jet propulsion laboratory (JPL) via satellite data and the chosen ENSO event to adjust the function was the one from years 1982/1983 which is considered as one of the most intense ENSO events recorded in the recent decades, and which was responsible for severe climatic impacts such as extreme wet weather in parts of the Americas and, concurrently, severe drought in parts of Australia and Asia [24].

The global ocean model was integrated for 36 months in this experiment. The annual mean values for volume transports in each section were subsequently estimated and analyzed. The main idea in the perturbed experiment is not getting the balance, as in the control experiment, but understand how volume transports will vary over the three years in response to a transient wind stress field.

3.3. Volume Transport Estimates

The position and geometry of each zonal section was chosen to cover the BC flow, considering its meridional and zonal extension between 23°S and 30°S. Each section had its own geometry based on the surface annual mean BC meridional velocity profile and flow width obtained in the climate experiment. Three zonal sections were chosen (Figure 1). The choice of these sections is justified by the existence of some historical volume transport estimates near them [1, 2]. Table 1 shows the exact coordinates of each section (latitude and longitude limits).

The volume transport estimates for each zonal section were performed using (3) and are expressed in Sverdrup (Sv = 106 m3·s−1). Consider

$$v_t = \int_0^L \int_{-h(x)}^0 v(x,z,t)\, dz\, dx. \tag{3}$$

is the volume transport (Sv), is the section width, is the depth, is the meridional velocity (m/s), is the depth (m) and is the zonal distance delimited exactly by the meridional limit of each section presented in Table 1.

Table 1: Coordinates and thickness of each monitored section. The longitude values represent the exact offshore limit of the section.

Section	Latitude	Longitude	Thickness (meters)
Section 1	23°S	38°W	298.3
Section 2	25°S	38°W	335.7
Section 3	30°S	44°W	524.2

The volume transport components were estimated over the meridional direction because of the mainly north-south direction of the BC flow in the studied region. Equation (1) was vertically integrated between the surface and the climate experiment annual mean level, where the meridional velocities were zero in each section. This level was chosen following Zemba (1991, in [2]), who considered kinematic criteria to define the current. Following this approach, the current can be defined as the total continuous flux flowing in the same direction as the surface current. In the studied region, this flux would represent the water volume transported south by the BC. The thickness of each section that was considered in the volume transport estimates was extracted from these profiles and represented the meridional volume transport zero-motion level (Table 1). The climate meridional volume transport vertical distribution in each section demonstrates a deepening of the meridonal volume transport no-motion level to the south (Figure 2).

The meridional volume transport anomaly fields for each section were also estimated. The anomaly values represent the difference between the annual mean perturbed experiment values and the same values obtained for the climate experiment. Importantly, negative (positive) meridional volume transport values represent southward (northward) flow direction and the occurrence of negative (positive) anomaly values in these regions represent an increase of the southward (northward) velocity field. This methodology was also applied to produce the volume transport anomaly vector fields.

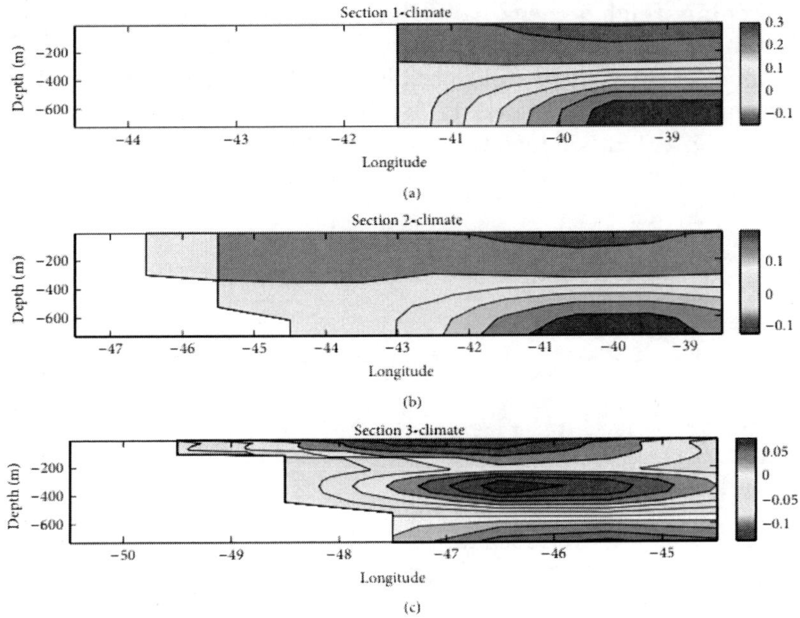

Figure 2: Annual mean vertical meridional volume transport distribution in Sections 1 (a), 2 (b), and 3 (c) obtained by the climate experiment. Sections are showed in Figure 1.

4. RESULTS

4.1. Control Experiment

The annual mean volume transport fields for the entire SBB area at the three different vertical layers defined in Table 1 are presented in Figure 3. It is possible to observe the most intense volume transport fields in the upper layer. It is also possible to observe the strengthening of southward flow toward the higher latitudes. This fact could be explained, in part, by the deepening of the no-motion level toward the same direction and a consequently greater southward water volume transported by the BC. It is also possible to observe the presence of the bifurcation zones associated with the anticyclonic circulation that occurs in different water layers along SBB region feeding BC southward flow in higher latitudes and the northward NBC flow in lower latitudes (Figure 3). This behavior was observed in Figure 2 where the deepening of the zero-motion level associated with the deepening and strengthening (higher negative values indicate stronger southward flow) of the southward BC flow toward higher latitudes was found [25]. In Sections 1 and 2, it is still possible to observe positive meridional velocity values below 300 meters depth, indicating the presence of the northward flow that will feed the NBC flow at lower latitudes.

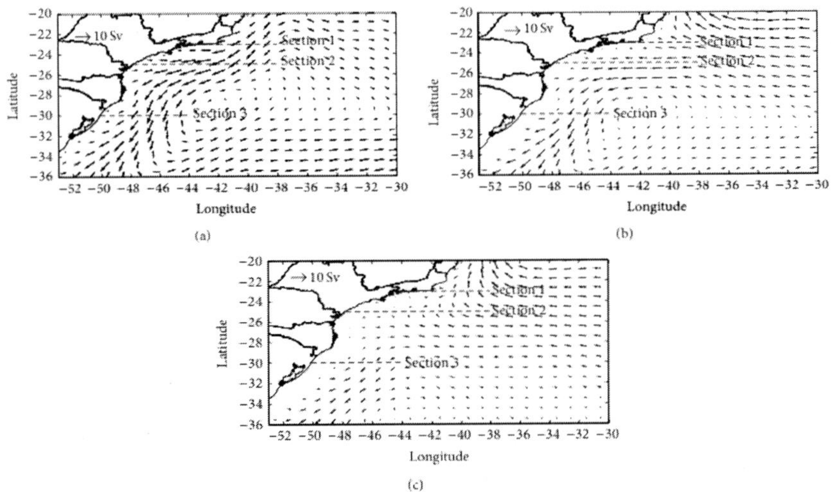

Figure 3: Climate annual mean integrated volume transport for 0–298.3 meters (a), 298.3–335.7 meters (b), and 335.7–524.2 meters (c).

Generally, classical seasonal behavior was observed, with enhanced southward transports during austral summer months (December–January–February) and decreasing southward volume transport during austral winter months (June–July–August), as shown in Figure 4. The analysis of these time series plots revealed again an increase of the ocean volume transports toward the higher latitudes. Section 1 presented a minimum magnitude of −4.0 Sv in July and a maximum magnitude of −5.0 Sv in February. Section 2 presented a minimum volume transport magnitude of −5.0 Sv in June and a maximum magnitude of −6.7 Sv in February. Section 3 presented a minimum magnitude of −7.0 Sv in June and a maximum magnitude of −9.0 Sv in January. This analysis also revealed a representative climatological time lag between the seasonal cycles of the ocean volume transports in each section. For example, a time lag of one month between the minimum southward transport magnitude was observed in Section 1 compared to Sections 2 and 3. Section 1 reached the minimum magnitude during July. Sections 2 and 3 reached their minimum southward magnitudes during June. Additionally, the maximum southward magnitude (more negative) of Sections 1 and 2 occur in February, whereas Section 3 reaches the largest magnitudes in January.

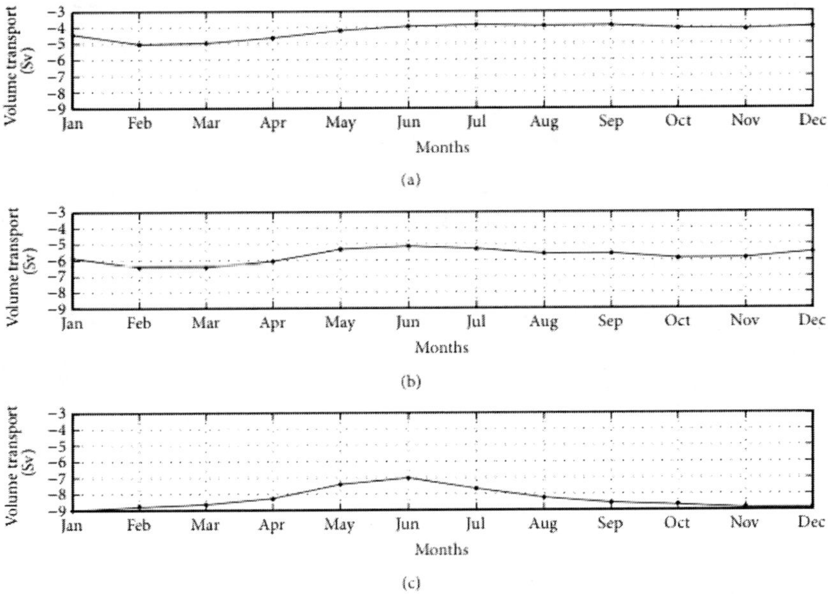

Figure 4: Time series of monthly mean meridional volume transport vertically integrated for Sections 1 (a), 2 (b), and 3 (c).

In Section 1, it is possible to observe the dominance of the southward flow at the surface (upper layers). The annual mean of the southward meridional transport reaches an absolute value of −4.26 Sv, with a standard deviation of ±0.16. In Section 2, as in Section 1, it is possible to observe the strong influence of a southward flow in the upper layer transports. The annual mean ocean volume transport had values of −5.74 Sv (±0.15). In Section 3, the highest values of annual mean transports volumes into upper layers reached −8.32 Sv (±0.36). These values are close to the estimations obtained by others studies in neighboring regions presented by da Silveira et al. [2] (Table 2).

Table 2: Volume transports estimates obtained from hydrographic data sets in the monitored section latitudes. Adapted from da Silveira et al. [2].

Latitude	Reference level (meters)	Volume transport (Sv)	Reference
23°	550	2.2/2.7	[26]
23°	500–1300	4.4	[27]
25°	750	7.3	[28]
28–30°	1550–1600	11.4	[29]

Although the comparisons between the ocean volume transport values obtained in this work and the values estimated from observed data found in the scientific literature are close, it is important to emphasize the differences between them. The vertical water column extensions used to estimate the transports in this work (Table 1) are shallower when compared to historical estimates based on observed data sets for near ocean regions (Table 2). This fact could explain the differences observed between the ocean volume transport estimates presented in both cited tables.

4.2. Perturbed Experiment

The imposition of an interannual perturbed global wind stress field generated interannual variability in the global ocean kinetic energy (OKE) cycle during the three simulated years. Figure 5 presents the time series of global OKE and wind stress anomaly energy for the entire period of the perturbed experiment. Only the imposition of an interannual ENSO-type global wind stress field was able to introduce significant global OKE anomalies. The second year presented the most energetic months, and the global ocean kinetic energy maximum peak occurred at the 23rd month of integration. The global ocean work performed by the wind stress field over the sea surface had an energy peak in the 21st month. This result suggests a time lag of approximately two months between the peaks in global wind energy that led to the global OKE (Figure 5).

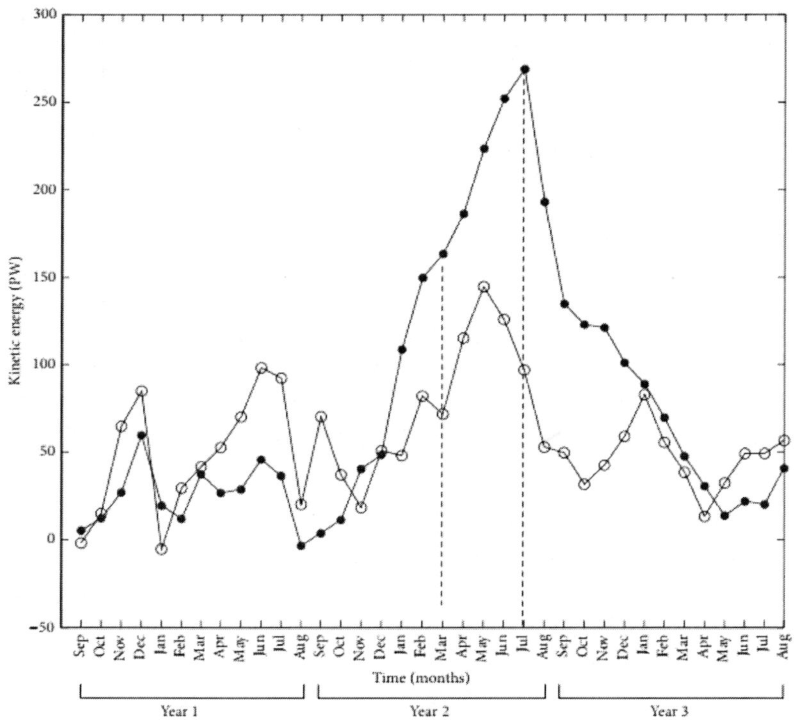

Figure 5: Global kinetic energy anomaly time series plot (filled circles for the ocean kinetic energy and empty circles for the global wind energy over the sea surface). The values of the work performed by the wind were multiplied by a factor of 103 [13].

The annual mean vertically integrated meridional volume transport values were estimated by each integrated perturbed year, being year 1 the first perturbed year, year 2 the second perturbed year and year 3 the third perturbed year. Interannual variations of BC meridional vertically integrated volume transport were observed in the three sections. Section 3 presented strongest variability with an increase of approximately 2 Sv from its climatological meridional volume transport value during the third perturbed year (Table 3). The three sections presented a weakening of the meridional volume transport during the first perturbed integrated year. This variability is apparently locally related to intensity variations of the South Atlantic atmospheric subtropical high (SASH). This fact can be inferred from the annual mean local anomaly wind stress fields shown in Figure 6 for each perturbed simulated year, which shows the weakening of the wind stress field during the first year and the strengthening of this property during the second and the third years of integration. In fact, the influence of local wind stress field over upper layer BC flux had already been identified by other works. Matano [30], studying the causes of the BC separation from the coast, concluded that when the MC flow is weak, the wind stress curl governs the upper layer dynamics of the BC flux.

In Figure 7, the first layer (0–298.3 m) annual mean difference vector of integrated volume transport for the three simulated years is shown. The upper layer volume transport decreases for the first year and the BC upper layer flux is enhanced for the next two simulated years in direct response to the imposition of the interannual global wind stress field. The last two years presented the most intense BC volume transport in Sections 2 and 3. In the final year, it is possible to observe a decrease in the southward BC flux near Section 1.

The intermediate (298.3–335.7 m) level of integrated volume transport difference vector field showed a behavior similar to that observed in the upper level, with a strong decrease in the southward transport between 25°S and 30°S for the first perturbed year. During this period, the model also represented significant changes in the bifurcation zone near 23°S with a decreasing of the southward BC flow over higher latitudes and also a decrease in the northward intermediate flow over lower latitudes. The second and third years showed a return to the climate circulation behavior with an increasing in the southward volume transport, below 23°S, associated with the southward BC flow and also in the northward flow above 23°S in lower latitudes.

The volume transports in the lower monitored level (335.7 m–524.2 m), similarly to the other layers, presented significant interannual variability associated with the imposition of the perturbed wind field. The volume transport difference vector field showed a decrease in the northward flow over 23°S and lower latitudes in the first and second year with an enhancement of this transport during the third year. The bifurcation zone that, on the climate experiment, is centered in 28°S was slightly moved to the north in the third year, which also presents an intensification of the southward volume transport over 30°S and higher latitudes.

Table 3: Annual mean (and the standard deviation) volume transport values for each monitored section in the climate experiment and for the three perturbed simulated years.

Sections	Climate	Perturbed year (1)	Perturbed year (2)	Perturbed year (3)
Section 1	−4.26 (±0.16)	−3.86 (±0.68)	−4.63 (±0.76)	−3.82 (±0.44)
Section 2	−5.74 (±0.15)	−5.27 (±0.97)	−6.345 (±0.89)	−6.08 (±0.75)
Section 3	−8.32 (±0.36)	−8.27 (±0.52)	−8.81 (±1.51)	−10.29 (±0.59)

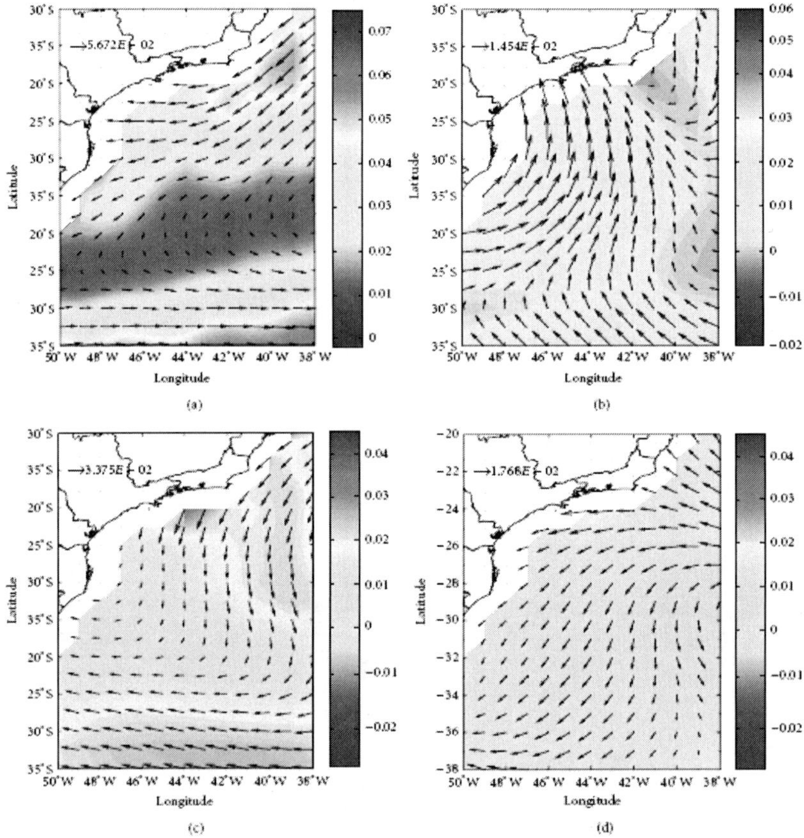

Figure 6: Annual mean wind stress vector field superimposed by the magnitude field for the climate experiment (a). Annual mean wind stress anomaly field superimposed by the wind stress difference vector field for year 1 (b), year 2 (c), and year 3 (d).

Figure 8 shows the annual mean meridional volume transport anomaly vertical distribution over the three sections for year 1. Positive values of the meridional velocity are observed in the first 100 meters of all three sections. These positive values indicate a decrease in the southward flow associated with the BC. Section 3 presents annual mean negative anomaly values in depths below 300 meters, indicating an increase in the southward flow that was already observed in Figure 9. Section 1 presented meridional volume transport negative anomaly values for depths below 300 meters, which is associated with a northward volume transport decrease in this region for the first perturbed integration year.

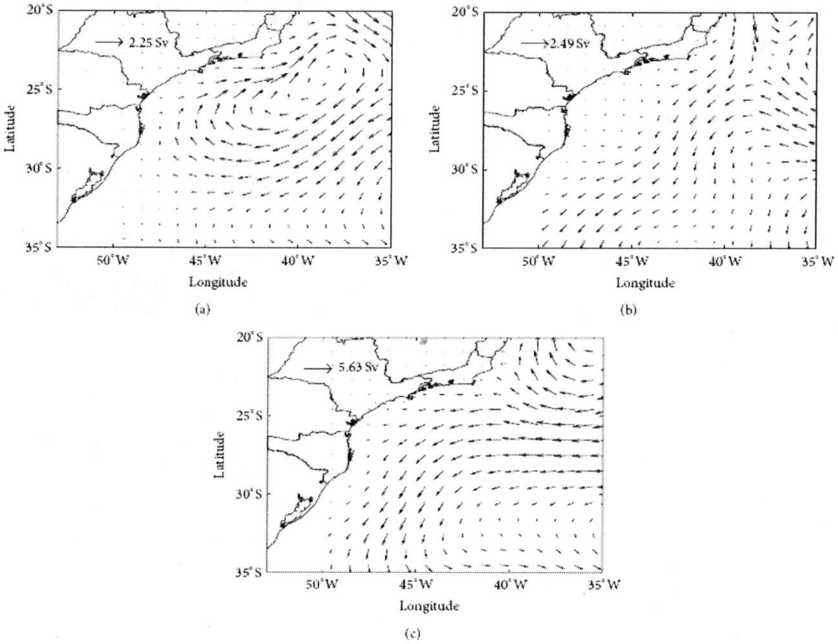

Figure 7: Annual mean integrated volume transport difference vector (units of Sv) field for 0–298.3 meters for the 3 perturbed years: year 1 (a), year 2 (b), and year 3 (c).

Section 3 presented an increase in the climate southward flow observed in Figure 2 by the presence of a negative anomaly meridional velocity component in the entire water column. These positive anomaly meridional volume transport values are stronger predominantly below 400 meters in the third perturbed year.

Table 3 shows the annual mean values (and the associated standard deviation) of meridional volume transport vertically integrated along the three zonal sections for the climate and the perturbed experiments. The analysis of these values reveals common characteristics for the three sections, such as the volume transport decreasing over the first perturbed year and a relative increase over the second integration year. Section 2 showed an enhancement of the annual mean volume transport for the last two perturbed years. In the first year, the volume transport values decreased. Section 3 showed an enhancement of the annual mean volume transport for the third perturbed year that also presented the highest volume transport value. In fact, in Section 3, the southward anomaly volume transport vector field at this latitude contributed to the southward enhancement over all of the layers.

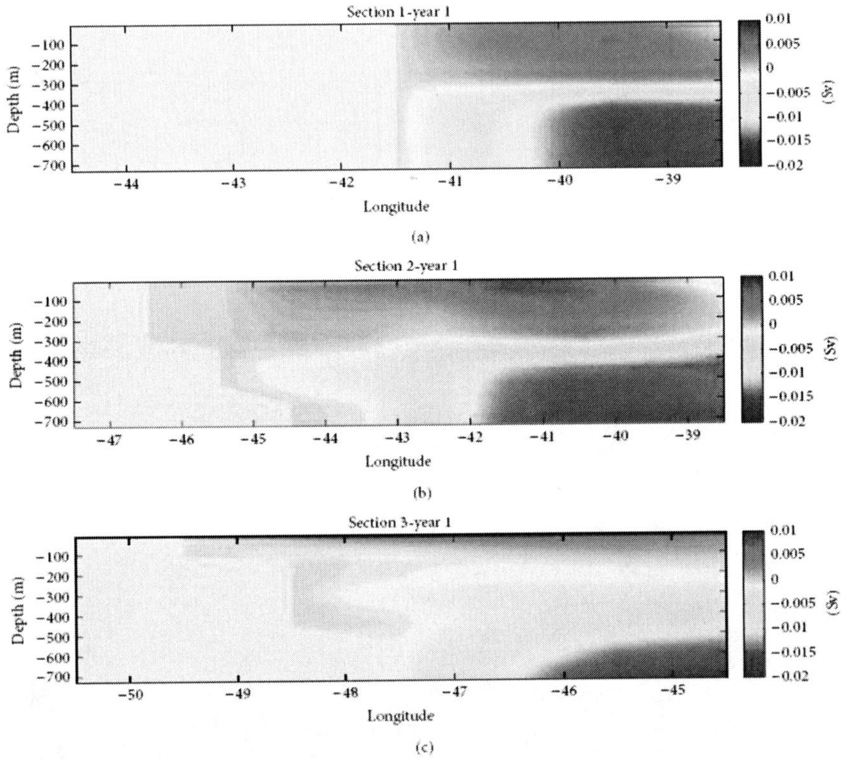

Figure 8: First perturbed integration year annual mean meridional volume transport anomaly field (units of Sv) for Sections 1 (a), 2 (b), and 3 (c). Sections are showed in Figure1.

Monthly mean meridional volume transport vertically integrated fluctuations were analyzed in the ocean layer between the surface and 524.2 meters. Sections 1 and 2 are in good agreement with the local wind modulation. During the second perturbed integration year, an intensification of the annual mean northeast wind fields was observed, as a consequence, accelerated the integrated volume transport field obtained for Sections 1 and 2 (Figure 9). These two sections presented the most intense monthly mean volume transport values in July of the second integration year, which was also when the global OKE peak occurred (Figure 5). Section 1 presents less intense values that are related to its shallower southward transport configuration and with a northward transport presence below it. Section 3 presents the most intense southward volume transports for the three perturbed simulated years (Figure 10). This intensity may be related to its deeper southward transport configuration and the contributions of an anticyclonic circulation cell transport that flows in this level and bifurcates near 30°S (Figure 3). The most intense monthly mean integrated volume transport value for this section is observed in November of the third perturbed simulated year, just four months after Sections 1 and 2 reach their maxima (Figure 11). This result indicates that the ocean volume transports in the studied region have different response time scales to the local surface wind field. This fact could be related to the interaction in deeper layers with termohaline oceanic transports fluctuations that are not directly influenced by the interannual variability of the local SBB wind and have more importance in feeding the southward volume transport field associated with BC in higher latitudes.

5. DISCUSSION

This work identifies and quantifies the occurrence of significant interannual variability in the BC volume transport during the three simulated years. In spite of its coarse resolution, the model is capable of representing Sverdrup-style response to large-scale wind stress curl changes. When considering the studies on climate the low frequency flow can supply significant information on the impact of heat distribution in the ocean thus contributing to future studies on its influence on the climate over near continental areas. However, the coarse horizontal resolution of the grid (1° latitude × 1° longitude) in the studied region could also influence the results obtained. In fact, a significant portion of coastal volume transport values and eddy flow interactions are not represented in this model grid configuration and consequently are not considered in transports estimates. This fact may contribute to the low intensity of the volume transports. It is also important to note that the thickness of BC flow presented for the three sections was always shallower than the scientific literature observed values over neighboring regions [1, 2]. This fact could be related also to the monthly frequency of the velocity field that would underestimate the volume transport values. The short time of model integration for the perturbed experiment could underestimate the observed interannual variability in the SBB region, particularly those related to the BC flux. This fact can be more clearly observed in the OKE anomaly data in the SBB and near offshore region delimited area (60°W–30°W and 20°S–30°S) integrated for the first 650 meters (Figure 12). The behavior of this curve is different from the global integrated OKE (Figure 5). The time series analysis revealed that for SBB region, the OKE is increasing even in the third perturbed integration year, when the highest integrated anomaly energy values occurred. This result suggests a direct association with the wind stress decrease observed during the first perturbed integrated year when negative OKE anomaly values were observed (Figure 6). During the second and especially in the third perturbed integrated year, the SBB local southward meridional wind stress component increased, and as a consequence, it is possible to observe the occurrence of positive anomaly OKE values in the SBB region.

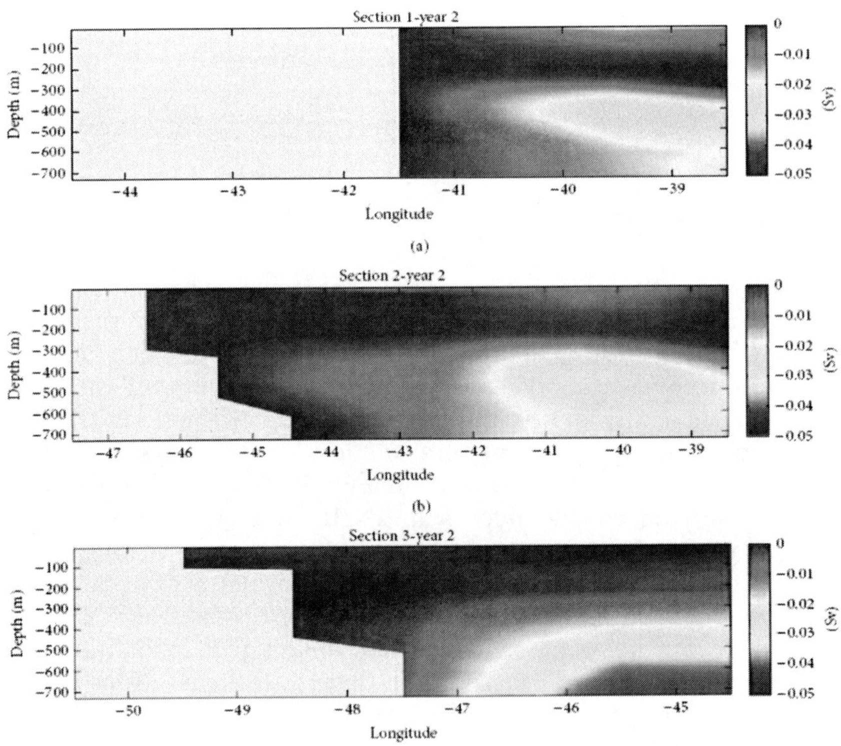

Figure 9: Second perturbed integration year annual mean meridional volume transport anomaly field (units of Sv) for Sections 1 (a), 2 (b), and 3 (c). Sections are showed in Figure1.

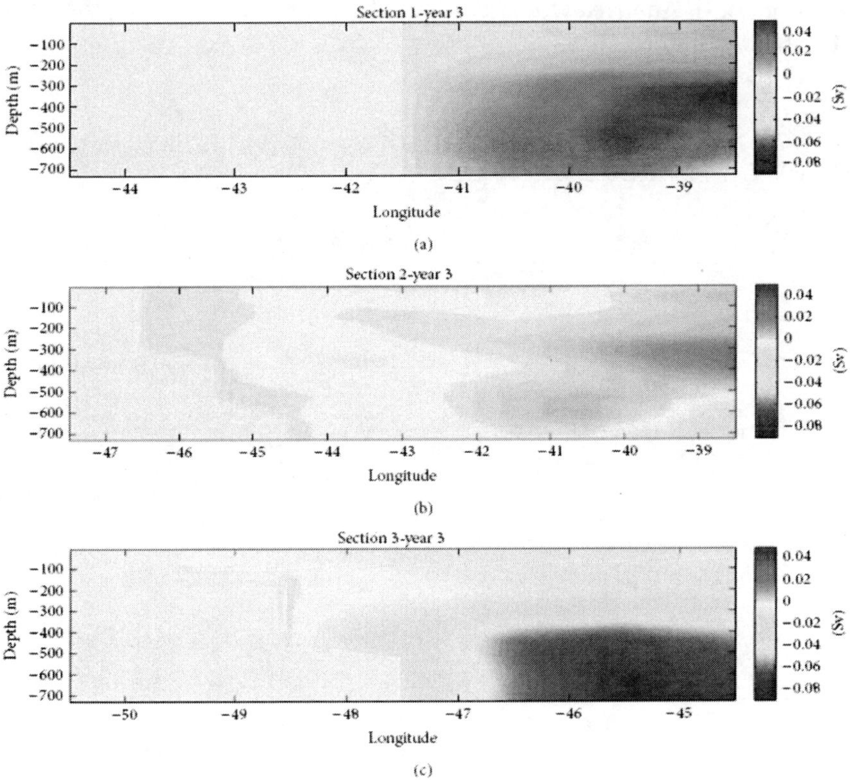

Figure 10: Third perturbed integration year annual mean meridional volume transport anomaly field (units of Sv) for Sections 1 (a), 2 (b), and 3 (c). Sections are showed in Figure1.

The remote or local (South Atlantic basin) wind-generated oceanic processes could also be affecting the SBB volume transport three-dimensional field. One way to better investigate the time evolution of the SBB OKE time distribution would be to increase the time of the model integration, restoring the global wind stress conditions that were used in the control experiment. This methodology would also be useful to identify the influence of different potential ocean variability modes over the studied region generated by the global ENSO-type wind stress field, such as Rossby, Kelvin or continental shelf waves that would be remotely generated. These waves would disturb the density fields distributions and consequently affect the ocean volume transports. Vivier et al. [10], based on observed data, suggested that Kelvin waves originating from the Equatorial Pacific and propagating along the west coast of South America would contour the continent, reaching the South Atlantic basin until 40°S.

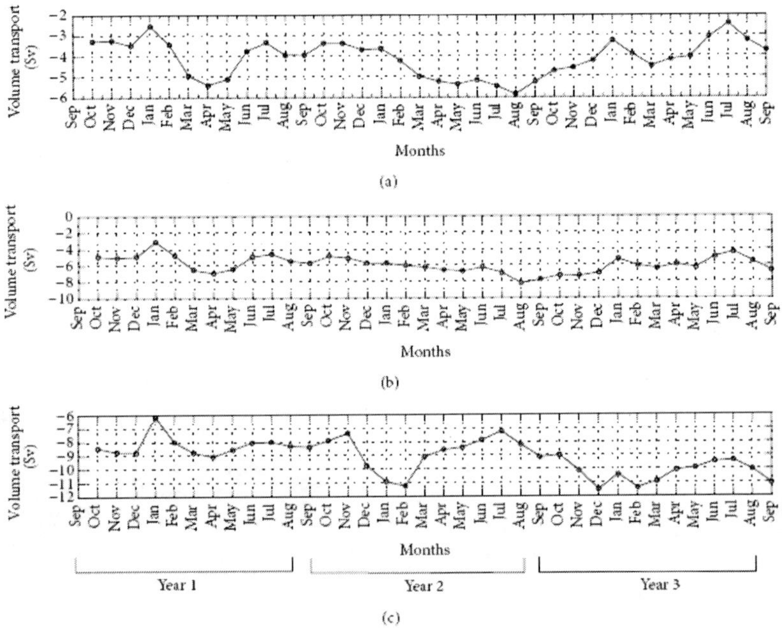

Figure 11: Time series plot of perturbed monthly mean meridional volume transport vertically integrated for Sections 1 (a), 2 (b), and 3 (c).

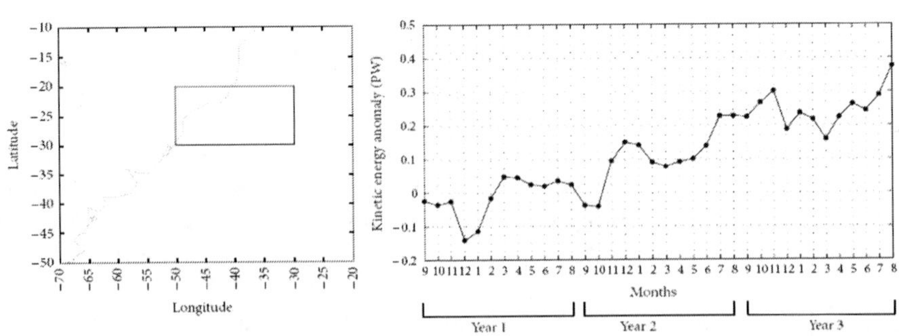

Figure 12: Regional SBB OKE anomaly time series plot. The OKE values are integrated for the first 650 meters.

The interannual variability of the meridional volume transports zero-motion levels that were used to estimate the vertical extension of each section are presented in Figure 13. It is possible to find significant change in these levels for each section. Sections 1 and 2 presented a similar behavior, becoming deeper during the end of the second integration year and shallower in the end of the third integration year. Section 3 presented a continuously deepening process during the entire perturbed experiment. Although significant no-motion level changes were observed in the three section, it is important to note that even for Sections 1 and 2, these changes were very small compared with Section 3. Section 3 presented the most significant changes, which occurred predominantly at the end of the third integration year.

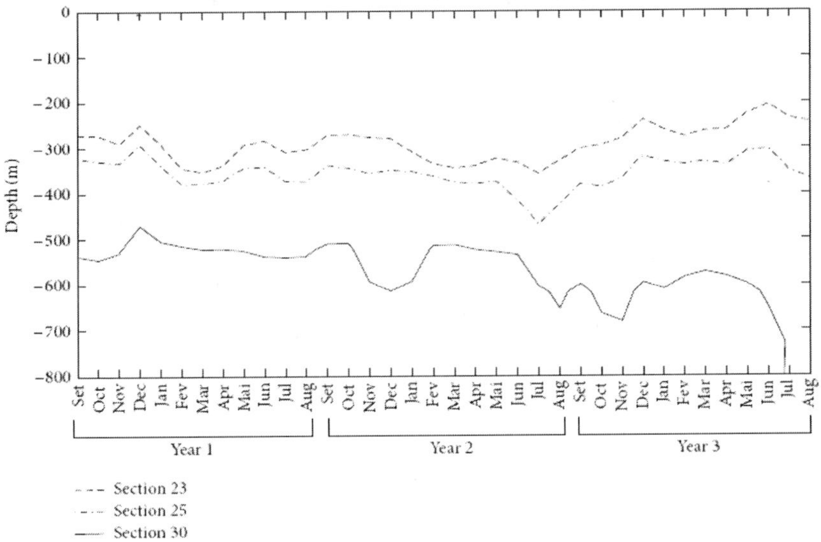

Figure 13: Meridional volume transport zero-motion level time series plot for the three integrated years and zonal averaged for each monitored section.

6. CONCLUSIONS

The imposition of an ENSO-type global monthly mean wind stress field in an OGCM revealed important interannual variability fluctuations in the tridimensional interannual volume transport fields over the SBB and near offshore region. Interannual and annual fluctuations over the ocean volume transports in the studied region are mainly detected over the region where the BC flux is located. The BC upper and medium layer volume transports present a decrease in their intensity during the first year, while an increase is observed for the last two integrated perturbed years. The deep layer ocean volume transport exhibited a continuous intensification process, which indicates a minor direct dependence of this layer to the local wind stress field and possibly indicates a dominance of remote ocean process influences.

It was also possible to observe a small time lag between the integrated volume transports changes in each section during the climate and perturbed experiment. Integrated volume transport time series indicate that Sections 1 and 2 are almost in phase with the occurrence of the most intense southward volume transport that occurred in July of the second perturbed integrated year. Volume transport over Section 3 presented its most intense value just in November of the third perturbed year (four months after Sections 1 and 2).

This work contributed to the knowledge of seasonality and interannual variability of BC system volume transport field associated with changes in the global wind stress field. It was concluded that an anomalous ENSO type global wind stress field could affect locally the BC velocity field and significantly alter its seasonal dynamic behavior. These results would contribute to the understanding of some ocean properties variability over SBB such as temperature field.

CONFLICT OF INTERESTS

The authors declare that there is no conflict of interests regarding the publication of this paper.

REFERENCES

1. R. G. Peterson and L. Stramma, "Upper-level circulation in the South Atlantic Ocean," Progress in Oceanography, vol. 26, no. 1, pp. 1–73, 1991.

2. C. A. da Silveira, A. C. K. Schmidt, E. J. D. Campos, S. S. de Godoi, and Y. Ikeda, “A corrente do Brasil ao largo da costa leste brasileira,” Revista Brasileira de Oceanografia, vol. 48, no. 2, pp. 171–183, 2000.

3. F. A. Schott, M. Dengler, R. Zantopp, L. Stramma, J. Fischer, and P. Brandt, “The shallow and deep Western boundary circulation of the South Atlantic at 5°–11°S,” Journal of Physical Oceanography, vol. 35, no. 11, pp. 2031–2053, 2005.

4. L. Stramma, M. Rhein, P. Brandt, M. Dengler, C. Böning, and M. Walter, “Upper ocean circulation in the Western Tropical Atlantic in boreal fall 2000,” Deep Sea Research Part I: Oceanographic Research Papers, vol. 52, no. 2, pp. 221–240, 2005.

5. R. Lumpkin and S. Garzoli, “Interannual to decadal changes in the western South Atlantic’s surface circulation,” Journal of Geophysical Research C: Oceans, vol. 116, no. 1, Article ID C01014, 2011.

6. D. R. A. Veleda, M. Araujo, M. Silva, and R. Montagne, “Seasonal and interannual variability of the southern south equatorial current bifurcation and meridional transport along the eastern Brazilian edge,"Tropical Oceanography, vol. 39, no. 1, pp. 27–59, 2011. 7.

7. G. J. Goni, F. Bringas, and P. N. Dinezio, “Observed low frequency variability of the Brazil Current front,"Journal of Geophysical Research C: Oceans, vol. 116, no. 10, Article ID C007198, 2011.

8. M. Cataldi, L. P. F. Assad, A. R. Torres Júnior, and J. L. D. Alves, “Estudo da influência das anomalias da TSM do Atlântico Sul extratropical na região da Confluência Brasil-Malvinas no regime hidrometeorológico de verão do Sul e Sudeste do Brasil,” Revista Brasileira de Meteorologia, vol. 25, no. 4, pp. 513–524, 2010.

9. M. A. Alexander, I. Bladé, M. Newman, J. R. Lanzante, N.-C. Lau, and J. D. Scott, “The atmospheric bridge: The influence of ENSO teleconnections on air-sea interaction over the global oceans,” Journal of Climate, vol. 15, no. 16, pp. 2205–2231, 2002.

10. F. Vivier, C. Provost, and M. P. Meredith, “Remote and local forcing in the Brazil-Malvinas region,"Journal of Physical Oceanography, vol. 31, no. 4, pp. 892–913, 2001.

11. C. A. D. Lentini, G. G. Podestá, E. J. D. Campos, and D. B. Olson, “Sea surface temperature anomalies on the western south atlantic from 1982 to 1994,” Continental Shelf Research, vol. 21, no. 1, pp. 89–112, 2001.

12. A. R. Torres Jr., Estudo numérico sobre tele-conexão atmosférica entre fenômenos oceânicos do Pacífico Equatorial e do Atlântico Sul [Ph.D. thesis], Federal University of Rio de Janeiro, 2005.

13. L. P. F. Assad, A. R. Torres Jr., R. N. Candella, and A. S. Mascarenhas Jr., "Brazil-Malvinas Confluence upper ocean temperature anomalies induced by an ENSO wind forcing," Ciencias Marinas, vol. 36, no. 3, pp. 267–284, 2010.

14. B. M. Castro and L. B. Miranda, "Physical oceanography of the Western Atlantic continental shelf located between 4 N and 34 S," in The Sea, A. R. Robinson and K. H. Brink, Eds., vol. 11, pp. 209–251, John Wiley & Sons, New York, NY, USA, 1998.

15. R. G. Soutelino, I. C. A. da Silveir, A. Gangopadhyay, and J. A. Miranda, "Is the Brazil current eddy-dominated to the North of 20°S?" Geophysical Research Letters, vol. 38, no. 3, Article ID L03607, 2011.

16. I. C. A. Silveira and L. B. Miranda, "On the origins of the North Brazil Current," Journal of Geophysical Research, vol. 99, no. 11, pp. 22501–22512, 1994.

17. L. B. Miranda and B. M. Castro, "Geostrophic flow conditions of the Brazil Current at 19°S," Ciencia Interamericana, vol. 22, pp. 44–48, 1982.

18. L. Stramma and M. England, "On the water masses and mean circulation of the South Atlantic Ocean,"Journal of Geophysical Research C: Oceans, vol. 104, no. 9, pp. 20863–20883, 1999.

19. R. C. Pacanowsky and S. M. Griffies, The MOM3 Manual, Geophysical Fluid Dynamics Laboratory/NOAA, Princeton, NJ, USA, 1999.

20. L. P. F. Assad, A. R. Torres Júnior, W. A. Zumpichiatti, A. S. Mascarenhas Júnior, and L. Landau, "Volume and heat transports in the world oceans from an ocean general circulation model," Brazilian Journal of Geophysics, vol. 27, no. 2, pp. 181–194, 2009.

21. F. Röeske, "An atlas of surface fluxes based on the ECMWF re-analysis—a climatological dataset to force global ocean general circulation models," Tech. Rep. 323, Max-Planck Institut für Meteorologie, Hamburg, Germany, 2001.

22. C. Sun, M. M. Rienecker, A. Rosati et al., "Comparison and sensitivity of ODASI ocean analyses in the Tropical Pacific," Monthly Weather Review, vol. 135, no. 6, pp. 2242–2264, 2007.

23. L. P. de Freitas Assad, "ENSO—type wind stress field influence over global ocean volume and heat transports," in Climate Variability—Some Aspects, Challenges and Prospects, Abdel Hannachi, Ed., pp. 99–120, InTech, Rijeka, Croatia, 2011.

24. E. M. Rasmusson and J. M. Wallace, "Meteorological aspects of the El Niño/Southern oscillation,"Science, vol. 222, no. 4629, pp. 1195–1202, 1983.

25. R. R. Rodrigues, L. M. Rothstein, and M. Wimbush, "Seasonal variability of the south equatorial current bifurcation in the atlantic ocean: a numerical study," Journal of Physical Oceanography, vol. 37, no. 1, pp. 16–30, 2007.

26. L. B. de Miranda and B. M. Castro, "Condições do movimento geostrófico das águas adjacentes a Cabo Frio (RJ)," Boletim do Instituto Oceanográfico, vol. 28, no. 2, pp. 79–93, 1979.

27. S. R. Signorini, "On the circulation and the volume transport of the Brazil Current between the Cape of São Tomé and Guanabara Bay," Deep-Sea Research, vol. 25, no. 5, pp. 481–490, 1978.

28. E. J. D. Campos, J. E. Goncalves, and Y. Ikeda, "Water mass characteristics and geostrophic circulation in the South Brazil Bight: summer of 1991," Journal of Geophysical Research, vol. 100, no. 9, pp. 18537–18550, 1995.

29. A. Fisher, The circulation and stratification of the Brazil Current [Dissertation], New York University, New York, NY, USA, 1964.

30. R. P. Matano, "On the separation of the Brazil Current from the coast," Journal of Physical Oceanography, vol. 23, no. 1, pp. 79–90, 1993.

10

VERTICAL DISTRIBUTION OF MEIOFAUNA ON REFLECTIVE SANDY BEACHES

Mariana de Oliveira Martins1 *, Tito Cesar Marques de Almeida1, Maikon Di Domenico1

1Centro de Ciências Tecnológicas da Terra e do Mar-CTTMar, Universidade do Vale do Itajaí - UNIVALI. (Cx. Postal 360, Itajaí, SC, CEP: 88302-202, Brazil)

ABSTRACT

Extreme physical conditions usually limit the meiofauna occurrence and distribution in highly hydrodynamic environments such as reflective beaches. Despite sediment grains of the upper layers being constantly resuspended and deposited, the high energy of the swash zone besides depositing coarse sediments allows an ample vertical distribution of meiofaunal organisms. The effect of physical, chemical and sediment variables on the vertical distribution of meiofaunal organims and nematodes was analysed on two reflective exposed beaches. Sampling was conducted at three sampling points on each beach in the swash zone. The sediment collected was divided into four 10-cm strata (0-10 cm, 10-20 cm, 20-30 cm, 30-40 cm). The statistical differences between strata due to factors previously established (i.e. meiofaunal composition, density of most abundant taxa) were tested using a hierarchical PERMANOVA applied under similarity and euclidian distances. An inverse relation among average grain size, content of organic matter and sediment sorting was evident. Coarser sediment characterized the upper layers, while at deeper layers the sediment was very poorly sorted and presented a higher content of organic matter. A similar pattern in the vertical distribution of meiofaunal and nematofaunal composition and density was detected. The lowest densities were associated with the first stratum (0-10 cm), highly affected by hydrodynamics. The vertical distribution of organisms was statistically different only when the interaction among factors was considered. This result suggests that zonation and vertical distribution of meiofaunal organisms are determined by the within-beach variability.

KEYWORDS

Reflective beach; Vertical distribution; Spatial distribution; Meiofauna; Nematoda; Sediment

1. INTRODUCTION

Variations in horizontal and vertical distributions are the major focus for understanding the causes of spatially structured ecological patterns of meiofauna (CHAPMAN et al., 2010; VIEIRA; FONSECA, 2013). On sandy beaches, the distribution is primarily controlled by the physical and chemical extremes of interstitial spaces (MCLACHLAN; BROWN, 2006; SCHLACHER et al., 2008). These extremes may be attributed to the degree of sorting, size and shape of the grains, primary production and local hydrodynamic energy (MCLACHLAN; BROWN, 2006; GRAY; ELLIOT, 2009).

Capillary forces, water retention, low infiltration and oxygenation, dominate the small interstitial spaces in fine and mud sediment habitats (FENCHEL; RIEDL, 1970; VIEIRA; FONSECA, 2013). In these interstitial habitats dominated by chemical extremes, the densities, number of species and species composition change immediately below the sediment surface (MCLACHLAN, 1989; STEYAERT et al., 2003; LEVIN, 2003). This is mainly to be explained in terms of food sources and redox potential (VIEIRA; FONSECA, 2013). Although the vertical distribution of mud and fine sediment is well understood (STEYAERT et al., 2003; JOINT et al., 1982; HEIP et al., 1985; PALMER, 1986), and the variability in the horizontal dimension is known to be as large on the centimeter scale as on that of meters and kilometers (GALLUCI et al., 2009; FONSECA et al., 2010), few tests have been carried out to correlate the vertical and horizontal dimensions (VIEIRA; FONSECA, 2013). This is specially true for exposed reflective beaches, which represent the opposite geomorphologic extreme controlled by physical limits (MCLACHLAN, 1989; DEFEO et al., 2001). The frequent incidence of large waves, with rising and falling tides in the intertidal zone, favors the deposition of coarse and asymmetrical grains (MASSELINK; RUSSEL, 2006; MASSELINK; PULEO, 2006), which creates a system with high sediment permeability and oxygenation. The grain size characteristics and the energy and hydrodynamic conditions result in the formation of a system with extensive interstitial spaces, which in some cases comprise as much as 40% of the total sediment volume (MCLACHLAN; BROWN, 2006). The three- dimensional spaces formed between the grains of sand create an extensive vertical layer that favors an extensive vertical distribution of meiofauna (MCLACHLAN; BROWN, 2006).

On exposed reflective beaches, meiofauna organisms such as Nematoda, Copepoda, Harpacticoida, Tardigrada, Gastrotricha, Annelida and Mystacocarida, exhibit high abundance (MCLACHLAN; BROWN, 2006; GIERE, 2009). These organisms inhabit different sediment layers; however, few studies have attempted to investigate their vertical distribution in reflective sandy beach sediments (PLATT, 1977; SHARMA; WEBSTER, 1983; NICHOLAS; HODDA, 1999;ALBUQUERQUE et al., 2007; MARIA et al., 2012; DI DOMENICO et al., 2014). The sediment oxygen concentration in reflective beaches is assumed to be the environmental variable that directly influences the vertical distribution of meiofauna communities (BRAECKMAN et al., 2011), as it is a non- limiting factor due to the great hydrodynamic force of the waves that break on the beach face (FENCHEL; RIEDL, 1970; RODRIGUEZ et al., 2001). The consequence of this assumption is the non-stratified distribution of meiofauna and Nematoda in reflective beach sediments. However, this assumption is frequently disregarded in studies on the distributions of sandy beach organisms (GHESKIERE et al., 2004; GHESKIERE et al., 2005; GHESKIERE et al., 2006; KOTWICKI et al., 2005). Other abiotic variables, such as grain size, salinity, redox potential, organic matter, temperature, water content, and depth of the water table (REISE; AX, 1979; STEYAERT et al., 1999), as well as biological interactions, may also determine the stratification patterns in reflective beaches.

The goal of this study was to test and correlate the vertical and horizontal distributions of meiofauna and Nematoda in response to the physical, chemical and grain size characteristics of two reflective beaches on the southeast/south Brazilian coast. Their overall geomorphological description was first made 15 years ago (KLEIN; MENEZES, 2001; PETTERMANN et al., 2006; DI DOMENICO et al., 2009) and they were then characterized by their large waves. The following hypotheses were tested: (H1) Nematoda and meiofauna are not stratified and (H2) are randomly distributed in the intertidal zone of reflective beaches.

2. MATERIAL AND METHODS

2.1 Study site

Two beaches - Estaleiro Beach and Estaleirinho Beach - previously characterized as exposed reflective beaches (KLEIN; MENEZES, 2001; PETTERMANN et al., 2006; DI DOMENICO et al., 2009), located on the centre-north coast of the state of Santa Catarina (Figure 1), were examined in June (i.e. late autumn) 2008. Both beaches exhibit a north-south orientation and are composed of sediments that are primarily characterized by coarse and medium sand, steep beach faces and the absence of a surf zone during the greater part of the year. When storm surges occur, these beaches exhibit narrow surf zones formed by sand banks and the beaches' morphodynamic profile becomes temporarily intermediate (KLEIN; MENEZES, 2001).

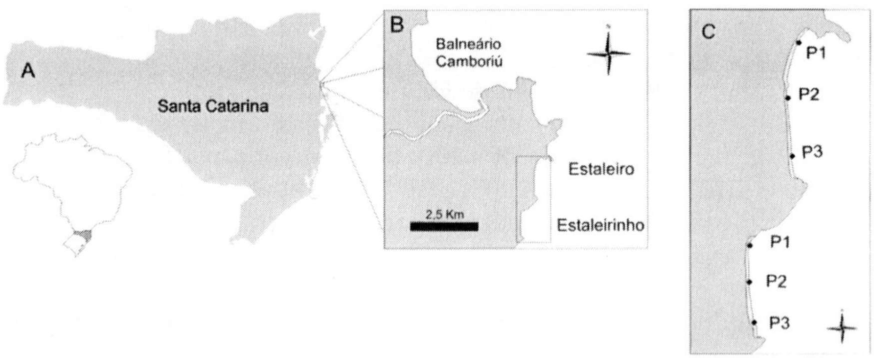

Figure 1 Location map of the studied sites and distribution of the sampling points at Estaleiro Beach and Estaleirinho Beach.

2.2 Sampling

Three points located more than 100 meters apart were sampled in the intertidal zone of each beach during low spring tide. Five replicates were collected at each sampling point for the meiofauna analysis and one replicate was collected for the grain size distribution analysis. The sampling was carried out on the same day.

Samples were collected using a PVC tube of 3.5 cm diameter and 1 m length, which was buried 40 cm into the sediment. The sediment collected was divided into four 10-cm strata, beginning with the most superficial layer (0-10 cm, 10-20 cm, 20-30 cm and 30-40 cm), and fixed in a 4% formaldehyde-saline solution.

2.3 Laboratory methods

The meiofauna was separated from the sediment through a floatation technique using colloidal silica LUDOX TM 50 (Sigma Aldrich, St. Louis, USA). The suspended material was separated using a 0.045 mm mesh sieve and stored in 4% formaldehyde. The collected organisms were quantified and identified by preparing permanent slides in accordance with SOMERFIELD and WARWICK (1996).

Samples for the grain size distribution analysis were processed according to SUGUIO (1973), and the grain size parameters were obtained using SIGA (*Sistema Gerenciador de Amostras Sedimentológicas* [Manager System of Sedimentologic Samples]) software. The organic matter concentration was determined using the weight difference after they had been burned in a muffle furnace at 800ºC for eight hours. The CaCO3 concentration was determined by acid dissolution in 10% hydrochloric acid.

Chlorophyll a extraction from the sediment was performed by adding 90% acetone for 12 hours. Chlorophyll a concentrations were calculated according to the equation by LORENZEN (1967), using absorbances at 665 and 750 μm and measured in a spectrophotometer before and after acidification with 2 N HCl.

The sediment permeability was determined using a device developed by CAPUTO (1980). The salinity was measured using a HORIBATM analyzer and the temperature with a thermometer placed in contact with the sand in a hole dug with a shovel. The bathymetric profile was established using a tape measure and a transparent hose as a level. The measurements were performed crosswise to the beach line from the beginning of the sand bank to the intertidal zone.

2.4 Data analysis

The spatial variation of the environmental variables was analyzed using principal component analysis (PCA). The PCA was performed for the following variables: calcium carbonate (%), average grain diameter, sorting, skewness and organic matter and chlorophyll a concentrations. The selection of variables assumed non-collinearity and included the measurements of variables at all depths. The data matrix was standardized for the application of the PCA (LEGENDRE; LEGENDRE, 1998). The analysis of the abiotic factors to check for collinearity revealed a correlation between the sand and gravel proportions; therefore, these parameters were excluded from the analysis.

Hypotheses 1 and 2 (H1 and H2), which were tested on the specific abundances and total densities of organisms, were used to calculate the similarity and distance matrices, respectively. The similarities in the compositions of the meiofauna and Nematoda were measured by the Bray-Curtis coefficient and the Euclidean distance of total meiofauna density. A permutational multivariate analysis of variance (PERMANOVA) was performed using the resulting matrices by considering the following factors: strata (four fi levels including 0-10, 10-20, 20-30, 30-40 cm), beaches (two random levels - Estaleiro Beach and Estaleirinho Beach - nested within the strata), and sampling points (three random levels - 1, 2 and 3 - nested within the beaches).

To identify the associations tested by the PERMANOVA, a proximity analysis (multidimensional scaling, MDS) was performed. The contributions of different taxa to the similarities amongst the strata were evaluated by the average similarity between the tested and significant factors, using a similarity percentage analysis (SIMPER). All statistical multivariate analyses were performed using the PRIMER 6.1.6. software (2008) (CLARKE; WARWICK, 2001).

3. RESULTS

3.1 Beach morphodynamics

Both beaches exhibited a large subaerial volume - 75-50 m wide at Estaleiro Beach and 40-50 m at Estaleirinho Beach - pronounced steepness and reflective beach characteristics. Estaleiro Beach exhibited a larger variation in the morphodynamic profile of the sampling points (Figure 2). The measured temperatures varied between 21 and 2ºC. Higher chlorophyll a concentrations were found in the more superficial strata; the highest concentration (215.29 µg/l) was observed in the 0-10 cm stratum of sampling point 3 on Estaleiro Beach and the lowest concentration (53.90 µg/l) was observed in the 30-40 cm stratum of sampling point 1 on Estaleirinho Beach (Table 1).

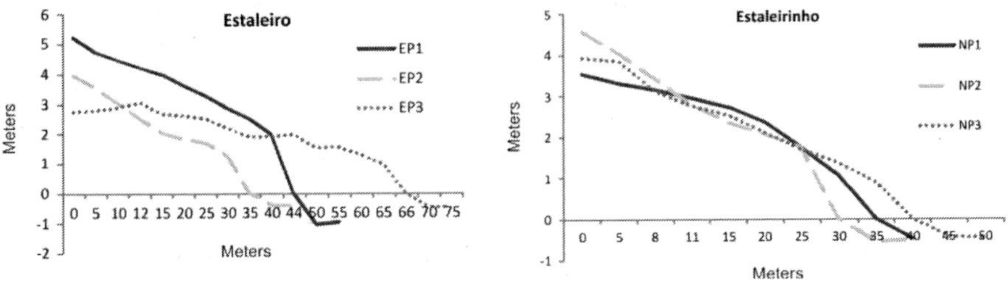

Figure 2 Topographic profiles of Estaleiro Beach and Estaleirinho Beach. The solid lines represent sampling point 1, the dashed lines represent sampling point 2 and the dotted lines represent sampling point 3.

The beaches were characterized by a predominance of moderately sorted coarse sand, with modal peaks of grains between 1.0 and 2.0 phi. The grain-size distribution was similar amongst the strata, sampling points and beaches. Sampling point 3 on Estaleiro Beach exhibited higher heterogeneity amongst the strata, with the average grain size varying between 0.34 phi in the 20-30 cm stratum to 1.4 phi in the 0-10 cm stratum.

The principal component analysis (PCA) resulted in the extraction and interpretation of two factorial axes, which explained 61.63% of the total variation. Axis 1 separated Estaleiro Beach, which was positioned on the positive side of the axis, with a higher organic matter concentration and coarser sediments, from Estaleirinho Beach, which was positioned on the negative side of the axis and was characterized by smaller average grain sizes and a lower organic matter concentration. Axis 2 separated the sampling points of Estaleiro Beach from the 0-10 cm and 10-20 cm strata with higher chlorophyll a concentrations. The 20 to 40 cm strata were significant due to a greater quantity of sorted grains on Estaleiro Beach and a higher carbonate concentration on Estaleirinho Beach (Figure 3).

Table 1 Sediment grain size and chlorophyll *a* concentration: (AGS) average grain size (phi), (Sort) sorting, (Skew) skewness (phi), (%CaCO3) carbonate percentage, (%OrgMat) organic matter percentage and (Chl-*a*) chlorophyll a concentration

Point	Depth	°C	MGS	Sort	Skew	%$CaCO_3$	%OrgMat	Chl-a (µg/l)
Estaleiro - P1	10	20.8	0.8717	0.7107	-0.0560	3.03	0.15	132.336
	20		0.9891	0.6712	-0.0780	2.73	0.12	88.9239
	30		0.8845	0.7212	-0.0713	2.79	0.12	86.4936
	40		0.7651	0.7375	-0.0735	3.69	0.12	79.9762
Estaleiro - P2	10	22	1.1816	0.6414	-0.1871	2.73	0.14	101.296
	20		1.088	0.6593	-0.1015	2.93	0.09	149.237
	30		0.9498	0.7261	-0.0872	3.63	0.11	115.435
	40		0.8074	0.8027	-0.0729	3.42	0.14	84.7262
Estaleiro - P3	10	21.3	1.0208	0.6998	-0.0825	3.52	0.18	215.295
	20		1.4341	0.5808	-0.2718	2.95	0.12	125.046
	30		0.3476	0.9616	-0.0442	3.73	0.18	181.051
	40		1.1935	0.6374	-0.2331	3.36	0.09	111.569
Point	**Depth**	**°C**	**MGS**	**Sort**	**Skew**	**%$CaCO_3$**	**%OrgMat**	**Chl-a (µg/l)**
Estaleirinho - P1	10	21.7	1.3041	0.6453	-0.1947	4.33	0.1	97.761
	20		1.3051	0.6656	-0.0823	4.43	0.1	85.7204
	30		1.1851	0.651	-0.0027	4.85	0.09	74.7844
	40		0.9623	0.684	0.0668	4.57	0.1	53.9066
Estaleirinho - P2	10	21.7	0.9105	0.8893	0.0201	4.58	0.11	92.6796
	20		1.2904	0.7141	-0.1382	4.72	0	128.47
	30		1.1866	0.7487	-0.0511	4.29	0.1	102.842
	40		1.2487	0.6784	-0.0385	4.97	0.11	76.9937
Estaleirinho - P3	10	21.3	1.2023	0.6846	0.0646	4.46	0.11	152.772
	20		1.2696	0.6484	-0.0049	4.61	0.1	132.668
	30		1.4509	0.6819	-0.1706	5.31	0.1	104.555
	40		1.3807	0.6006	-0.0990	5.03	0.09	128.139

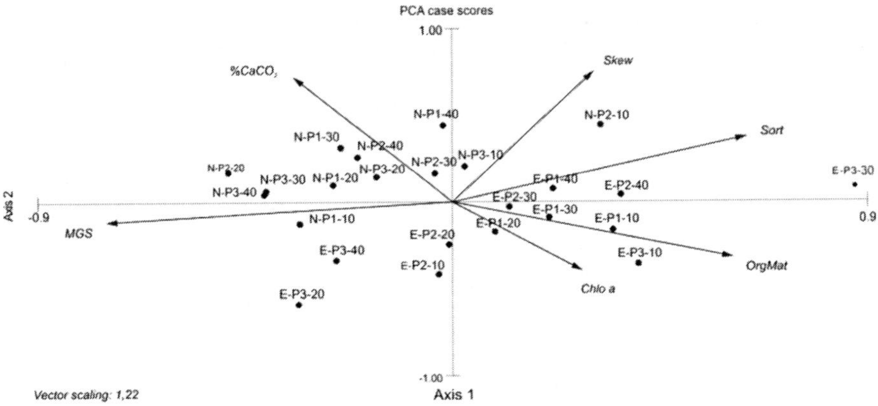

Figure 3 PCA for Estaleiro Beach and Estaleirinho Beach for different depths. The vectors represent skewness (Skew), sorting (Sort), organic matter (OrgMat), chlorophyll *a* (Chl-*a*), average grain size (AGS) and carbonate percentage (%CaCO3). At the points, the first letter represents Estaleiro (E) Beach and Estaleirinho (N) beach, followed by the indication of the sampling points (P1, P2 and P3) and strata (0-10, 0-20, 0-30 and 0-40).

3.2 Meiofauna distribution

The meiofauna was composed of 15 taxonomic groups. Nauplius, which in this case belongs to the copepod, was the most abundant followed by adult Copepoda, Nematoda, Gastrotricha, Tardigrada, Platyhelminthes, Ostracoda and Polychaeta. The remaining taxa exhibited a total abundance lower than 5% (Table 2). The highest meiofauna densities were found in the intermediate strata (10-20 and 20-30 cm) and the lowest meiofauna densities were observed in the 0-10 cm stratum. The highest average meiofauna density (620.4 ind.10 cm-2) was observed in the 20-30 cm stratum of sampling point 3 on Estaleirinho Beach. The lowest density (65.8 ind.10 cm-2) was observed in the 0-10 cm stratum of sampling point 3 of Estaleiro Beach (Figure 4). Significant differences were observed at the lowest interaction point amongst the factors (beach (stratum)) (P-MC = 0.001>) (Table 3), i.e., the vertical distribution patterns of the meiofauna depended on the sampling points and the beaches.

Table 2. Total abundance and frequency of meiofauna taxa

Taxa	Abundance	(%)
Nauplius	8076	30.63
Copepoda	4443	16.85
Nematoda	3640	13.81
Gastrotricha	3151	11.95
Tardigrada	1642	6.23
Turbellaria	1604	6.08
Ostracoda	1440	5.46
Polychaeta	1129	4.28
Mystacocarida	776	2.94
Halacoroidea	235	0.89
Oligochaeta	163	0.62
Sarcomastigophora	48	0.18
Cnidaria	10	0.04
Bivalvia	9	0.04
Total	26366	100

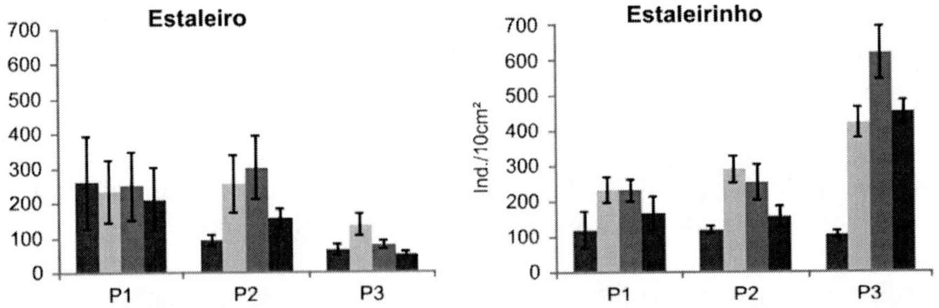

Figure 4 The density of meiofauna (ind./10 cm2) at Estaleiro Beach and Estaleirinho Beach for different sampling points (P1, P2 and P3) and four sediment strata (0-10, 10-20, 20-30, 30-40 cm).

Table 3 Results of the PERMANOVA applied to the Euclidean distance matrix for the total density of meiofauna

Source	df	SS	MS	Pseudo-F	P (perm)	perms	P (MC)
Stratum (St)	3	582.37	194.12	1.6875	0.319	105	0.319
Beach (St)	4	460.13	115.03	1.774	0.158	999	0.188
Point (Beach (St))	16	1037.5	64.842	3.8649	0.001	998	0.001
Res	96	1610.6	16.777				

The multidimensional scaling (MDS) proximity analysis applied to the meiofauna groups showed the separation of the 0-10 cm stratum from the remaining strata (Figure 5). The analysis of the contributions of different species to the similarities amongst strata showed that the nematodes and turbellarians were the most representative in the 0-10 cm stratum by contributing more than 50% of the similarities amongst the samples. Below the 10-20 cm stratum, the copepods and Nauplius yielded the largest contributions to the similarities amongst the samples (Table 4).

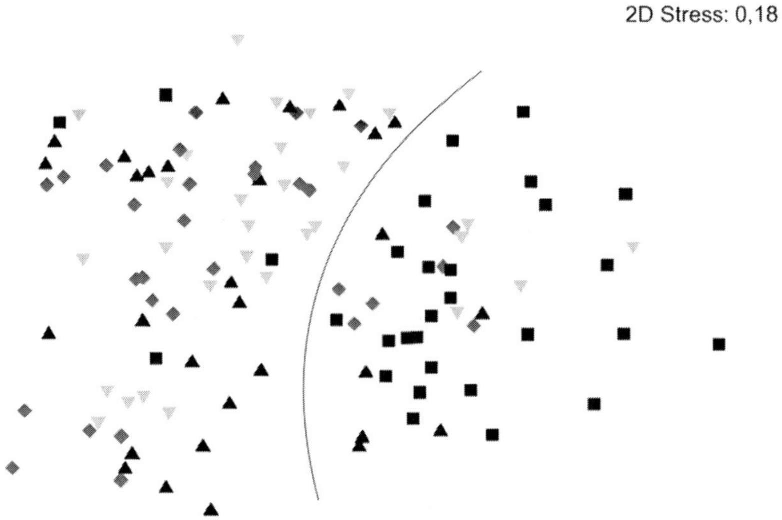

Figure 5 Representation of the similarities using a nonparametric MDS plot. Similarities calculated with the Bray-Curtis coefficient using the abundance values for the meiofauna groups observed at depths from 0 to 40 cm (stress = 0.18). 0-10 cm strata = black squares, 0-20 cm strata = black triangles, 0-30 cm stratum = dark grey diamonds and 0-40 cm stratum = grey triangles. Line separates superficial strata.

3.3 Nematoda distribution

Nematoda belongs to 24 genera, the Epsilonema, Axonolaimus, Microlaimus, Theristus, Enoploides, Daptonema, Bathylaimus, Monoposthia and Sphaerolaimus being the most abundant, with 96% of the total number of organisms counted. The remaining genera exhibited a total abundance lower than 0.1% (Table 5).

Table 4 Contribution to the average similarity (SIMPER). Average density and contribution (%) to the taxa similarity for each depth strata of Estaleiro Beach and Estaleirinho Beach

	Estaleiro				Estaleirinho		
depth	Taxa	Average density	Cont. (%)	Depth	Taxa	Average density	Cont. (%)
	Nematoda	21.87	38.45		Nematoda	29	29.12
10	Turbellarians	17.47	32.42	10	Turbellarians	16.07	22.58
	Nauplius	61.67	8.16		Copepoda	29.4	22.55
	Copepoda	27.33	23.64		Nauplius	68.2	25.13
	Nauplius	78	23.05		Gastrotricha	54.6	19
20	Nematoda	16.13	14.31	20	Nematoda	49.67	17.73
	Ostracoda	25.4	13.6		Copepoda	40.27	15.08
	Nauplius	112.53	32.79		Nauplius	80.27	31.48
	Copepoda	30.13	26.13		Nematoda	63.33	18.54
30	Nematoda	9.6	10.41	30	Gastrotricha	56.13	16.74
	Nauplius	71.2	36.38		Nauplius	49.67	35.4
	Copepoda	29.6	23.46		Nematoda	46.2	15.94
40	Turbellarians	8.07	11.75	40	Gastrotricha	56.4	15.2

Table 5 Total abundance and frequency of nematode genera

Taxa	Abundance	(%)
Epsilonema	1096	30.53
Axonolaimus	666	18.55
Microlaimus	461	12.84
Theristus	333	9.28
Enoploides	238	6.63
Bathylaimus	236	6.57
Daptonema	233	6.49
Monophostia	136	3.79
Sphaerolaimus	60	1.67
Oncholaimus	27	0.75
Camacolaimus	16	0.45
Viscosia	15	0.42
Eurystomina	14	0.39
Thichotheristus	12	0.33
Rhabdodemania	12	0.33
Terschellingia	11	0.31
Neotonchus	7	0.19
Chromaspirina	6	0.17
Enoplolaimus	5	0.14
Ascolaimus	2	0.06
Paracyatholaimus	1	0.03
Rhynchonema	1	0.03
Desmoscolex	1	0.03
Metachromadora	1	0.03
Total	3590	100

The highest average nematode density (104.6 ind. cm-2) was observed in the 20-30 cm stratum of sampling point 3 of Estaleirinho Beach, and the lowest (3.6 ind.cm-2) was observed in the 20-30 cm stratum of sampling point 1 of Estaleiro Beach (Figure 6). Significant differences were observed for the interaction point (beach (stratum)) (P-MC =

0.001) (Table 6), i.e., the vertical distribution patterns for the nematodes depended on the sampling points and beaches.

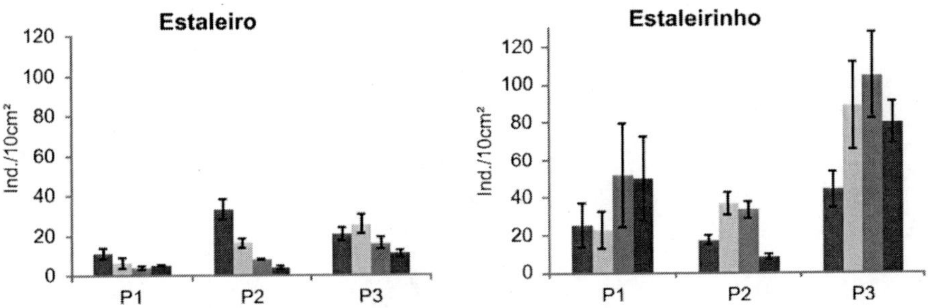

Figure 6 Nematode average density (nº ind./10 cm2) at Estaleiro Beach and Estaleirinho Beach for different sampling points (P1, P2 and P3), and depths from 10 cm to 40 cm.

Table 6 Results of PERMANOVA regarding the Nematoda taxonomic composition

Source	df	SS	MS	Pseudo-F	P (perm)	perms	P (MC)
Stratum (St)	3	16300	5433.5	0.65686	0.684	105	0.802
Beach (St)	4	33088	8271.9	2.9776	0.003	999	0.002
Point (Beach (St))	16	44449	2778.1	3.5471	0.001	998	0.001
Res	96	75187	783.2				

The proximity analysis (MDS) performed for the Nematoda genera matrix showed a trend towards a grouping of the 0-10 cm strata (Figure 7). The individual contributions of the different genera were similar for each stratum and beach. At Estaleiro Beach, the genus Daptonema contributed to the similarities in the most superficial layer (0-10 cm). The genus Theristus contributed to the similarities in the 30-40 cm stratum. On Estaleirinho Beach, Epsilonema and Microlaimus contributed to the similarities in the strata below 10 cm. The genus Axonolaimus contributed to the similarities in almost all strata of both beaches (Table 7).

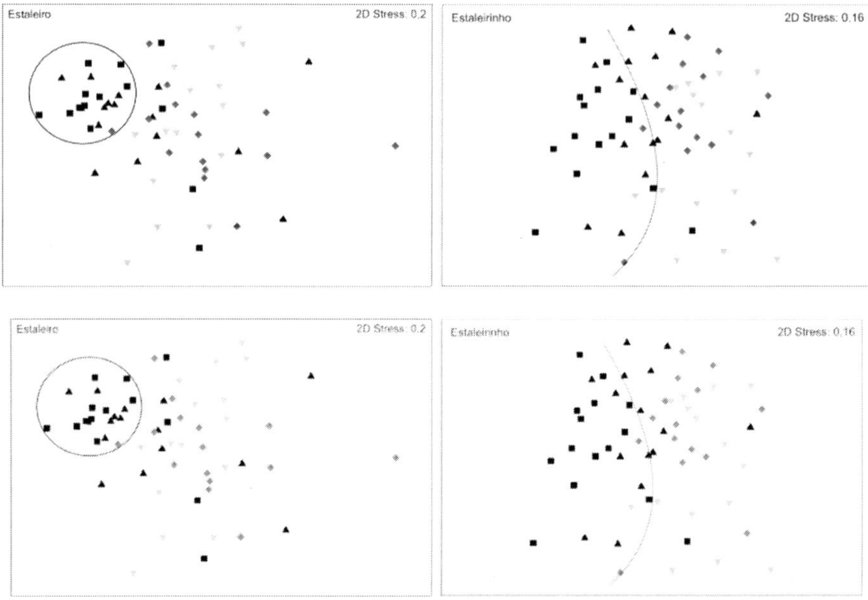

Figure 7 Nonparametric MDS plot. Similarities calculated with the Bray-Curtis coefficient using the nematode abundance values for Estaleiro Beach and Estaleirinho Beach. 0-10 cm strata = black squares, 0-20 cm strata = black triangles, 0-30 cm stratum = dark grey diamonds and 0-40 cm stratum = grey triangles. Circle and line separates superficial strata.

4. DISCUSSION

The spatial variability of the composition and densities of both meiofauna and Nematoda, were higher amongst the sampling points located on the same beach face, than amongst the different vertical strata. Different studies on meiofauna suggest that the meiofauna's variability on the horizontal dimension is large on the centimeter scale when compared to the meter and kilometer scales for mud and fine sediments (GALLUCI et al., 2009;FONSECA et al., 2010; VIEIRA; FONSECA, 2013). The horizontal and vertical distribution patterns were related to the beach profiles at each sampling point, which resulted from different areas of erosion and deposition. The large heterogeneity of the intertidal zones of reflective sandy beaches is primarily due to the formation of turbulent regions, which are caused by the direct impact of waves (MASSELINK; RUSSEL, 2006). A higher deposition of coarse grains was evident at the erosion points, and the accumulation of gravel composed primarily of shell fragments resulted in a lower stratification. These characteristics were more pronounced at sites where rip currents are formed, as the force of the water carries the finer sediment particles to those sites (SHORT, 1999; MASSELINK; RUSSEL, 2006). Associated with this physical factor, the patchy distribution characteristic of the meiofauna may be essential for the determination of vertical and horizontal distribution patterns (MCLACHLAN; BROWN, 2006). Similar results were found for the horizontal distribution of interstitial annelids on Estaleiro beach (DI DOMENICO et al., 2009), where differences among transects (i.e. points) on the same beaches are higher than on similar beaches.

In this study, the highest meiofauna densities were observed in the 10 to 30 cm strata.This contrasts with previous reports, according to which the highest meiofauna densities were generally observed in the top 10 cm of sediment (MCLACHLAN, 1978; OLAFSSON; ELMGREN, 1997; DE JESUS-NAVARRETE; HERRERA-GOMEZ, 2002; KAPUSTA et al., 2002; KOTWICKI et al., 2005). High densities at the sediment surface are generally related to the existence of an anoxic layer (chemical gradient) that forms immediately below the top centimeters of sediment in dissipative beaches (KOTWICKI et al., 2005), ultra- dissipative beaches (MARIA et al., 2012) and tidal flats (VIEIRA; FONSECA, 2013). However, on reflective beaches, such as Estaleiro and Estaleirinho Beaches, the low densities in the 0-10 cm stratum resulted from the re-suspension of the top centimeters of sediment by the wave energy (MCLACHLAN, 1989; MCLACHLAN; BROWN, 2006), which conditioned the meiofauna to live in intermediate strata.

High chlorophyll a concentrations below the top layers of the sediment may somehow support microscopic chains and enable the proliferation of meiofauna and Nematoda in reflective beaches (JESUS et al., 2006). The combination of high values of skewness and average grain size produce larger interstitial spaces, greater water infiltration and penetration of sunlight below the top millimeters of sediment (JESUS et al., 2006).

Nauplius was the most abundant group. The high densities of these organisms are common in intertidal zones of coarse sand reflective beaches (RODRIGUEZ et al., 2001). The high oxygen concentration characteristic of these environments is optimal for these organisms as they are highly sensitive to reductions in interstitial oxygen (GHESKIERE et al., 2005; MCLACHLAN; BROWN, 2006).

Table 7 Contribution to average similarity (SIMPER). Average abundance (ind./10 cm2), average similarity, contribution (%) and cumulative contribution (%) of the Nematoda genera for each stratum of Estaleiro Beach and Estaleirinho Beach

	Estaleiro						Estaleirinho				
Depth	***Genus***	**Average abundance**	**Similarity**	**Cont. (%)**	**Cum. (%)**	**Depth**	**Genus**	**Average abundance**	**Similarity**	**Cont. (%)**	**Cum. (%)**
10	*Daptonema*	8.8	18.2	38.48	38.48	10	*Axonolaimus*	10	19.97	49.06	49.06
	Axonolaimus	5.33	11.88	25.11	63.59		*Theristus*	5.67	6.7	16.47	65.53
	Enoploides	2.47	5.96	12.59	76.19		*Enoploides*	3.2	5.64	13.86	79.39
20	*Axonolaimus*	4.33	10.88	28.24	28.24	20	*Axonolaimus*	10.73	14.03	37.14	37.14
	Enoploides	3.07	7.98	20.72	48.95		*Theristus*	4.33	4.27	11.32	48.46
	Theristus	2.4	7.38	19.14	68.09		*Microlaimus*	14.2	4.27	11.31	59.77
30	*Epsilonema*	2.13	14.87	38.9	38.9		*Monophostia*	3.07	4.25	11.26	71.03
	Theristus	2.27	9.97	26.08	64.99	30	*Epsilonema*	30.07	18.79	46.98	46.98
	Enoploides	1.93	5.92	15.49	80.48		*Axonolaimus*	6.8	8.32	20.81	67.79
40	*Enoploides*	1.53	9.84	24.46	24.46		*Microlaimus*	12.07	3.81	9.52	77.31
	Axonolaimus	1	9.32	23.19	47.65	40	*Epsilonema*	29.87	23.56	67.49	67.49
	Theristus	1.13	9.28	23.09	70.75		*Axonolaimus*	4.73	4.78	13.68	81.16

Nematoda were responsible for the similarity patterns in the top 10 cm of sediment. Both beaches' concentrations of chlorophyll a and organic matter were slightly higher in the more superficial layers. Copepoda and Nauplius were the organisms most represented at the remaining depths.

Only tiny, well-adapted species such as Epsilonematid and Draconematid nematodes (GHESKIERE et al., 2006) were expected to inhabit these highly dynamic zones (MCLACHLAN; BROWN, 2006). Amongst the nematodes, the genus Epsilonema was the most abundant as it existed at the deepest layers between 30 and 40 cm. The organisms belonging to this genus were small and can respond opportunistically given their high reproduction rates; thus they can attain high densities when food was available (VANAVERBEKE et al., 2004). The lower mobility of small organisms such as Epsilonema, hinder their efficient return to the surface after they have been washed away due to the high permeability in the zones of reflective beaches (GHESKIERE et al., 2005). Meanwhile, our horizontal distribution showed that the heterogeneity of the substrate of the same beach, as found at point 3 on Estaleiro, might show the ideal condition for Nematoda abundance.

The separate analysis of the vertical distribution of the different genera in the sediment column revealed that they are not homogeneously distributed. Each genus displays a preference for a given depth range. The preference of some species for given depths as suggested by the occurrence of exclusive genera in each stratum determines the vertical distribution patterns of Nematoda, which can occupy different ecological niches (WIESER 1952). In this study, different genera contributed differently to each stratum, which indicated a possible division of ecological niches.

Daptonema, Theristus and Axonolaimus, which are classified as non-selective deposit feeder nematodes, were responsible for the similarity between superficial strata, whereas the genus Enoploides (predators) made the largest contribution to the intermediate strata. The non-selective deposit feeder genera are confined to the superficial layers of the sediment due to the larger availability and variety of food at the surface and the presence of predators and copepod species (STEYAERT et al., 2003; MARIA et al., 2012).

The community distribution and species abundance in reflective sandy beaches is primarily determined by physical factors, such as sediment characteristics and hydrodynamic stress (MCLACHLAN et al., 1993). The low densities found in the superficial layer were a result of the direct impact of waves in the intertidal zone, which resulted in the vertical migration of organisms to escape the physical stress and the drying characteristics of these zones (URBAN-MALINGA et al., 2004). However, the existence of the Nematoda genera indicated a possible partitioning of the habitat, with preferential zones for non-selective deposit feeder species in superficial layers and predator species in deeper strata.

5. CONCLUSIONS

The meiofauna distribution in reflective beaches is primarily controlled by physical factors. In this scenario, our hypotheses regarding the absence of stratification, and its independence of the horizontal distribution patterns along the intertidal zone in the sediments of reflective beaches were partially rejected. We observed low densities in the top 10 cm of the sediment and that the higher concentration of organisms in intermediate strata (20 and 30 cm) were potentially caused by the migration of the organisms in an attempt to escape the physical stress caused by the wave impact and desiccation characteristic of the intertidal zone. Similarly, the lower density observed below a depth of 30 cm may be attributed to predators. Then the biological interactions assume a secondary role with predator species in deeper strata. However, significant differences in the vertical distribution of organisms were only detected when the lowest interaction between factors was considered, which indicates that the spatial variation on the beaches is a determinant of the vertical distribution of organisms.

ACKNOWLEDGEMENTS

We wish to thank UNIVALI/University of the Itajaí Valley, for providing sampling equipment and assistance. This study was financially and logistically supported by the Itajaí Harbor authorities and also by the Brazilian National Council for Technological and Scientific Development (CNPq - Process 140611/2008-8), which provided MDD's PhD fellowship, and the São Paulo Research Foundation (FAPESP - Process 2012/08581-0) which provided MDD's postdoctoral fellowships and grants.

REFERENCES

1. ALBUQUERQUE, E. F.; PINTO, A. P. B.; PEREZ, A. D. Q.; VELOSO, V. G. Spatial and temporal changes in interstitial meiofauna on a sandy ocean beach of South America. Braz. J. Oceanogr., v. 55, n. 2, p. 121-131, 2007.

2. BRAECKMAN, U.; VAN COLEN, C.; SOETAERT, K.; VINCX, M.; VANAVERBEKE, J. Contrasting macrobenthic activities differentially affect nematode density and diversity in a shallow subtidal marine sediment. Mar. Ecol. Prog. Ser., v. 422, p. 179-191, 2011.

3. CAPUTO, H. P. Mecânica de solos e suas aplicações. 4 ed. Rio de Janeiro: Livros Técnicos e Científicos, 1980. 219 p.

4. CHAPMAN, M. G.; TOLHURST, T. J.; MURPHY, R. J.; UNDERWOOD, A. J. Complex and inconsistent patterns of variation in benthos, micro-algae and sediment over multiple spatial scales. Mar. Ecol. Prog. Ser., v. 398, p. 33-47, 2010.

5. CLARKE, K. R; WARWICK, R. M. Change in marine communities: an approach to statistical analysis and interpretation. 1ed. Plymouth: Plymouth Marine Laboratory, 2001. 144 p.

6. DE JESÚS-NAVARRETE, A.; HERRERA-GÓMEZ, J. Vertical distribution and feeding types of nematodes from Chetumal Bay, Quintana Roo, Mexico. Estuaries, v. 25, n. 6, p. 1131-1137, 2002.

7. DEFEO, O.; GOMEZ, J.; LERCARI, D. Testing the swash exclusion hypothesis in sandy beach populations: the mole crab Emerita brasiliensis in Uruguay. Mar. Ecol. Prog. Ser., v. 212, p. 159-170, 2001.

8. DI DOMENICO, M.; LANA, P. D. C.; GARRAFFONI, A. R. S. Distribution patterns of interstitial polychaetes in sandy beaches of southern Brazil. Mar. Ecol., v. 30, n. 1, p. 47-62, 2009.

9. DI DOMENICO, M.; MARTÍNEZ, A.; ALMEIDA, T. C. M.; MARTINS, M. O.; WORSAAE, K.; LANA, P. C. Response of the meiofaunal annelid Saccocirrus pussicus (Saccocirridae) to sandy beach morphodynamics. Hydrobiologia, v. 734, n. 1, p. 1-16, 2014.

10. FENCHEL, T. M.; RIEDL, R. J. The sulfide system: a new biotic community underneath the oxidized layer of marine sand bottoms. Mar. Biol., v. 7, n. 3, p. 255-268, 1970.

11. FONSECA, G.; SOLTWEDEL, T.; VANREUSEL, A.; LINDEGARTH, M. Variation in nematode assemblages over multiple spatial scales and environmental conditions in Arctic deep seas. Prog. Oceanogr., v. 84, n. 3, p. 174-184, 2010.

12. GALLUCCI, F.; MOENS, T.; FONSECA, G. Small-scale spatial patterns of meiobenthos in the Arctic deep sea. Mar. Biodivers., v. 39, n. 1, p. 9-25, 2009.

13. GHESKIERE, T.; HOSTE, E.; VANAVERBEKE, J.; VINCX, M.; DEGRAER, S. Horizontal zonation patterns and feeding structure of marine nematode assemblages on a macrotidal, ultra-dissipative sandy beach (De Panne, Belgium). J. Sea Res., v. 52, n. 3, p. 211-226, 2004.

14. GHESKIERE, T.; MAGDA, V.; GREET, P.; STEVEN, D. Are strandline meiofaunal assemblages affected by a once-only mechanical beach cleaning? Experimental findings. Mar. Environ. Res., v. 61, n. 3, p. 245-264, 2006.

15. GHESKIERE, T.; VINCX, M.; URBAN-MALINGA, B.; ROSSANO, C.; SCAPINI; F.; DEGRAER, S. Nematodes from wave-dominated sandy beaches: diversity, zonation patterns and testing of the isocommunities concept. Estuar. Coast. Shelf Sci., v. 62, p. 365-375, 2005.

16. GIERE, O. Meiobenthology: The microscopic motile fauna of aquatic sediments. 2. ed. Berlin: Springer-Verlag, 2009.

17. GRAY, J. S.; ELLIOTT, M. Ecology of marine sediments: from science to management. 2. ed. New York: Oxford University Press, 2009. 225 p.

18. HEIP, C.; VINCX, M.; VRANKEN, G. The ecology of marine nematodes. Oceanogr. Mar. Biol. Annu. Rev., v. 23, p. 399-489, 1985.

19. JESUS, B.; MENDES, C. R.; BROTAS, V.; PATERSON, D. M. Effect of sediment type on microphytobenthos vertical distribution: Modelling the productive biomass and improving ground truth measurements. J. Exp. Mar. Biol. Ecol., v. 332, n. 1, p. 60-74, 2006.

20. JOINT, I.; GEE, J.; WARWICK, R. Determination of fine-scale vertical distribution of microbes and meiofauna in an intertidal sediment. Mar. Biol., v. 72, p. 157-164, 1982.

21. KAPUSTA, S. C.; WÜRDIG, N. L.; BEMVENUTI, C. E. Distribuição vertical da meiofauna, inverno e verão, no estuário de Tramandaí-Armazém, RS-Brasil. Acta Limnol. Bras., v. 14, n. 1, p. 81-90, 2002.

22. KLEIN, A. H.; MENEZES, J. T. Beach morphodynamics and profile sequence for headland bay coast. J. Coatal Res., v. 17, n. 4, p. 198-215, 2001.

23. KOTWICKI, L.; DE TROCH, M.; URBAN-MALINGA, B.; GHESKIERE, T.; WESLAWSKI, J. M. Horizontal and vertical distribution of meiofauna on sandy beaches of the North Sea (The Netherlands, Belgium, France). Helgoland Mar. Res., v. 59, n. 4, p. 255-264, 2005.

24. LEGENDRE, P.; LEGENDRE, L. Numerical ecology. 2. ed. Amsterdam: Elsevier Science, 1998. 853 p.

25. LEVIN, L. A. Oxygen minimum zone benthos: adaptation and community response to hypoxia. Oceanogr. Mar. Biol. Annu. Rev., v. 41, p. 1-45, 2003.

26. LORENZEN, C. Determination of chlorophyll and phaeopigments: spectrophotometric equations. Limnol. Oceanogr., v. 12, n. 2, p. 343-346, 1967.

27. MARIA, T. F.; VANAVERBEKE, J.; ESTEVES, A. M.; DE TROCH, M.; VANREUSEL, A. The importance of biological interactions for the vertical distribution of nematodes in a temperate ultra-dissipative sandy beach. Estuar. Coast. Shelf Sci., v. 97, p. 114-126, 2012.

28. MASSELINK, G.; PULEO, J. A. Swash-zone morphodynamics. Cont. Shelf Res., v. 26, n. 5, p. 661-680, 2006.

29. MASSELINK, G.; RUSSEL, P. Flow velocities, sediment transport and morphological change in the swash zone of two contrasting beaches. Mar. Geol., v. 227, n. 3, p. 227-240, 2006.

30. MCLACHLAN, A. A quantitative analysis of the meiofauna and the chemistry of the redox potential discontinuity zone in a sheltered sand beach. Estuar. Coast. Mar. Sci., v. 7, p. 275-c, 1978.

31. MCLACHLAN, A. Water filtration by dissipative beaches. Limnol. Oceanogr., v. 34, p. n. 4, 774-780, 1989.

32. MCLACHLAN, A.; BROWN, A. C. The ecology of sandy shores. Burlington: Academic Press, 2006. 392 p.

33. MCLACHLAN, A.; JARAMILLO, E.; DONN, T. E.; WESSELS, F. Sandy beach macrofauna communities and their control by physical environment: a geographical comparison. J. Coatal Res., v. 15, p. 27-38, 1993.

34. NEVES, I. F.; ROCHA, O.; ROCHE, K. F.; PINTO, A. A. Zooplankton community structure of two marginal lakes of the river Cuiabá (Mato Grosso, Brazil) with analysis of Rotifera and Cladocera diversity. Braz. J. Biol., v. 63, n. 2, p. 329-343, 2003.

35. NICHOLAS, W. L.; HODDA, A. M. The free-living nematodes of a temperate, high-energy, sandy beach: faunal composition and variation over space and time. Hydrobiologia, v. 394, p. 113-127, 1999.

36. OLAFSSON, E.; ELMGREN, R. Seasonal dynamics of sublittoral meiobenthos in relation to seasonality in a coastal marine ecosystem. Oecologia, v. 67, p. 157-168, 1997.

37. PALMER, M. Water flow and the vertical distribution of meiofauna: a fl experiment. Estuaries, v. 9, p. 225-228, 1986.

38. PETTERMANN, R. M.; AMIN JR, A. H.; BEAUMORD, A. C.; STREFLING, L. Geology, Land Uses and Quaternary Paleogeographic Evolution of Estaleiro Beach in Santa Catarina State, Southern Brazil, J. Coatal Res., v. 39, p. 970-975, 2006.

39. PLATT, H. M. Vertical and horizontal distribution of free-living marine nematodes from Strangford Lough, Northern Ireland. Cah. Biol. Mar., v. 18, p. 261-c, 1977.

40. REISE, K.; AX, P. A meiofaunal "thiobios" limited to the anaerobic sulfide system of marine sand does not exist. Mar. Biol., v. 54, p. 225-237, 1979.

41. RODRIGUEZ, J. G.; LOPEZ, J.; JARAMILLO, E. Community structure of the intertidal meiofauna along a gradient of morphodynamic sandy beach types in southern Chile. Rev. Chil. Hist. Nat., v. 74, p. 885-897, 2001.

42. SCHLACHER, T. A.; SCHOEMAN, D. S.; DUGAN, J.; LASTRA, M.; JONES, A.; SCAPINI, F.; MCLACHLAN, A. Sandy beach ecosystems: key features, sampling issues, management challenges and climate change impacts. Mar. Ecol., v. 29, n. 1, p. 70-90, 2008.

43. SHARMA, J; WEBSTER, J. M. The abundance and distribution of free-living nematodes from two Canadian Pacific beaches. Estuar. Coast. Shelf Sci., v. 16, p. 217-227, 1983.

44. SHORT, A. D. Handbook of beach and shoreface morphodynamics. New York: Wiley, 1999. 379 p.

45. SOMERFIELD, P. J; WARWICK, R. M. Meiofauna in marine pollution monitoring programmes: a laboratory manual. Lowestoft: Directorate of Fisheries Research (MAFF), 1996. 71 p.

46. STEYAERT, M.; GARNER, N.; VAN GANSBEKE, D.; VINCX, M. Nematode communities from the North Sea: environmental controls on species diversity and vertical distribution within the sediment. J. Mar. Biol. Ass. U. K., v. 79, p. 253-264, 1999.

47. STEYAERT, M.; VANAVERBEKE, J.; VANREUSEL, A.; BARRANGUET, C.; LUCAS, C.; VINCX, M. The importance of fine-scale, vertical profiles in characterising nematode community structure. Estuar. Coast. Shelf Sci., v. 58, n. 2, p. 353-366, 2003.

48. SUGUIO, K. Introdução à sedimentologia. São Paulo: Edgard Blücher/EDUSP, 1973. 317 p.

49. URBAN-MALINGA, B.; KOTWICK, L.; GHESKIERE, T.; JANKOWSKA, K.; OPALINSKI, K.; MALINGA, M. Composition and distribution of meiofauna, including nematode genera, in two contrasting Arctic beaches. Polar. Biol., v. 27, n. 8, p. 447-457, 2004.

50. VANAVERBEKE, J.; SOETAERT, K.; VINCX, M. Changes in morphometric characteristics of nematode communities during a spring phytoplankton bloom deposition. Mar. Ecol. Prog. Ser., v. 273, p. 139-146, 2004.

51. VIEIRA, D. C.; FONSECA, G. The Importance of Vertical and Horizontal Dimensions of the Sediment Matrix in Structuring Nematodes Across Spatial Scales. PLoS ONE, v. 8, n. 10, p. e77704, 2013.

52. WIESER, W. Die Beziehung zwischen Mundhöhlengestalt, Ernährungsweise und Vorkommen bei freilebenden marinen Nematoden: eine ökologisch-morphologische Studie. Ark. Zool., v. 4, p. 439-484, 1952.

11

ULTRASOUND-ASSIST EXTRUSION METHODS FOR THE FABRICATION OF POLYMER NANOCOMPOSITES BASED ON POLYPROPYLENE/MULTI-WALL CARBON NANOTUBES

Carlos A. Ávila-Orta 1,*, Zoe V. Quiñones-Jurado 2, Miguel A. Waldo-Mendoza 2, Erika A. Rivera-Paz 2, Víctor J. Cruz-Delgado 1, José M. Mata-Padilla 1, Pablo González-Morones 1 and Ronald F. Ziolo 1

1 Centro de Investigación en Química Aplicada, Blvd. Ingeniero Enrique Reyna Hermosillo. 140, Col. San José de los Cerritos, Saltillo, Coahuila CP 25294, Mexico

2 Innovación y Desarrollo en Materiales Avanzados A. C., Grupo POLYnnova, Carr. San Luis Potosí-Guadalajara 1510, Nivel 3, Local 12, Lomas del Tecnológico, San Luis Potosí, SLP, CP 78211, Mexico

ABSTRACT

Isotactic polypropylenes (iPP) with different melt flow indexes (MFI) were used to fabricate nanocomposites (NCs) with 10 wt % loadings of multi-wall carbon nanotubes (MWCNTs) using ultrasound-assisted extrusion methods to determine their effect on the morphology, melt flow, and electrical properties of the NCs. Three different types of iPPs were used with MFIs of 2.5, 34 and 1200 g/10 min. Four different NC fabrication methods based on melt extrusion were used. In the first method melt extrusion fabrication without ultrasound assistance was used. In the second and third methods, an ultrasound probe attached to a hot chamber located at the exit of the die was used to subject the sample to fixed frequency and variable frequency, respectively. The fourth method is similar to the first method, with the difference being that the carbon nanotubes were treated in a fluidized air-bed with an ultrasound probe before being used in the fabrication of the NCs with no ultrasound assistance during extrusion. The samples were characterized by MFI, Optical microscopy (OM), Scanning electron microscopy (SEM), Transmission electron microscopy (TEM), electrical surface resistivity, and electric charge. MFI decreases in all cases with addition of MWCNTs with the largest decrease observed for samples with the highest MFI. The surface resistivity, which ranged from 1013 to 105 Ω/sq, and electric charge, were observed to depend on the ultrasound-assisted fabrication method as well as on the melt flow index of the iPP. A relationship between agglomerate size and area ratio with electric charge was found. Several trends in the overall data were identified and are discussed in terms of MFI and the different fabrication methods.

KEYWORDS

polypropylene; carbon nanotubes; melt flow index; extrusion; ultrasound; surface resistivity; electric charge

1. INTRODUCTION

Carbon nanotube/polymer nanocomposites [1,2,3,4,5,6] have been studied for over a decade due to their outstanding properties, which result from the combination of a soft phase (polymer matrix) and a functional phase (dispersed nanomaterials). Among other properties and attributes, carbon nanoparticles can provide mechanical strength [7,8] to polymers, act as gas barriers in polymers [9,10], and effect electron transport in polymers [11]. Nevertheless, obtaining outstanding properties requires an effective dispersion of the carbon nanotubes (CNTs) in the polymer matrix, which is easily hindered by particle-particle attraction to form agglomerates and by the lack of affinity between the nanoparticles and the matrix.

To address and solve this problem, a number of fabrication methods, based on *in situ* polymerization [12,13], melt blending [14,15], and solution mixing [16,17] have been developed in order to have an adequate dispersion that optimizes the targeted properties. On the other hand, polymers, as well as CNTs, are often subjected to treatments prior to nanocomposite fabrication in order to reduce the free energy difference between the components. In this sense, the application of high-frequency sound waves during pretreatment, as well as during fabrication, could provide an effective energy source to improve the dispersion of the functional CNTs in the polymer matrix.

A few examples of this approach can be found in the literature. For example, prior to fabrication using a solution mixing method, nanoparticles in solution are often pretreated ultrasonically [18]. On the other hand, nanoparticles can be de-agglomerated, say in a beaker, using an ultrasound probe in an open air system that creates a fluidized air bed [19,20]. Thereafter, the treated nanoparticles can be used in melt-blending nanocomposite fabrication.

Another example is the direct application of ultrasound waves during melt blending in an extrusion process [21,22,23]. This is typically done at a single or fixed acoustic frequency. However, other fabrication methods use a number of different, sweep, or variable frequencies instead of a single frequency, in an expansion chamber, which seems to improve the dispersion even more [24,25]. As we noted earlier, this behavior might be related to the fact that there is a resonant frequency or frequencies that might be associated with a given length of polymer chain (vibrational motion of chains mechanism) rather than to cavitation effects [26]. Bearing in mind that the polymer chain lengths have a wide distribution, we note that a number of different frequencies is necessary to change the chain conformation from random to elongate. Once the given acoustic frequency is no longer applied, the chain conformation relaxes to the random conformation. This process or mechanism might improve chain mobility and polymer access by the CNTs and, therefore, lead to more efficient dispersion of the nanoparticles.

Up to this point, it is has not been possible to compare each of the ultrasound-assist methods since they have been used in different polymer systems or under different conditions. Given the paucity of data and difficulty in comparing different ultrasound extrusion fabrication studies, we focused on a model system to determine the effects of ultrasound-assist methodology on the properties of nanocomposites formed by the extrusion process. For this purpose, we choose a system based on isotactic polypropylene (iPP) with three different melt flow index as the matrix and on multi-walled carbon nanotubes (MWCNTs) as the functional additives. Such polymer/CNT systems have been widely studied during the last decade due to their industrial application and scientific interest. In particular, MWCNTs are excellent electric conductors, which can improve the electrical properties of non-conductive iPP, and polymer chain mobility can be monitored and controlled by using iPP with different melt flow index.

2. RESULTS AND DISCUSSION

2.1. Melt Flow Index

The melt flow index (MFI) for iPP homopolymers and iPP/MWCNT nanocomposites fabricated with different methods are shown in Table 1. The four fabrication methods are designated as W-U for fabrication without ultrasound, F-U for fixed-frequency ultrasound fabrication, V-U for variable-frequency ultrasound fabrication and PT for pretreated carbon nanotubes, where the carbon nanotubes were treated in a fluidized air-bed with an ultrasound probe before being used in the fabrication of the nanocomposites (NCs), with no ultrasound assistance during the extrusion. The usual designation for multi-wall carbon nanotubes (MWCNTs) was shortened to MNT and used hereafter.

The results are explained bearing in mind the power law relationship between MFI and molecular weight, where a steep decay in molecular weight is observed at low MFI, while minor changes in molecular weight are observed at intermediate and high MFI [27,28,29]. MFI for the as-received iPP homopolymers was evaluated prior to nanocomposite fabrication. The values obtained in this study were the average value of measurements taken on five different portions of each suppliers' iPP and agree well with the suppliers' value with only minor deviations. In the case of iPPMFI = 2.5, a slight increase in MFI is obtained after processing in normal conditions, *i.e.*, fabrication without ultrasound-assist (W-U), probably due to chain scission. Nonetheless, the use of ultrasound waves seems to decrease the MFI. This can be due to the fact that ultrasound waves can either break or bond some of the polymer chains. In the latter case, an increase in the chain length would indeed diminish homopolymer fluidity. For samples with a higher MFI, iPPMFI = 34, the extrusion processing effect and the use of ultrasound waves do not seem to affect, in a marked manner, the fluidity of the sample. Finally, melt extrusion of iPPMFI = 1200 seems to reduce its fluidity probably due to the bonding of short polymer chains. Furthermore, in this particular case, ultrasound application seems to reduce chain length thus increasing MFI.

Regardless of the fabrication method, the MFI for iPP/MNT nanocomposites is much lower compared to the pure iPP homopolymers. In the case of nanocomposites fabricated with homopolymers iPPMFI = 2.5 and iPPMFI = 34 the MFI values obtained are close to half of the values obtained for the pure homopolymers, while in the case of iPPMFI = 1200 the values for the nanocomposites are up to ten times smaller. In a first approach, it can be considered that MNTs act as mobility barriers hindering the transport of polymer chains, thus increasing viscosity [28]. Thus, MFI tends to decrease with the presence of a large amount (10 wt %) of MNTs. Nevertheless, the complexity of processing imposes barriers for a full understanding of the system, regarding the effect of the ultrasound-assist fabrication method.

Table 1. Melt Flow Index (g/10 min) for isotactic polypropylenes (iPP) homopolymers, and iPP/multi-wall carbon nanotubes (MNT) nanocomposites fabricated by different methods.

Sample	Fabrication Method				
	As Received	W-U	F-U	V-U	PT
$iPP_{MFI = 2.5}$	2.70	3.00	2.91	2.85	3.00
$iPP_{MFI = 34}$	34.34	35.60	37.57	35.73	35.60
$iPP_{MFI = 1200}$	1208.87	1013.27	1250.85	1281.86	1013.27
$iPP_{MFI = 2.5}$/MNT	-	1.49	1.59	1.55	1.42
$iPP_{MFI = 34}$/MNT	-	17.53	18.63	18.50	18.40
$iPP_{MFI = 1200}$/MNT	-	179.40	159.12	171.61	160.01

2.2. Morphology

The morphology of iPP/MNT polymer nanocomposites was analyzed in terms of the MNT agglomerate size, the area ratio of the MNT agglomerates, and dispersion of the MNTs in the iPP matrix. Although MNTs have a diameter in the scale of nanometers, their length is typically on the order of 1–20 μm and they tend to agglomerate due to van der Waals forces. Thus, it is important to characterize the size of agglomerates to determine the effect of ultrasound-assist fabrication.

Optical microscopy micrographs are shown in Figure 1 for samples of iPPMFI = 2.5 prepared using different fabrication methods. The black spots correspond to MNT agglomerates, while the white/greyish colors correspond to the polymer.

The figure shows that the MNT agglomerates are markedly larger for the methods W-U and F-U than for methods PT and V-U; nonetheless, the direct observation of large agglomerates can lead to erroneous conclusions. Thus, in order to have a numerical comparison, micrographs for each method were analyzed, and the results are shown in Table 2. The agglomerate size of the MNTs in the polymer nanocomposites was close to 10 μm with a few deviations. Menzer [30] obtained similar values for as-received and ball-milled shortened carbon nanotubes. The lowest agglomerate size, 8.56 μm, was determined for the sample iPPMFI = 2.5 prepared using the V-U method, while the highest value of 12.95 μm was obtained for iPPMFI = 34using the PT method. In general, for the different homopolymers, agglomerate size is practically the same when no ultrasound is used, independent of the MFI. However, in the case of samples prepared with ultrasound-assist fabrication, either with fixed or variable frequency, the lower values were obtained for samples with the lowest MFI. Thus, the shear force of long polymer chains generated during melt extrusion seem to help agglomerate breaking, whereas the shear force of short polymer chains does not. Additionally, it can be noticed that a combination of high MFI (iPPMFI = 1200) using the V-U method results in a large agglomerate size. On the other hand, the MNT area ratio does not show a clear tendency (Table 3). Nonetheless, we note that the samples prepared using V-U show the lowest values, while iPPMFI = 2.5 prepared using PT show the highest value. It is also worth noting that increasing the fluidity of the sample the MNT area ratio does not change markedly in the cases of samples prepared with W-U and F-U. However, in the case of samples prepared using V-U, it can be noticed that increasing the MFI, the area ratio increases, while in the case of PT the values decrease. This behavior suggests that both, the combination of low fluidity homopolymer plus variable ultrasound in the former, and the use of high fluidity homopolymer with de-agglomerated MNT (without ultrasound), are effective ways to reduce the MNT area ratio.

On the other hand, the morphology of polymer nanocomposites was observed on the surface of fractured samples prepared by W-U (Figure 2). Agglomerates of 1 μm as well as individual MNTs are seen for iPPMFI = 2.5/MNT, while for iPPMFI = 1200/MNT only individual MNTs are observed. This behavior suggests that extrusion of high viscosity samples is not enough to break apart agglomerates, while the use of a low viscosity polymer helps to disentangle MNT bundles [28].

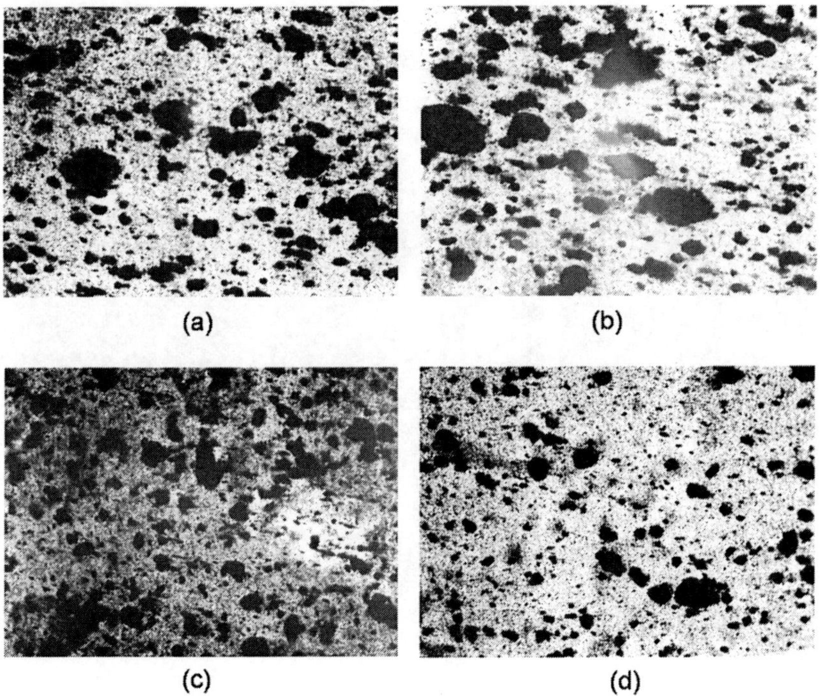

Figure 1. Optical micrographs for nanocomposites for isotactic polypropylenes/multi-wall carbon nanotubes (iPP/MNT) (10 wt %) fabricated by melt extrusion with and without ultrasound-assist fabrication for iPPMFI = 2.5. (**a**) W-U; (**b**) F-U; (**c**) PT; (**d**) V-U.

Table 2. Agglomerate size (μm) for iPP/MNT nanocomposites fabricated by different methods.

iPP_{MFI}/MNT MFI (g/10 min)	Fabrication Method			
	W-U	F-U	V-U	PT
2.5	10.14 (±2.12)	9.19 (±2.51)	8.56 (±5.40)	9.47 (±2.73)
34	9.60 (±1.02)	9.70 (±2.15)	10.63 (±1.21)	12.95 (±0.80)
1200	10.09 (±0.76)	10.35 (±1.46)	12.02 (±2.69)	10.11 (±2.98)

Table 3. Area ratio for iPP/MNT nanocomposites fabricated by different methods.

iPP_{MFI}/MNT MFI (g/10 min)	Fabrication Method			
	W-U	F-U	V-U	PT
2.5	37.96 (±9.89)	33.97 (±10.01)	11.18 (±6.78)	57.34 (±7.95)
34	41.16 (±7.69)	35.22 (±8.30)	17.45 (±2.07)	31.46 (±13.06)
1200	33.79 (±5.42)	35.48 (±6.27)	22.65 (±4.24)	24.29 (±2.41)

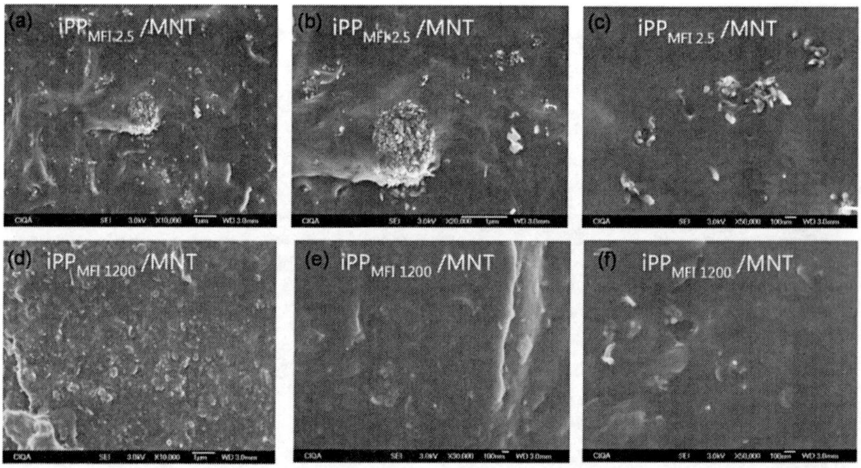

Figure 2. Scanning electron microscopy (SEM) micrographs of the surface fracture for nanocomposites of iPP/MNT (10 wt %) fabricated by melt extrusion without ultrasound-assist fabrication. (**a**) iPPMFI2.5/MNT at 10K magnification; (**b**) iPPMFI2.5/MNT at 20K magnification; (**c**) iPPMFI2.5/MNT at 50K magnification; (**d**) iPPMFI1200/MNT at 10K magnification; (**e**) iPPMFI1200/MNT at 30K magnification; (**f**) iPPMFI1200/MNT at 50K magnification.

2.3. Electrical Behavior

Electrical surface resistivity (SR) and electrostatic charge of the iPP homopolymers and polymer nanocomposites were measured for the three different MFIs and fabrication methods and evaluated within the limits of experimental measurement procedures.

2.3.1. Surface Resistivity

Electrical surface resistivity (SR) values of the iPP homopolymers and polymer nanocomposites are presented in Table 4. SR values for the iPP polymers appear independent of the MFI and show the polymers to be insulators at ≥1013 Ω/sq as expected. The SR values for the iPPMFI = 2.5/MNT samples, however, are at least an order of magnitude less, at 1012 Ω/sq, regardless of the fabrication method. Although the samples are insulators, they appear to be able to transport more charge than the iPP homopolymers.

Table 4. Surface resistivity (Ω/sq) for iPP homopolymers, and iPP/MNT nanocomposites fabricated by different methods.

Sample	Fabrication Method			
	W-U	F-U	V-U	PT
$iPP_{MFI = 2.5}$	$\geqslant 10^{13}$	-	-	-
$iPP_{MFI = 34}$	$\geqslant 10^{13}$	-	-	-
$iPP_{MFI = 1200}$	$\geqslant 10^{13}$	-	-	-
$iPP_{MFI = 2.5}$/MNT	10^{12}/(I)	10^{12}/(I)	10^{12}/(I)	10^{12}/(I)
$iPP_{MFI = 34}$/MNT	10^{12}/(I)	10^{9}/(SD)	10^{6}/(SD)	10^{5}/(C)
$iPP_{MFI = 1200}$/MNT	10^{5}/(C)	10^{6}/(SD)	10^{8}/(SD)	10^{5}/(C)

I = Insulator material; SD = static dissipative material; C = conductive material.

A drastic change in SR is observed for the samples with higher MFI, where a marked decrease in resistivity is observed. For example, for samples iPPMFI = 34/MNT, a decrease of up to seven orders of magnitude is observed when the sample is prepared by the PT fabrication method (105 Ω/sq) to give a slightly more conductive material. Application of ultrasound waves in the melt also decreases the SR, which is more pronounced when a variable frequency is applied (106 Ω/sq) than when a fixed frequency is used (109 Ω/sq). In this case, however, the materials produced are static-dissipative. On the contrary, the sample prepared without ultrasound assist shows an SR of 1012 Ω/sq compared with the sample with lower MFI and is also an insulator.

For samples of iPPMFI = 1200/MNT, a remarkable reduction in electrical resistivity is observed (105 Ω/sq) for the sample prepared without ultrasound-assist as compared to samples with lower MFI using the same fabrication method. The same result was obtained for samples prepared by the PT method, which already showed this resistivity level for samples with MFI of 34 g/10 min. In the case of ultrasound-assisted methods, slightly higher values were obtained, *i.e.*, for static dissipative materials.

A wide range of electrical surface resistivity values is found in the literature for iPP/MNT systems prepared using melt blending. Since reported electrical resistivity or conductivity values can refer to volume or surface resistivity, we converted the reported data to electrical surface resistivity for comparison purposes. For example, Steinmann [31] reported a value of 102 Ω/sq for MNT loadings of 7.5 wt % using an iPP with MFI of 25 g/10 min, while Wang [32] used an iPP with an MFI of 1.2 obtaining a value of 109 Ω/sq.

Although it might be argued that MFI plays an important solitary role in determining SR, it is known that a number of factors contribute to the final surface resistivity of the material, such as (1) the intrinsic properties of the iPP such as the molecular weight, which is intimately linked to MFI; (2) the MNT type, outer diameter, length, purity, agglomerate size, *etc.*; and (3) the fabrication parameters including type of machine, temperatures, shear rates, *etc.* In this regard, different authors [33,34,35] reported values of 105 Ω/sq for MNT concentrations *ca.* 5–6 wt % and each of them used different MFI, types of nanotubes and processing parameters. Finally, Pan [36] used a 10 wt % concentration of MNT in iPP with an MFI of 35 g/10 min that showed an electrical resistivity of 106 Ω/sq. The materials used in Pan's study are at least comparable to the ones used here, as are the reported SRs. Thus, the experimentally-observed SR values in this study compare favorably with SR values reported by other authors.

From the SR results obtained in this study, we note some trends that merge regarding the fabrication method and MFI of iPP NCs. First of all, insulator-like materials are obtained when a low MFI of 2.5 g/10 min is used, regardless of the fabrication method. The fabrication method, however, plays a key role in decreasing the electrical resistivity when higher MFI polypropylenes (MFI = 34 and 1200 g/10 min) are used, where static dissipative and mildly-conductive materials are obtained. For a discussion of the transition from insulating to conducting materials for samples prepared using solution blending, see [18].

Static-dissipative materials were obtained for samples fabricated with ultrasound-assist methods in the melt, while insulator and conductive materials were obtained when samples were fabricated without the presence of ultrasound waves in the melt. In the case of samples fabricated without ultrasound-assist, an insulator-like material is obtained for an MFI of 34 g/10 min, while a slightly conductive material is obtained when the MFI is 1200 g/10 min. Finally, mildly-conducting materials were obtained when the MNTs were de-agglomerated in the fluidized air bed for both MFIs.\

2.3.2. Electrical Charge

It is well know that vinyl polymers, such as isotactic polypropylene, can be easily charged via friction rubbing, for example, with a wool fabric. Following this simple methodology, we used 56 mm by 0.1 mm square plaques of iPP homopolymers and iPP/MNT NCs and rubbed them following the empirical procedure described in the experimental section to test if they could acquire and sustain an electrical charge. Charge results in kV are shown in Table 5. It is seen that the iPP homopolymers are charged negatively from their initial charged state. These values represent an average of three individual measurements on each sample and appear to trend more negatively for higher MFI.

Table 5. Electric charge (kV) for iPP homopolymers, and iPP/MNT nanocomposites fabricated by different methods.

Sample	Fabrication Method			
	W-U	F-U	V-U	PT
$iPP_{MFI = 2.5}$	0.00/ −0.40	-	-	-
$iPP_{MFI = 34}$	0.33/ −0.48	-	-	-
$iPP_{MFI = 1200}$	−0.02/ −1.22	-	-	-
$iPP_{MFI = 2.5}$/MNT	0.00/ −0.25	0.04/ −0.10	−0.01/ −1.04	0.00/ −0.03
$iPP_{MFI = 34}$/MNT	0.00/0.00	0.00/0.00	0.00/0.00	0.00/0.00
$iPP_{MFI = 1200}$/MNT	0.00/0.00	0.00/0.00	0.00/0.00	0.00/0.00

The slash (/) indicates charge before and after friction rubbing.

On the other hand, samples fabricated using iPPMFI = 2.5/MNT appear to charge negatively after rubbing, with the final charge depending on the fabrication method. The initial charge ranges from 0 to +40 V, while the final charge ranges from −30 to −1040 V. The lowest final charge was obtained for samples prepared using the PT method, while the V-U method rendered samples with the highest charge. The values obtained for samples prepared using the W-U and F-U methods were −250 and −100 V, respectively.

Interestingly, the electric charge measured for all of the iPPMFI = 34/MNT and iPPMFI = 1200/MNT NCs was zero within the time frame of the measurement, irrespective of the fabrication method. These observations can be explained with the aid of the so-called MNT area ratio, which is the ratio of the area of the MNT agglomerates to that of the sample as determined by optical microscopy, and the surface resistivity results.

As expected, iPP can be easily charged since they are insulating materials. However, the electric charge of samples prepared using iPPMFI = 2.5/MNT appears to depend on the fabrication method. The lowest value was obtained using de-agglomerated MNT in the gas phase (PT) while the highest value was obtained for the method that used variable frequencies in the molten state (V-U). Intermediate values are found for the two remaining methods. This behavior can be associated with the MNT area ratio, where the highest value was obtained for PT samples, and the lowest for the V-U samples, while intermediate values were found for the other two methods. Thus, it appears that a correlation emerges between MNT area ratio and the electric charge at low MFI, *i.e.*, a large area ratio means that the surface of the sample is covered with a greater amount of carbon nanotubes that are not easily charged, while a small value suggests that large portions of iPP can be charged. It is not surprising to note that even though the SR values for these samples are all 1012 Ω/sq, indicating their insulator nature, the materials behave differently in terms of electrical charge, which relates to triboelectric effects. To support this idea, TEM images were collected for the sample iPPMFI = 2.5/MNT prepared using V-U and the one prepared using W-U (see Figure 3). In terms of the extrusion process, we believe that both samples have no difference right before the influence of the ultrasound waves. Once the sample enters into the influence of ultrasound, a change in behavior is expected, as seen in Section 2.2. Many images were collected and showed a trend that untreated samples (W-U) are more loosely packed than the treated ones (V-U). We speculate that during the ultrasound irradiation, individual MNT are displaced from the agglomerates by erosion and end up in the polymer phase where there is a minimal concentration of CNTs, but at the same time, increasing the density of packaging. A larger distance separates these densely packaged agglomerates (see area ratio in Table 3), then showing a charge behavior. However, the agglomerates that were untreated (W-U) were not prone to show this effect and maintained a large agglomerate size with minor distance between agglomerates (see area ratio in Table 3).

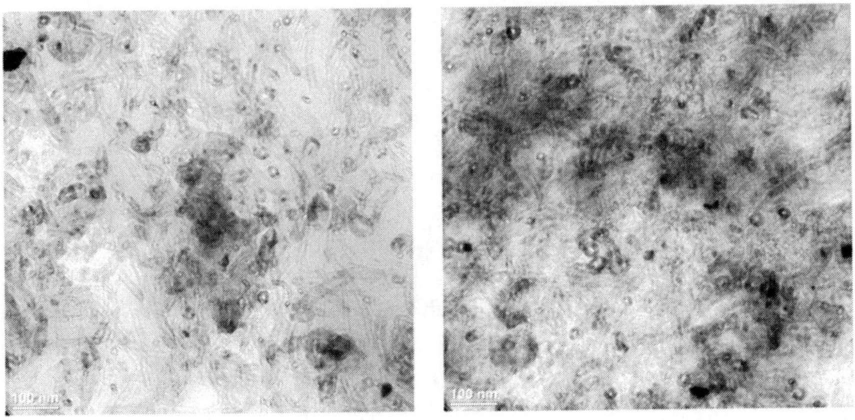

(a) (b)

Figure 3. TEM micrographs of a section for nanocomposites of iPPMFI = 2.5/MNT (10 wt % fabricated by melt extrusion (**a**) without, and (**b**) with, V-U ultrasound-assist fabrication.

Finally, we note that regardless of the MFI used and the fabrication method, samples prepared with iPP with MFI of 34 and 1200 g/10 min showed no initial and no final electrical charge, even for the sample iPPMFI = 34/MNT prepared using W-U that show insulating behavior, and for samples with static-dissipative values. The results obtained in this study suggest that both electrical resistivity and electrical charge are necessary to determine the electrical behavior for hydrophobic materials such as iPP/MNT NCs.

3. EXPERIMENTAL SECTION

Materials: Three iPP homopolymers with different MFI were used. The homopolymers were provided by Exxon (Houston, TX, USA), grade PP4712 E1 (MFI = 2.5 g/10 min), Indelpro grade PL835N (MFI = 34 g/10 min) and, finally, Indelpro grade Profax PL 505 (MFI = 1200 g/10 min). These three materials were designated as iPPMFI = 2.5, iPPMFI = 34 and iPPMFI = 1200, according to their manufacturer reported MFI. Multiple-wall carbon nanotubes grade IGCNTs were purchased from Cheaptubes, Cambridgeport, VT, USA, with an outer diameter of 20–40 nm, length 10–30 μm, and 90% purity.

Methods: iPP/MNT nanocomposites with 10 wt % of MNT were fabricated using a laboratory twin-screw extruder model 24MC (Thermo Scientific, Stone, UK) with an L/D = 40:1, a speed of 100 rpm, and a plain temperature profile of 220 °C along the barrel. iPP pellets and MNT powder were fed at the main feeder of the extruder at a feed rate of 2 kg/h and 200 g/h, respectively. A specially-designed expansion chamber for ultrasonic treatment was adapted at the die zone and kept at 220 °C, which allowed ultrasound application in the melt. Details of this device were reported elsewhere [23]. The 750 W ultrasound generator was made in-house and can perform a sweep function to scan a frequency range with discrete increments [24].

Four different fabrication methods were used to fabricate the polymer nanocomposites. In the first method, the processing of the materials was carried out with no ultrasound-assist and materials were used as received. This method was designated W-U for "without ultrasound". In the second and third methods, the materials were used as received, however the ultrasound generator was switched on in both cases, and a commercial transducer from Advanced Sonic Processing Systems with a main frequency of 20 ° 0.2 kHz was used. A fixed frequency of 20 ± 0.1 kHz was used in the second method, while a frequency scan of 20–50 kHz (with 0.1 scan step and 250 cycles per scan) was used in the third method. In both cases the generator output was 500 W. These methods were designated as F-U for fixed frequency ultrasound-assist fabrication and V-U for variable-frequency ultrasound-assist fabrication. The fourth method was similar to the first method in the sense that the materials were not subjected to ultrasound during melt extrusion. However, the MNTs were treated in a fluidized air-bed with an ultrasound probe before being used in the fabrication of the NCs. This method promotes the size reduction of the MNT agglomerates and was designated as PT for pretreatment.

Molded plaques 15 cm × 15 cm × 0.1 mm were obtained in a mold using a hydraulic compression press model Q230H-X4A (PHI, Industry, CA, USA) at 200 °C. Molded plaques were cut into 56 mm squares and were conditioned at 23 °C and 50% RH 50% during 48 h to homogenize the samples prior to evaluation at ambient conditions.

3.1. Characterization Techniques

Melt Flow Index (MFI). Melt flow of materials was evaluated using an extrusion plastometer in accordance with the ASTM D 1238 standard method [37]. For this purpose a Kayeness Melt Indexer was used at 230 °C, with a weight of 2.16 kg. Five measurements were made and the average value is reported.

Optical Microscopy (OM). This technique was used to determine the size (mean diameter, *D*m) of agglomerate in iPP/MNT nanocomposites made using the four different fabrication methods. Image Pro Plus software from Media Cybernetics (Rockville, MD, USA), was used to analyze the images. OM observations were performed in an optical microscope, Olympus BX53 (Tokyo, Japan).

Scanning Electron Microscopy (SEM). SEM observations were used to determine the nanostructure within the matrix of the polymer nanocomposites. Pieces of compression molded plaques were cryo-fractured and coated with gold. The SEM observations were realized directly on the external surface and cryo-fractured surface of the iPP/MNT nanocomposites fabricated without ultrasound (W-U). SEM micrographs were obtained using a field emission scanning electron microscope, JSM-74101F- JEOL VR (JEOL, Tokio, Japan) with a secondary electron detector (SEI) using a voltage of 4.0 kV.

Electrical Surface Resistivity. This property was measured on homopolymers and polymer nanocomposites using the 56 mm square molded plaques described above and a Surface Resistance Meter from Antistat (Ipswich, UK) model ANT093-0050. The apparatus was placed on the surface of the molded plaques for 30 s; each sample was measured three time and the values averaged.

Electrostatic Charge. Electrostatic charge was evaluated with a portable electrostatic field meter Simco-Ion model FMX-004 (Simco Ion, Industrial Group, Hatfiel, PA, USA), in the low range mode (0–3.00 kV). The same samples employed for electrical resistance were used in this test and three independent measurements were carried out on each sample and averaged. The samples were left to equilibrate for 2 h in a small open chamber protected from dust and draft at ambient conditions (23 °C, 40% RH and an O2 partial pressure of ~135 mm Hg in Saltillo). Each sample was clamped to a grounded steel ring and friction rubbed edge to edge once with a clean wool fabric cloth applying a steady pressure.

Transmission Electron Microscopy (TEM). TEM was used to analyze the inner morphology of MNT agglomerates in the PP/MNT nanocomposites. Thin sections (~90 nm) of compression molded plaques of PPMFI=2.5/MNT, prepared without ultrasound (W-U) and variable frequency ultrasound (V-U), were cryo-microtomed. TEM micrographs were obtained using a TEM Titan 89-300 (FEI, Hillsboro, OR, USA).

3.2. Complementary Characterization Techniques.

Polypropylene homopolymers with different melt flow indexes (2.5, 34 and 1200 g/10 min) and iPP/MNT nanocomposites prepared by different methodologies were analyzed by complementary techniques as follows. (Results shown in theSupplementary Materials).

Thermogravimetric Analysis (TGA). TGA was used to analyze the thermal behavior. Analyses were performed using a Q500 analyzer (TA Instruments, Newcastle, DE, USA) with a heating rate of 10 °C/min and gas flow rate of 50 mL/min. Sample runs were carried out from 40 to 600 °C under an N2 atmosphere and under an O2 atmosphere above 600 °C. The results are shown in Figure S1 and Tables S1, S2 and S3 of Supplementary Materials.

Differential Scanning Calorimetry (DSC). A DSC model Discovery from TA Instruments was used to analyze the melting and crystallization behavior. The analyses were performed at a heating and cooling rate of 10 °C/min in the range of 25–200 °C. Pieces of each material were weighed (8 ± 1 mg) and sealed within aluminum sample pans before experiments. The equipment was calibrated with an indium standard in an N2 atmosphere. The results are shown in Figure S2 and Tables S4, S5 and S6 of Supplementary Materials.

Dynamic Mechanical Analysis (DMA). Dynamic mechanical properties (storage modulus (E'), loss modulus (E"), and damping, tan δ, which is the ratio of E"/E'), were determined using a dynamic mechanical analyzer model Q800 (TA Instruments). Rectangular sections of compression-molded parts (3 cm × 1.2 cm × 0.1 cm) were used to perform the analysis. Each analysis was carried out at a heating rate of 3 °C/min from −50 to 120 °C, an oscillation amplitude of 0.3 to 0.5 mm, and a frequency of 1 Hz. The results are shown in Figure S3 and Tables S7, S8 and S9 of Supplementary Materials.

4. CONCLUSIONS

The present study showed that methods used for ultrasound-assist extrusion fabrication influence the morphology and electrical properties of isotactic polypropylenes (iPP) with different MFI and of nanocomposites (NCs) with 10 wt % loadings of multi-wall carbon nanotubes using the same polymers. Four fabrication methods were studied and compared, two with variable and fixed frequency ultrasound and two with no ultrasound assist during extrusion. The MFI of the NCs decreased to half the value for low (2.5 g/10 min) and medium (34 g/10 min) MFI polypropylene, and an order of magnitude for high (1200 g/10 min) MFI resin, and appeared independent of the NC fabrication method. The lowest agglomerate size was found for the NC prepared with iPPMFI = 2.5 using variable frequency assist; at the same time, this sample showed the lowest agglomerate area ratio, consistent with the observed electrical behavior where large polymer areas exist with a minimal presence of carbon nanotubes. Surface resistivity values for these samples were in the insulator range; thus, the samples were able to sustain electric charge by friction rubbing. The value of the electric charge appeared dependent on the fabrication method. The NC prepared with iPPMFI = 2.5, using the variable frequency assist fabrication method, showed the highest charge, 1040 V. For NCs with higher MFI, the resistivity decreased up to seven orders of magnitude (105 Ω/sq) and was dependent on the fabrication method. Surface resistivity values were ≥1013 Ω/sq for the pure polymers and 1012 Ω/sq for iPPMFI=2.5 NCs independent of the fabrication method. Pure polymers were easily charged negatively by friction rubbing with wool. Nanocomposites prepared with the higher MFI values showed no charging regardless of fabrication method, consistent with their more conductive nature. The observed trends in the electrical behavior are consistent with the nanocomposite morphology.

ACKNOWLEDGMENTS

We wish to thank A Schulman de México SA de CV for the financial support of this study. This work was partially supported by Consejo Nacional de Ciencia y Tecnología (CONACYT) through Project 250848 Laboratorio Nacional de Materiales Grafénicos.

AUTHOR CONTRIBUTIONS

Carlos Avila conceived and wrote the paper and revised the final version. Zoe Jurado, Miguel Waldo and Erika Rivera performed experiments and analyzed the electrical data. Víctor Cruz and Pablo Gonzalez prepared the polymer nanocomposites and ultrasound pretreatments. Jose Padilla performed experiments and analyzed the flow and morphology. Ron Ziolo provided technical discussions and reviewed and contributed to the final revised manuscript.

CONFLICTS OF INTEREST

The authors declare no conflict of interest.

REFERENCES

1. Fairley, P. The start of something big. *Chem. Week* **2001**, *163*, 23–26.

2. Breuer, O.; Sundararaj, U. Big returns from small fibers: A review of polymer/carbon nanotube composites. *Polym. Compos.* **2004**, *25*, 630–645.

3. Spitalsky, Z.; Tasis, D.; Papagelis, K.; Galiotis, C. Carbon nanotube-polymer composites: Chemistry, processing, mechanical and electrical properties. *Prog. Polym. Sci.* **2010**, *35*, 357–401.

4. Min, C.; Shen, X.; Shi, Z.; Chen, L.; Xu, Z. The electrical properties and conducting mechanisms of carbon nanotube/polymer nanocomposites: A review. *Polym. Plast. Technol. Eng.* **2010**, *49*, 1172–1181.

5. Byrne, M.T.; Gun'ko, Y.K. Recent advances in research on carbon nanotube-polymer composites. *Adv. Mater.* **2010**, *22*, 1672–1688.

6. Gulrez, S.K.H.; Mohsin, M.E.A.; Shaikh, H.; Anis, A.; Pulose, A.M.; Yadav, M.K.; Qua, E.H.P.; Al-Zahrani, S.M. A review on electrically conductive polypropylene and polyethylene. *Polym. Compos.* **2014**, *35*, 900–914.

7. Mičušík, M.; Omastová, M.; Krupa, I.; Prokeš, J.; Pissis, P.; Logakis, E.; Pandis, C.; Pötschke, P.; Pionteck, J. A comparative study on the electrical and mechanical behaviour of multi-walled carbon nanotube composites prepared by diluting a masterbatch with various types of polypropylenes. *J. Appl. Polym. Sci.* **2009**, *113*, 2536–2551.

8. Haque, M.A.; Mina, M.F.; Alam, A.; Rahman, M.J.; Bhuiyan, M.A.; Asano, T. Multiwalled carbon nanotubes-reinforced isotactic polypropylene nanocomposites: Enhancement of crystallization and mechanical, thermal, and electrical properties. *Polym. Compos.* **2012**, *33*, 1094–1104.

9. Kashiwagi, T.; Grulke, E.; Hilding, J.; Groth, K.; Harris, R.; Butler, K.; Shields, J.; Kharchenko, S.; Douglas, J. Thermal and flammability properties of polypropylene/carbon nanotube nanocomposites. *Polymer* **2004**, *45*, 4227–4239.

10. Kashiwagi, T.; Mu, M.F.; Winey, K.; Cipriano, B.; Raghavan, S.R.; Pack, S.; Rafailovich, M.; Yang, Y.; Grulke, E.; Shields, J.; *et al.* Relation between the viscoelastic and flammability properties of polymer nanocomposites. *Polymer***2008**, *49*, 4358–4368.

11. Seo, M.K.; Park, S.J. Electrical resistivity and rheological behaviors of carbon nanotubes-filled polypropylene composites. *Chem. Phys. Lett.* **2004**, *395*, 44–48.

12. Jafariesfad, N.; Ramazani, S.A.; Azinfar, B. Property investigation of polypropylene/multiwall carbon nanotube nanocomposites prepared via *in situ* polymerization. *Polym. Int.* **2014**, *63*, 689–694.

13. Kaminsky, W. Metallocene based polyolefin nanocomposites. *Materials* **2014**, *7*, 1995–2013.

14. Mayoral, B.; Garrett, G.; McNally, T. Influence of screw profile employed during melt mixing on the micro-scale dispersion of MWCNTS in poly(propylene). *Macromol. Mater. Eng.* **2014**, *299*, 748–756.

15. Mayoral, B.; Lopes, J.; McNally, T. Influence of processing parameters during small-scale batch melt mixing on the dispersion of MWCNTS in a poly(propylene) matrix. *Macromol. Mater. Eng.* **2014**, *299*, 609–621.

16. Avila-Orta, C.A.; Medellín-Rodríguez, F.J.; Dávila-Rodríguez, M.V.; Aguirre-Figueroa, Y.A.; Yoon, K.; Hsiao, B.S. Morphological features and melting behavior of nanocomposites based on isotactic polypropylene and multiwalled carbon nanotubes. *J. Appl. Polym. Sci.* **2007**, *106*, 2640–2647.

17. Xu, D.; Wang, Z. Role of multi-wall carbon nanotube network in composites to crystallization of isotactic polypropylene matrix. *Polymer* **2008**, *49*, 330–338.

18. Ávila-Orta, C.A.; Raudry-López, C.E.; Dávila-Rodríguez, M.V.; Aguirre-Figueroa, Y.A.; Cruz-Delgado, V.J.; Neira-Velázquez, M.G.; Medellín-Rodríguez, F.J.; Hsiao, B.S. Morphology, thermal stability, and electrical conductivity of polymer nanocomposites of isotactic polypropylene/multi-walled carbon nanotubes. *Int. J. Polym. Mater. Polym. Biomater.* **2013**, *62*, 635–641.

19. Ávila-Orta, C.A.; Valdéz-Garza, J.A.; Neira-Velázquez, M.G.; Borjas-Ramos, J.J.; González-Morones, P.; Espinoza-Gonzalez, C.J.; España-Sánchez, B.L. Proceso de desaglomeración, fragmentación y reducción de tamaño de los aglomerados de nanopartículas mediante la aplicación de ultrasonido en fase gas. Mx. Patent MX/a/2010/014326, 20 December 2010.

20. España-Sánchez, B.L.; Ávila-Orta, C.A.; Padilla-Vaca, F.; Neira-Velázquez, M.G.; González-Morones, P.; Rodríguez-González, J.A.; Hernández-Hernández, E.; Rangel-Serrano, A.; Barriga, E.D.; Yate, L.; *et al.* Enhanced antibacterial activity of melt processed poly(propylene) Ag and Cu nanocomposites by argon plasma treatment. *Plasma Process. Polym.* **2014**, *11*, 353–365.

21. Isayev, A.I.; Kumar, R.; Lewis, T.M. Ultrasound assisted twin screw extrusion of polymer-nanocomposites containing carbon nanotubes. *Polymer* **2009**, *50*, 250–260.

22. Yang, L.; Zhang, Z.; Wang, X.; Chen, J.; Li, H. Effect of ultrasonic irradiation on the microstructure and the electric property of PP/CPP/MWNT composites. *J. Appl. Polym. Sci.* **2013**, *128*, 1510–1520.

23. Mata-Padilla, J.M.; Ávila-Orta, C.A.; Medellín-Rodríguez, F.J.; Hernández-Hernández, E.; Jiménez-Barrera, R.M.; Crúz-Delgado, V.J.; Valdéz-Garza, J.; Solís-Rosales, S.G.; Torres-Martínez, A.; Lozano-Estrada, M.; *et al.* Structural and morphological studies on the

deformation behavior of polypropylene/multi-walled carbon nanotubes nanocomposites prepared through ultrasound-assisted melt extrusion process. *J. Polym. Sci. Part B Polym. Phys.* **2015**,*53*, 475–491.

24. Ávila-Orta, C.A.; Martínez-Colunga, J.G.; Bueno-Baqués, D.; Raudry-López, C.E.; Cruz-Delgado, V.J.; González-Morones, P.; Valdéz-Garza, J.A.; Esparza-Juárez, M.E.; Espinoza-González, C.J.; Rodríguez-González, J.A. Proceso continuo asistido por ultrasonido de frecuencia y amplitud variable, para la preparación de nanocompuestos a base de polímeros y nanopartículas. Mx. Patent MX/a/2009/003842, 19 September 2014.

25. Avila-Orta, C.; Espinoza-Gonzalez, C.; Martinez-Colunga, G.; Bueno-Baques, D.; Maffezzoli, A.; Lionetto, F. An overview of progress and current challenges in ultrasonic treatment of polymer melts. *Adv. Polym. Technol.* **2013**, *32*, E582–E602.

26. Espinoza-Gonzalez, C. Ultrasound effects mechanism in molten state polymer matrices and in the dispersion of nanostructures. Model system: Nylon-n and multi-walled carbon nanotubes. Ph.D. Thesis, Centro de Investigación en Química Aplicada, Saltillo, Coahuila, Mexico, December 2012.

27. Bremner, T.; Cook, D.G.; Rudin, A. Further comments on the relations between melt flow index values and molecular weight distributions of commercial plastics. *J. Appl. Polym. Sci.* **1991**, *43*, 1773–1773.

28. Besco, S.; Lorenzetti, A.; Hrelja, D.; Boaretti, C.; Roso, M.; Ferri, D.; Modesti, M. Influence of melt viscosity on the structure and properties of electrically conductive nanocomposites produced by masterbatch process. *Macromol. Mater. Eng.* **2014**, *299*, 814–824.

29. Ferg, E.E.; Bolo, L.L. A correlation between the variable melt flow index and the molecular mass distribution of virgin and recycled polypropylene used in the manufacturing of battery cases. *Polym. Test.* **2013**, *32*, 1452–1459.

30. Menzer, K.; Krause, B.; Boldt, R.; Kretzschmar, B.; Weidisch, R.; Pötschke, P. Percolation behaviour of multiwalled carbon nanotubes of altered length and primary agglomerate morphology in melt mixed isotactic polypropylene-based composites. *Compos. Sci. Technol.* **2011**, *71*, 1936–1943.

31. Steinmann, W.; Vad, T.; Weise, B.; Wulfhorst, J.; Seide, G.; Gries, T.; Heidelmann, M.; Weirich, T. Extrusion of CNT-modified polymers with low viscosity-influence of crystallization and CNT orientation on the electrical properties.*Polym. Polym. Compos.* **2013**, *21*, 473–482.

32. Wang, B.; Sun, G.; Sun, G.; He, X.; Liu, J. The thermal characterization and rheology behavior of PP/MWNTs nanocomposites. *Acta Polym. Sin.* **2006**, *1*, 408–413.

33. Xin, F.; Li, L. Decoration of carbon nanotubes with silver nanoparticles for advanced CNT/polymer nanocomposites.*Compos. Part A Appl. Sci. Manuf.* **2011**, *42*, 961–967.

34. Wang, Z.; Fan, X.; Wang, K.; Deng, H.; Chen, F.; Fu, Q. Fabrication of polypropylene/carbon nanotubes composites via a sequential process of (rotating solid-state mixing)-plus-(melt extrusion). *Compos. Sci. Technol.* **2011**, *71*, 1397–1403.

35. Hom, S.; Bhattacharyya, A.R.; Khare, R.A.; Kulkarni, A.R.; Saroop, M.; Biswas, A. Blends of polypropylene and ethylene octene comonomer with conducting fillers: Influence of state of dispersion of conducting fillers on electrical conductivity. *Polym. Eng. Sci.* **2009**, *49*, 1502–1510.

36. Pan, Y.; Cheng, H.K.F.; Li, L.; Chan, S.H.; Zhao, J.; Juay, Y.K. Annealing induced electrical conductivity jump of multi-walled carbon nanotube/polypropylene composites and influence of molecular weight of polypropylene. *J. Polym. Sci. Part B Polym. Phys.* **2010**, *48*, 2238–2247.

37. American Society for Testing Materials. *Standard Test Method for Melt Flow Rates of Thermoplastics by Extrusion Plastometer; ASTM D1238-13*; ASTM International: West Conshohocken, PA, USA, 2013.

INDEX

O

P

R

S

T

U

V